Cottonfields *to* Copters

Fifty-Five Years of Flying on the
EDGE OF THE BLADE

Dwayne Williams

outskirts
press

Cottonfields to Copters
Fifty-Five Years of Flying on the Edge of the Blade
All Rights Reserved.
Copyright © 2022 Dwayne Williams
v6.0

The opinions expressed in this manuscript are solely the opinions of the author and do not represent the opinions or thoughts of the publisher. The author has represented and warranted full ownership and/or legal right to publish all the materials in this book.

This book may not be reproduced, transmitted, or stored in whole or in part by any means, including graphic, electronic, or mechanical without the express written consent of the publisher except in the case of brief quotations embodied in critical articles and reviews.

Outskirts Press, Inc.
http://www.outskirtspress.com

Paperback ISBN: 978-1-9772-4717-9
Hardback ISBN: 978-1-9772-4770-4

Cover Photo © 2022 Bell Textron Inc. All rights reserved - used with permission.

Outskirts Press and the "OP" logo are trademarks belonging to Outskirts Press, Inc.

PRINTED IN THE UNITED STATES OF AMERICA

Dedication

This book is dedicated to Lynnette Williams, my loving wife of 55 years. She was there at the beginning of my aviation career and we are still together. She is, and always will be, the wind beneath my wings.

In Appreciation

Bob Dillenback is currently the CFO of AeroDynamix, Inc. and earned his MBA from Boston College. Even though he wasn't drafted during the Vietnam war, numerous family members have served in America's wars throughout history. He has also written a book under a pseudonym, and from the time we first met, it quickly became obvious he was very interested in my military and civilian aviation career. Later, when I expressed an interest in writing about my career and having it published, he quickly offered his assistance. In fact, without his assistance and encouragement, I might not have attempted to write this book. I cannot express my appreciation enough for Bob's assistance.

Brad Dingus holds a Mechanical Engineering degree from the University of Oklahoma, an MBA from the University of Texas at Arlington and currently serves as a Staff Engineer at Bell Textron, Inc. When Brad became aware I was writing a book about my aviation career, he generously offered his time to spend countless hours reading, and editing, my original manuscript. His timely assistance spurred me to complete this book, and I sincerely appreciate all the hours Brad spent, not only editing this book, but also for his belief in me, and his encouragement.

Table of Contents

Prologue		Page i
Chapter 1	My Early Years	Page 1
Chapter 2	The Military Comes Calling	Page 32
Chapter 3	Attending Flight School as Warrant Officer Candidate (WOC)	Page 41
Chapter 4	Welcome to Vietnam	Page 54
Chapter 5	Memorable Battles and Events	Page 70
Chapter 6	From Vietnam to Fort Wolters, Texas	Page 139
Chapter 7	Petroleum Helicopters, Inc. (PHI)	Page 154
Chapter 8	Mining in the Clouds	Page 178
Chapter 9	Back Home in Louisiana	Page 208
Chapter 10	Bell Helicopter International (BHI)	Page 228
Chapter 11	Goodbye BHI, Hello Bell Helicopter March, 1979	Page 264
Chapter 12	Interesting Assignments at Bell Helicopter	Page 277
Chapter 13	Lost in The Bermuda Triangle	Page 300
Chapter 14	Interesting Times in Colombia	Page 316
Chapter 15	Flying the 222/680 Helicopter	Page 338
Chapter 16	The Tiltrotors	Page 354
Chapter 17	Interesting Flight Tests	Page 381
Chapter 18	First Flight of (Bell Agusta) BA-609	Page 417
Chapter 19	Staying Busy with Other Companies	Page 428
Chapter 20	First Flight of Marenco Helicopter SKYe SH09	Page 447
Epilogue		Page 455

Prologue

Throughout my aviation career, I have often been asked why I became a pilot, and I wish I had an answer for that question. I mean a good, positive answer where I could say without hesitation, "I became a pilot because . . . ", then go on to provide a reason full of details about why I became a pilot, but I cannot do that. I can tell you that as a child, some of my fondest memories are of stopping work in the middle of a cotton field to watch B-25 bombers fly training missions out of Reese AFB, located just west of Lubbock, Texas. If that was the cake in my life, convincing my wonderful mother to take my brothers and me to airshows held annually at Reese AFB, was most certainly the icing! Those airshows always brought in wonderful, strange looking airplanes, and quite often, even stranger looking helicopters! An airshow back in those days cannot compare to the mega-airshows we have today; but to a young child living in the vast flatlands of west Texas, it was definitely a happening event!

West Texas did not provide a lot of spectacular scenery, nor many diversions to break the monotony of huge expanses of totally flat countryside that produced cotton, tumbleweeds and dust storms. Such surroundings did, however, promote fertile minds, because with just a little bit of lumber salvaged from old vegetable crates, a few nails and a dab of imagination, I could conjure up my own version of some of the more famous World War II fighter planes. In my imaginative mind, my creations had been exact replicas of the most powerful aircraft in the world that had rid the skies of our enemies over Germany and Japan! In fact, some of my aircraft even managed to get airborne for short periods of time, thanks to a few lengths of rope and sturdy Elm trees that had been planted in the dust bowl days.

However, I must admit to suffering the indignities of numerous crash landings, not so much a result of my being shot down during my imaginary combat patrols, but primarily because of a shortage of good, strong ropes. New ropes cost more money than I could afford, so I was limited to whatever I could salvage in the alleys. Those well-worn ropes had been unable to keep me airborne with the consistency I desperately needed to completely rid the skies of my imaginary foes!

    Once I finally realized the structural integrity of my designs offered little aside from numerous scrapes, bruises and contusions, my next step was to build model airplanes. Fortunately, plastic models of that era were more affordable than those in model shops today. By chopping cotton in the summer, and pulling it in the fall, I could afford one model aircraft a week, painstakingly assemble it, then repeat the process the following week, assuming I had another hard-earned dollar to do so. That hobby aided me in identifying various types of aircraft amid the wide-open expanses that provided ample opportunities to observe airplanes. I was thrilled by large flights of B-25 bombers in formation, not to mention an occasional B-36 flying high overhead with its distinctive throbbing sound of six reciprocal engines slightly out-of-synchronization with each other. Once, I even saw an F-86 Sabre jet scooting around low level, and that had been exciting! Perhaps seeing me jumping up and down and waving from a large, open pasture, he had even made a wide turn and dived directly toward me before roaring over-head, then disappeared over the horizon! I could scarcely sleep that night.

    Those plastic models and high-flying bombers provided just enough snippets of aviation to keep me indelibly linked to aviation, and also harbored my dreams of someday becoming a pilot. One of my friends and I had even become "hangar rats" about once a month by walking several miles across an open prairie to visit the small airport located just outside the city limits of Littlefield. There we enjoyed the sight of multi-colored airplanes, plus the distinctive smell of dope used to repair brittle wing fabric and high-octane fuel that powered their engines. Our greatest wish was for some

pilot to offer us a ride someday, even if just around the pattern. Unfortunately, the day we so fervently wished for never came to be; nonetheless, it never dampened my enthusiasm of one day being able to fly.

Aside from an uncle who had flown as an engineer/gunner on a B-24 bomber in World War II and had been shot down and captured, no one in my family had ever flown. There had been no one to whom I could turn to offer guidance on what I had long dreamed of doing. Realizing I was never going to fly unless I made it happen, I simply drove out to the Littlefield Airport on November 11, 1964, and signed up with the Maner Flying Service for flying lessons. It was simple as that! A flight instructor by the name of Verrol Briggs took me up that same day in a Piper 140 airplane. Soaring in and around clouds that resembled huge cottonfields, I was hooked from the moment we became airborne!

Since that time, my journey has taken me around the world multiple times, during which I have amassed more than 16,000 flight hours. It has been an incredible journey; yet, as I look back, it feels like my 55-year journey of flying helicopters began only yesterday,

CHAPTER 1

My Early Years

I doubt many folks have heard of Greenville, Texas, but that's where my journey began. Located in east Texas, it's approximately 50 miles northeast of Dallas, Texas. Established way back in 1846, it was named after Thomas J. Green, a significant contributor to the establishment of the Republic of Texas. It became the largest community in Hunt County where it became widely known for producing cotton on the rich, black farm land that surrounds it. In fact, from the 1920's to the mid-60's, there had been a banner stretching across Main Street that proudly proclaimed: "Welcome to Greenville, the Blackest Land, the Whitest People". That same sentiment was also painted across the town's water tower. The original intent was to describe the color and richness of the area's soil, and the kindness of its citizens. Unfortunately, the unintended suggestion of racial overtones resulted in the slogan being changed in 1965 during the escalation of civil unrest rumbling throughout the south. "Whitest People" became "Greatest People", and I think it's still a place where great people of all races enjoy living and raising their families.

Prior to the war, life was an era of modest and simple living, when locals knew all their neighbors, and looked after each other. "Mass Media" consisted of radios, newspapers and the neighborhood's "grapevine" of gossip which provided most of the local information. Prices were much lower then, as were incomes. You could buy a Coca-Cola for five cents, a gallon of gas for fifteen cents, a new car for $900 and an average sized new house for a mere $3,600! That sounds wonderful until you realize the average annual salary back then was about $2,000, or $166.00 a month. Things became increasingly more difficult when World War II broke out, as most canned goods, cooking oil, meat, butter, cheese and gasoline were rationed. In fact, there were precious few commodities that weren't rationed in order to support the young men and women in uniform who were serving, fighting and dying on distant shores.

In that era of minimal wages, spanning from the Great Depression until the beginning of World War II, there was a class of workers often referred to as share-croppers. They were hard-working people, whose entire families often resided in crumbling, decrepit shacks on some piece of hard scrabble land owned by a landlord. Sharecroppers farmed the land and did odd jobs for the landowners, then shared in the profit when the crops were harvested and sold, assuming there had been any profit to share. Those were tough times, and being at the bottom rung of the farming business, those folks never came close to earning $2,000 a year. Barely earning enough money to feed their families, they were among the poorest of the poor in a nation struggling to emerge from the lean, hard years of the Great Depression.

Married on the evening of September 7, 1938 in Lone Oak, Texas, my parents were descendants of an extensive line of those same sharecroppers. They worked all day on the farm before getting married later that evening. I still have a photo of them on the day after their wedding that shows them dressed in old farm-style clothing, standing alongside two horses harnessed to a plow. Used to working long, hard days just to eke out a living, those two lovebirds took just enough time off to get married, have dinner, then return to the farm the next morning to plow fields. There had been no thoughts of a honeymoon. Money was simply too scarce to be wasted on such frivolities. They grew up in the Great Depression, survived all the rigors of those tough times, and eventually became charter members of what has become known as the "Greatest Generation". What an awesome achievement!

March 19, 1943 was a rather chilly morning on a day not unlike many endless days during that era of disturbing news, especially since the war had spread across the entire globe. However, I would like to think March 19th was a momentous day for Mr. and Mrs. Eliga

Daniel Williams, since that was when I became a member of their small family. Born at 3:30 am at Joe Becton Hospital in Greenville, Texas, I weighed in at 8 lbs., 4 oz.

I was named Eliga Dwayne Williams, and joined an older sister, Norma Gail Williams, who had been born on August 26, 1939. I would like to think my birth was a poignant event for my parents since they had lost their first son, Daniel Truett Williams, at childbirth on December 27, 1941. Born a mere three weeks after the bombing of Pearl Harbor, the death of their baby boy was overshadowed by world events; but for the grieving parents, the loss of their baby was a tragedy they would remember for the rest of their lives. In step with those desperate times, Daniel Truett's obituary in the local newspaper closed with the caption, "Keep 'Em Flying"!

The local population at that time was very concerned with the ebb and flow of World War II, as their thoughts and prayers were dedicated to the young men and women fighting for their country in foreign lands. In fact, one share-cropper's son from that area would go on to make quite a name for himself by becoming the most decorated soldier of the war. That same young man would also become a well-known and respected Hollywood cowboy star during the 50's and 60's. Although not too well known by today's generation, Audie Leon Murphy will never be forgotten in the annals of military history. He set an extraordinary example of a warrior's spirit, and is still fondly embraced as an honored son in Greenville, Texas.

Like everyone else during World War II, the Williams family provided their share of combatants since my dad was inducted into the U.S. Navy in April 1944. He left his young wife and two children at home to travel to San Diego, California, where he attended boot camp training. After completing training, he was assigned to the U.S.S. Sitkoh Bay, a small escort aircraft carrier commissioned on March 28, 1944. Although the ship's primary mission was to provide support to the much larger battle groups, Dad saw action on April 7, 1945 while delivering Marine Air Group 1 to Okinawa. On that day, a Japanese Yokosuka P1Y "Frances" aircraft dove toward

the carrier but was quickly shot down before splashing harmlessly into the sea, only 100 yards from the ship. My Dad still remembered that event with some horror since he had been below the flight deck when the action began. While other crewmen had been slamming water-tight doors shut, he was desperately fighting his way upward to the flight deck! He survived, and returned home to his beloved family in February, 1946.

I also had two uncles who fought in the European Theater of Operations (ETO). One of them, Milton R. Mitchell served in the infantry while Denton C. Powell flew as a flight engineer/gunner on a B-24 Liberator bomber before it was shot down on a bombing mission over Italy. He was quickly captured and spent 13 months in a German Prisoner-of-War camp. Fortunately, unlike so many others, he eventually made it home alive. As far as I know, he was the only member of my rather large family even remotely involved in aviation, and that ended following his discharge from the military. He returned to Greenville and spent the rest of his life as a professional painter, a good one at that. My Dad and uncles have always been heroes to me.

Relocating to West Texas

When my dad returned home to Greenville after being discharged from the Navy in 1946, it was a wonderful event; however, jobs were practically non-existent. It seemed every young man in Hunt County had served in the military, and when they began returning home from the war, there were very few jobs available. Since my aunt and uncle had relocated to the small town of Littlefield way out in west Texas, they invited my dad to visit them as he searched for a job. Boasting a population of approximately 5,000, it was named after a Mississippi native by the name of George Washington Littlefield, a former Confederate officer, cattleman and banker who had purchased the southern, or "Yellow House", division of the massive XIT ranch, one of the largest ranches in history.

How it came into being was rather interesting. It started in the early 1800's when a Major Long of the U.S. Army was ordered to lead a small party of officers and men to explore the vast, unexplored area where Littlefield was eventually founded. At the completion of his mission, he pronounced the land to be totally unfit for cultivation and furthermore, no one would ever want to live there. He proclaimed the land would remain forever the unmolested haunt of Indians, hunters, bison and jackals! Then, in 1876, the people of Texas voted to trade that same "worthless" piece of land to anyone who would build a state capitol building in Texas. Based on the harshness of the area, the swap involved the land being valued at only fifty cents an acre. The trade was eventually negotiated, and three million acres of harsh land became the XIT ranch that stretched more than 200 miles along the Texas/New Mexico border, but it quickly incurred a large amount of debt. After 16 unprofitable years, the XIT owners sold the first tract of land to George Littlefield in 1901 for $2.00 an acre. The massive XIT ranch lasted just a few years; the magnificent capitol building still stands in Austin, Texas.

Mr. Littlefield renamed his new ranch the Yellow House Ranch, and it became a successful working ranch. After learning a rail line would pass through the area, in 1912, he made the decision to sell the northeastern corner of his ranch to settlers who established the small town of Littlefield, Texas. Located on the new railway, the city was officially named on July 4, 1913 when the first Santa Fe train rolled into town. That same land deemed to be so barren and worthless so many years ago by an astute U.S. Cavalry Major, turned out to be my home for many years.

Living in Littlefield, Texas

My dad and sister had moved to Littlefield a few months before mom and me so he could find a job before uprooting his small family. He was lucky in his job search so in early 1948, the time came for us to move. I was about five years-old when my dad and older cousin arrived at our home in Greenville driving a large "bob-tail" truck my cousin used to haul cattle in; and it smelled

like it! Consequently, the first order of business was to use the water hose to clean it out before loading everything we owned into the back of the largest truck I had ever seen. All I can remember about the actual move itself was seeing nothing much more than signal lights and power lines while driving through many towns along the way. That's because I had to sit in the middle of the bench-style seat, tightly crammed between my dad and cousin. So much so, I could only look upwards through the windshield! I also remember it was a long, uncomfortable ride because in those days, traveling such long distances was pretty much an all-day affair. We didn't arrive in Littlefield until well after dark, so I had no idea where I was, or what the surrounding terrain looked like. It wasn't a fun trip for me!

    I vaguely recall waking up the next morning feeling lost and uncertain about my young future. Having no place to call home, we had spent the night at my aunt and uncle's house. Even though it was their home, it was hardly more than a roof over our heads since it was made from an old, well-used military style barrack literally wrapped in black tar paper held on by nails driven through shiny, silver discs. It was then lathered with a heavy coat of gray, crumbling stucco that did little to please the eye! There was no paint on the exterior at all, and the rooms were small and cramped.

    On that first morning, waking up and walking outside had been like taking a walk on a foreign planet! While east Texas had been lush and green, west Texas resembled a moonscape! A well-worn, broken-down picket fence stood guard around a yard closely resembling a giant adobe clay brick since there was no grass, bushes or trees! In fact, my aunt didn't mow the yard; she swept it! The only thing brightening the yard up were several sad-looking hollyhock flowers that had been lovingly nurtured by my aunt. She had planted them close to the house, hoping its shadow would provide some shade from the hot sun. Outside the yard, I saw a dry, dusty road that seemed endless as far as the eye could see. That U.S. Army Major hadn't been too far off in his assessment when he

deemed the land to be unfit for anyone but wild animals, Indians and hunters!

In addition to the shock of the barren landscape, the creature comforts were also rather stark. The bathroom was an outhouse sitting rather forlornly about 50-feet from the back of the house. I can tell you with complete sincerity, it was extremely cold in the winter! Fortunately, the toilet paper was a Sears and Roebuck catalog, which I thought was cool, especially at Christmas time. I could pore over all the neat selections of toys while using the bathroom! In fact, those old Sears catalogs, which always appeared in the mail just before Christmas, remained a highlight of my childhood for many years, long after we abandoned the outhouse for an indoor bathroom.

After staying what seemed like forever with my aunt and uncle, we finally managed to find a home and, much like my aunt and uncle's home, beauty was only in the eyes of the beholder, given the basic nature of the house. Our place was a small, two-bedroom structure, but it was home to our family of four, and we were happy to be settled. It was located very near the local schools which was a good thing since we had no car. My Dad walked to and from work every day at the Garland Motor Company located in downtown Littlefield. In fact, he often walked home during his lunch hour, since he could rarely afford lunch at the local cafe. His usual two daily round trips amounted to two or three miles each workday.

Similarly, every Saturday, our family would set off for grocery shopping with Dad pulling my sis and me behind him in my bright red American Flyer wagon. The wagon was only available for riding to the grocery store located many blocks away. On the return trip, my sis and I both had to walk, since the red wagon was brimming with groceries to last a week. It might have been a hardship to some, but for my sister and me, that's just the way it was. We did have an exciting event in late 1948, though; that's when my dad bought our first automobile, a well-used 1937 Chevrolet! Until that time, our primary mode of transportation was walking, except for our one-way trips in my American Flyer. We were all very happy to step up into the life of "high society".

About twice a week, an ice truck would stop by to deliver ice for our "ice box" to keep perishable items cool, since such things as milk, meat, eggs, etc. could be kept fresh a day or two longer than if they had been left outside. Milk was delivered twice a week, as well. Mom would set empty glass bottles out on the step that would be duly swapped out when the milkman made his deliveries. If she wanted anything extra such as butter, orange juice or buttermilk, she would write a note, roll it up and stick in an empty bottle. She never ordered more than what could be used in a couple of days since ice boxes were not that efficient.

Littlefield did have its share of stories to tell, though. One memorable event occurring on the night of October 26, 1943, involved the murder of a personable young physician by the name of Roy Hunt, and his lovely young wife, Mae. Founder of the popular Littlefield Medical Clinic in 1937, he had become a well-known and highly respected member of the community before he and his wife were discovered bound together on their bed. The doctor had been shot; his wife beaten to death. It was a horror scene made worse by the fact their bodies had been discovered by their two young daughters who had run screaming into the night for help. A reward was offered for the killer, or killers, but no one was arrested. I still remember Mom always making some comment about "how terrible it was" whenever we happened to pass by the Hunt's home. It wasn't fancy, nor was it sinister looking; it looked just like all the other houses on that street. It was notable only because it held an evil secret. A secret known only by the killers of the popular young couple, and they had chosen not to discuss it.

My First Helicopter

Shortly after settling into our new home, an exciting event occurred late on the afternoon of July 1, 1948. My neighborhood buddies and I had been playing outside when we heard a strange

noise overhead. Looking up, we saw one of the strangest looking machines we had ever seen, circling overhead before finally descending toward a large vacant field nearby. None of us had a clue about what it was, other than it flew! It also appeared to be landing just down the street from us! Clad only in shorts and T-shirts, being bare-footed didn't slow any of us down as we began running fast as we could down the street to see that incredible machine! We had no idea what it was and as soon as we arrived, we saw a few folks standing around it, plus a policeman or two. Then, two gentlemen stepped out of the machine and began talking. We had no idea what they were saying, nor did we care. We were too much in awe of the largest "propellers" we'd ever seen! Even more confusing, they were attached to a pole that came up through the middle of the machine, instead of being on the nose, like airplanes. We didn't know it, but we were looking at a helicopter. In fact, it was one of the first ever built, and we had no idea we were seeing history in the making.

I found out much later in life it had been none other than Lyndon Baines Johnson being flown around the Texas countryside while running for an open seat in the U.S. Senate in 1948. Having a medical problem just days before starting his campaign, he had been late in starting and was lagging far behind his opponent, the popular former Texas governor Coke Stevenson. Needing to make up for lost time, Johnson wanted to travel across the state in the most expeditious way possible. The helicopter was still a novelty since only a handful had been certified to carry passengers at that time. Putting his anxieties aside, he chose a helicopter to carry him around Texas to make up for lost time in his political runoff campaign.

What made that event rather unique was the fact that Johnson was running out of time and money; consequently, that stop in Littlefield, Texas was the last campaign stop he made in a helicopter. After making his speech, he had returned to Lubbock, Texas, then rode a train to Austin, Texas. That was the first usage of a helicopter for political activity, and my friends and I were eye-witness to it! Perhaps it was rather prophetic that the first helicopter I saw was made by Bell Helicopter, a legendary company where I would spend many happy years of my adult life.

Cotton Was King

That long, rectangular section that projects northward and defines the square shaped border with New Mexico and Oklahoma is the Panhandle Region of Texas. Littlefield sits directly south of that region, in the middle of a geographical area of Texas known as the South Plains. It sits at an elevation of approximately 3,600 feet with plenty of sun and low humidity, perfect conditions for growing cotton. In the 1940's and 50's, cotton was the economic engine that powered the entire region. Practically everything depended on cotton, and everyone was linked to it in some form or fashion. Whether they were driving tractors, operating gins, pulling cotton or chopping weeds, cotton fed their families.

My dad eventually worked his way up to become service manager at the Garland Motor Company, so he was spared having to toil in the cotton fields; but not so for Mom and me. Some of my earliest memories in the cotton fields were of playing at the end of the turn rows (an area where farmers turned their tractors around) while waiting for my mom, aunt and their friends as they pulled cotton. They would first work down a row to the end, then reverse and finish up at the turn-row where I spent the day playing. In fact, it was in the cotton patch where I met one of my life-long friends while playing and waiting for our moms to make their rounds in the cotton fields. Despite being long, hard days for our mothers, complaints were never heard from them. It was a necessity that could not be avoided.

The cotton patch was great fun when I was young, but soon as I was old enough, I was given my own cotton sack and sternly told to pick all the cotton I could after school, and all day on Saturdays! The pay was minimal, about sixty cents per 100 pounds, and that translated into about $2.00 after school, and maybe $6.00 on Saturdays. Big cotton trailers were parked at the end of the rows and once we made a "round" of pulling cotton, we had to weigh it, write the weight in the tally book, then someone had to climb into the trailer to shake cotton out of the sacks. Much to my dismay, I always seemed to have that unenviable task while everyone else enjoyed a longer break! Once we had pulled all the cotton in one area, the farmer relocated the trailer to another field of snow-white cotton ready for picking. That was systematically done until the trailer was either bulging with cotton, or we ran out of fields. The farmer then brought in another empty trailer and the process began all over again. It was a very repetitious and boring process, but at least it was cooler than hoeing weeds, since cotton was harvested in the fall months. Truth be told, I never cared for either option!

Chopping cotton takes very little skill. It does, however, require a very sharp hoe and a bit of stamina since it's normally accomplished in the heat of the summer, in July and August. Unfortunately, cotton wasn't the only thing that grew in such fertile ground. A wide variety of weeds also thrived, and they were always thirsty! They challenged cotton for any precious water that arrived in the form of rain or irrigation water pumped from underground reservoirs, if the farmer could afford it. There were always lots of weeds and most of them had deep roots, tough skins and some even had sharp thorns making them hard to hoe! Then, there was always the occasional rattlesnake or two that had taken up residence in some secluded area where field mice were plentiful.

Some rows of cotton were so long that the other end often seemed to shimmer in the heat, as if it were a mirage. It was like being in the desert! The sun's reflection looked like water, but it wasn't, and it only made me thirstier. It seems I was always thirsty working in the cottonfields! Thirsty for water with lots of ice! It was hot work that didn't require a lot of talent, but it did provide an opportunity to look around, and daydream. In those hot, blue Texas skies, there always seemed to be an airplane, or two. Appearing with great regularity, they flittered about before disappearing over the eastern horizon, probably returning to Reese Air Force Base, a very busy training base located just west of Lubbock, Texas.

Those cotton fields provided me with many opportunities to dream because I was continuously envisioning ways to escape the cotton fields and maybe, just maybe, one day I could even fly airplanes. Those planes were sometimes alone, sometimes in pairs, and on many occasions, there were as many as twelve of them in formation! Most of those aircraft were old World War II era Mitchell B-25 multi-engine bombers, the same aircraft in which James Doolittle had led the first attack on Japan in April, 1942. They were still plentiful in those days and just watching them filled me with such awe that I would spend most of my hard-earned money on purchasing model airplanes. If we weren't working on Saturdays, I would head straight toward the local hobby shop to buy the latest plastic model, even though it took most of my meager supply of money. Those same models filled me with hope that I would someday have an opportunity to pilot one of those magnificent machines.

All things considered, Littlefield wasn't a bad place to spend my childhood. We weren't always working in cotton fields and, even though we seemed to move from one rent house to another with great regularity, I managed to make a lot of friends. Moving about in a small town might change your neighborhood, but it really didn't change who your friends were. You simply had to walk a little further to visit them. The one house we lived in the longest was located on the west side of town, which was the poor side of town. But when everyone in the neighborhood is poor, how do you differentiate? There always seemed to be a lot of kids back then, and one family that lived just around the corner from us had a son that would go on to become very well known.

Back then, he was just another neighborhood boy who could play the guitar pretty well, plus sing some of the more popular songs, especially if they happen to be country western. He was closer in age to my sister, and they had become close friends, in a platonic way. Their link was they both loved country music made so popular by Hank Williams, and I still recall how upset they were when they heard he had died while traveling through West Virginia in 1952. Country music wasn't all that popular back then, but I still remember my sis listening to her "45" records on a small record player and writing down all the lyrics to a song. In the meantime, her friend was busy strumming and learning the chords to their latest favorite song. Pretty soon, he had it all down and sometimes even entertained us with his skill, artistry and songs. In those days, Waylon Jennings was a long way from being famous; he was just another neighborhood boy who lived around the corner, and dreamed of becoming a country-western star. We're all pretty proud of our old neighborhood friend who achieved his boyhood dream.

Even though the terrain was flat and virtually devoid of trees, I loved running with neighborhood buddies and enjoying clear skies, beautiful sunsets, cool mornings and pleasant evenings, even in the summertime. Cars weren't air conditioned, nor were most homes. Those that were used air conditioners called "swamp coolers", since they utilized ordinary house water pumped to the top of the unit, then allowed to drizzle down through straw padding on all sides. Typically mounted in the window, the fan "sucked" air through the wet straw, then pumped the cooler air into the room. It worked fairly well in the dry climate of west Texas, but not so well in areas of high humidity. For me, normal air conditioning was to simply place my bed close to an open window, then let the cool night air flow across my bed. If the weather is cool enough, that is still my favorite way to sleep.

We didn't live on the farm, but we lived in a house on the outskirts of town that had a very large pasture. Towns and cities didn't rule with an iron fist in those days, so in my younger years we always seemed to have a few chickens for eggs, plus a milk cow available for dad to milk, and mom to churn the butter. West Texas was famous for cotton, but farmers also planted big gardens and were willing to share, or sell their bounty at a cheap price. I can recall spending hours snapping and shelling peas, shucking corn or whatever else happened to be available for my mom and aunt to preserve, or can. After large potato and onion crops had been harvested, many farmers allowed families to take their own burlap bags and pick up the smaller ones for their own use, free of charge. We always had chickens for eggs and the "fryers" we had to chase down typically ended up as Sunday dinner! It didn't get much fresher than that.

People seemed to be more self-sufficient back then, and everyone was always willing to help each other whenever necessary. Flour used to be sold in cloth bags adorned with very colorful prints or objects. Each time we had to buy a bag of flour, the next one in line in my family of four kids would get to select their favorite flour bag, then Mom would use her sewing skills to make a beautiful shirt! I still have a photograph of me wearing my favorite shirt made from a flour sack that had rocket ships and planets on it. I guess even then I was reaching for the stars!

Washday

Monday was always wash day. The old wringer-style washer sat in the corner of the kitchen six days a week, every day except wash day. That's when it was moved into the center of the kitchen, plugged into the overhead socket the light bulb was screwed into, and put into action! I still recall when I was about seven-years old, when Freddie, my two-year old brother, and me decided to help my mom one fine Monday. We had watched Mom do the wash, then run the clothes through the two-roll rubber wringer to remove excess water. It looked like great fun! So, the next time Mom went out to hang the clothes to dry, we went into action! I got up on the chair and pulled the cord on the overhead socket to start the wringers going, then hopped down and proceeded to roll the clothes through. It worked like a charm! How great it was to be helping Mom with the laundry! She would be SO proud!

Freddie wanted to have some fun, too, so he hopped up on the chair. Pulling a soggy pair of pants out of the washer, I handed them to my little brother so he could feed them into the wringer. We were really having fun until he forgot to let go of the pants! Immediately, his arm was sucked through the wringer! The rubber wringer! Try as I might, I couldn't stop him from being rolled through the wringer! In my desperation, I could only imagine that first his hand, then the arm, his head and maybe his whole body being sucked through that wringer! What was I going to tell Mom when she came back inside and found her baby boy flat as a pancake? Desperate to stop the wringers from sucking him all the way through, I jumped back up on the chair, yanked the cord twice for a double click, and the wringers started rolling him back out! Thank God! I'm thinking maybe mom might not kill me after all! Then, I looked at his arm.

His arm was flat as a pancake, and the webbing between his thumb and finger had been ripped to such a degree his thumb now rested on his wrist. Mom was going to kill me after all! Knowing I suddenly had a problem I couldn't manage, I began screaming for Mom. She came rushing in, grabbed him up and we went racing to the hospital. I think it only cost about twenty dollars to stitch his thumb back into place, but they didn't do anything about his flat-as-a-pancake arm. They said he was so young it would soon fill back out on its own. Now, of course I'm staying quiet as a mouse through all that! I just knew I was in serious trouble for almost turning my little brother into a pancake! But all she said was not to ever help her with the laundry again, and we didn't! I've often wondered how that wonderful woman managed to raise three boys and somehow, keep her sanity.

Other Dangers

In those days, we had to endure and overcome a lot of dangers, but we just didn't know it. We survived such things as asbestos shingles and lead based paint, both sanctioned by the U.S. government. Polio, a horrifying disease that sickened and paralyzed the unlucky ones, was one of the scariest issues confronting the younger population. It was a time of fear for us, since several of our classmates contracted the disease and a couple of them ended up in an iron lung! Polio was an invisible, silent killer that seemed to lurk in the shadows. I still thank God for Dr. Jonas Salk, a dedicated man who plunged the stake into the heart of the disease!

Then there was Elvis. I mention Elvis because there were numerous parents and preachers who were convinced, and tried to convince us, that he was a sinful young man sent by the Devil to control all teenaged girls who were screaming, shaking, sobbing and literally passing out whenever he sang! Some disbelievers even burned his records. He was no big deal to the rest of us. Heck, we really liked his fresh rock and roll style of music and we watched all of his movies. In the end, we survived him, as we did all the other speed bumps we encountered, such as sandstorms. I suppose if you can survive West Texas sandstorms, you can survive just about anything.

My First Sandstorm

It's difficult to describe what sandstorms were like in those days. It wasn't that far removed from the memorable days of the Depression and the "dustbowls" that practically covered the Midwestern states in sand several feet thick in many places. The aftermath often resembled a moonscape! In the early fifties, we were in the midst of a terrible drought, and the first sandstorm I endured was horrifying! Even though it was noon, the sky went dark. I'm talking pitch black, even forcing cars to turn on headlights. My sis and I ran inside our house just before the wind hit the thin walls of our small house with such force that dust penetrated every crack, crevice, hole or slit, and blew directly into our house! A billowing cloud of suffocating dust forced us to grab blankets, sheets and whatever else that was available, then run into the kitchen. Using everything available, we plugged every hole in an effort to stop the powdery dust from entering the kitchen where we were cowering! Mom even passed out wet dishrags so we could hold

them to our faces to prevent inhaling the choking dust. It seemed like we stayed in our little "cave" for days, but it was probably no more than a day or so. Thank goodness we were in our kitchen because at least we had food. When the howling finally stopped, we stepped out through the doors of our kitchen sanctuary into the living room, and were stunned by what we saw!

Dust! Lots and lots of dust; it was everywhere. Piles of it! On the floor, on the furniture, on the walls, and even on the ceiling! It was unbelievable. Then, when we were brave enough to venture outside, we found the screen-door jammed tight, since sand had piled so high against it. Exiting through the back door, we entered a world of brown, as in brown sand, everywhere! It was hard to distinguish where roads, yards and ditches started or ended. Even cars had mounds of sand piled up against them! There wasn't much else to do except start shoveling. We shoveled sand out of the house, pitching it outside where all the other sand was piled. It took a lot of work for things to get back to normal. It was also easy to spot the cars that had been driving when that sandstorm hit, because the front part of their hoods and fenders were often shiny since the paint had been sandblasted off. That sandstorm was the first of many I endured while growing up in West Texas. "Oh well", the old-timers used to say, "this is West Texas!" Those storms still occur, just not as often, or as bad. I'm just glad I don't have to deal with them anymore!

Looking back, there were some tough times, but for mom and dad, both of whom had survived the Depression, life was probably pretty good. We might have been poor, but I can't recall ever missing a meal, even when we were running the streets with our buddies. We simply ate at whosever house we happened to be when dinner, or supper, was served, and there were never any complaints! The bread may have been stale and the soup a bit thin and watery, but we ate it with great relish! I remember we never knew who might show up for a meal at our house, but we always had plenty. Plus, mom was a great cook! Food was always served with a helping of love, and was accepted in the same manner.

I remember getting an allowance when I was still quite young. It was a whopping quarter. Twenty-five cents! Doesn't sound like much today, but it went a long way, especially on Saturdays. That was movie day! Mom would drop off my older sister and me at one of the movie theaters at noon. There was the Rio, the Ritz and Palace theaters from which to choose, and enjoy. A ticket cost nine cents. I could then get a handful of peanuts from the penny machine, and a coke for five cents. Popcorn was another five cents. That left an extra nickel I could save toward purchasing a model airplane after many weeks of saving. That is, if my sister didn't "borrow" it from me with promises to pay me back. I don't think she ever did, but on those wonderful Saturdays, we would watch the movie, cartoons, a weekly episode of a serial thriller, and then the main feature, which was usually a western, starring one our cowboy stars. Typically, we would watch the whole thing over and over. In those days, once you paid your money, you could stay until you wanted to leave. There were even two drive-in theaters for evening entertainment, the LFD and XIT. They had playgrounds for kids to enjoy until the movie started. At first, we had to roll the windows down to hear loudspeakers synched with the movie. Later, they became much more high-tech with speakers that hung on your car windows. It was wonderful!

Scotty, My Forever Friend

Life was good back then. I had two younger brothers, Freddie James and Larry Wayne, and there were probably ten or twelve neighborhood boys that ranged anywhere from four to ten years of age. Then there were the dogs. Every family had at least one dog, and they weren't your typical "house" dogs so common today. Those dogs were simply mutts of mixed pedigree with no pampered background, and were much like cats when it came to taking care of themselves. They lived outside and no one had to feed them, or otherwise take care of them. They lived on scraps thrown out after a meal, or whatever else they could drag up to dine on. My beloved

pet was one those mixed breeds. His name was Scotty, and I loved him dearly! Early on, I would try to make him stay home while my buddies and I went roaming and exploring various alleys in search of untold "treasures". That is, until the day he saved my life, or at least prevented a severe mauling.

It happened when I was heading to town on my bicycle and he tried to follow me. I scolded him and, with his tail tucked behind him, he lowered his head and sadly made his way home. About halfway to town, I saw two massive Dobermans racing across a yard and leaping a four-foot picket fence as if it weren't there! They had probably already eaten all the neighborhood kids, so I was fresh meat! They were growling, snarling, and already slobbering in anticipation of what would soon be their next meal! I tried to accelerate but it seemed as if my legs were slugging through molasses. It was no use; I was a goner!

That's when a white, explosive ball of fur hit them! Both of them! They never knew what hit them! Their retreat was faster than their attack! One was minus an ear; the other was bleeding profusely where an eye used to be! All I could do was fall to my knees and hug my hero. Sobbing and too weak-kneed to stand, I could only grab and hold him as if I were never letting go! I just kept telling him over and over again how much I loved him! We were soulmates after that and were constantly together until he disappeared without a trace some 12 years later. I walked everywhere, and searched for weeks. My mom would often drive me all over town, desperately trying to find him, but it was not to be. I never saw my best friend again, and I was beyond devastated. We had grown up together, and he will always be my forever friend of my childhood days.

Tunnels and Tree Houses

I cannot remember being bored since there were always neighborhood boys to pal around with. Being faced with a landscape devoid of mountains, rivers or lakes, we used our imagination and often went underground; as in digging tunnels! Lots of them! I don't know why we were so mesmerized by them, but we became quite adept at digging them. Maybe it was because they were normally a bit cooler than the ambient temperature. Plus, they were our secret hideaways. Whatever mischief we often got into, the caves and tunnels always provided safe havens. Only the neighborhood gang knew their whereabouts, and their locations were never shared with anyone outside the gang. It was always a well-kept secret, even to this day!

Then there were treehouses! Fortunately, there was a very nice stand of tired old elm trees dating back to the 1930's when they were planted by the Works Progress Administration (WPA) to combat severe dust storms that were so prevalent during the "Dust Bowl" era. They weren't what you would call massive, but they were certainly tall enough for building tree houses, plus they provided wonderful shade! If we weren't hiding out in the caves, we could always be found in our tree house, trying to mimic the chirping of numerous sparrows or other birds that flitted about.

Oddly enough, at age 10, I seemed to be the driving force, or "engineer", behind our unlimited number of projects. Using pen and paper, I would draw out how to best dig a tunnel, complete with secret compartments, or erect tree houses located some 15 – 20 feet high in whatever trees were available. Tree houses that would be constructed between big limbs and branches. We also tacked boards on the massive trunks from the ground up to the tree houses that served as ladders for everyone, including our younger four-year-old gang members, as well. I can still hear my mom, aunt and other moms yelling out us to us all the time: "Be careful! Take care of your little brothers! Don't let them get hurt!" And we did take care of them. No one was ever seriously hurt, although we certainly did provide a lot of opportunities to do so!

None more so than that one fine day when I announced to the neighborhood gang assembled in our back yard that we needed an escape route from our tree house. After all, we could be attacked at any time by a notorious gang from across town, so I had been doing some thinking about it. We could run a cable stretching from the tree house all the way out to the furthest tree, a distance of perhaps 50 feet, or so. It all sounded so simple! I mean, all l we needed was a small pipe, and a cable to slide it on. The search was on! One-half the gang went one way; the other half took off in another direction. Surprisingly, it wasn't long before everyone reassembled with a six-strand cable and small pipe that fit on the cable perfectly! We were on the verge of some real excitement! Something really cool!

Dragging the cable behind me, I quickly scaled the ladder up to the tree house and began lashing the cable to a very large, strong looking branch. Once that was done, one of my willing assistants on the ground stretched out the other end of the cable to the foot of the "anchor" tree. Scrambling down from my perch, we all then met to discuss how high up the tree we should tie the cable. After much discussion, we decided to lash it about 2 or 3 feet from the ground to achieve the proper angle for sliding down from the tree house. But first, we had to put the pipe on the cable. It fit fine, so we pulled as much tension out of the cable as possible, then lashed it down. Then came the next big step; sliding the pipe up the cable to the tree house! Unfortunately, try as we might, we couldn't propel the pipe all the way up; it would only go halfway. It needed oil, or grease! Or, how about lard! Oh yeah, that should work! And I knew exactly where to find some.

As soon as Mom left in the car, my brother and I went into stealth-mode and snuck into the kitchen, searching for that big bucket of lard stashed in the cupboard. Aha! We found it! That lard was going to work great! Filling both hands, we rushed back outside to accomplish the final task; grease the cable! And boy did we! We also stuffed the pipe with all the lard we could! I have to tell you, when we "threw" that pipe back up that cable, it nearly took the upper branch off! It was definitely zippy-doo-dah fast now! It was deemed ready for testing! Back up the tree I went, while everyone

else remained on the ground, and that seemed odd. That should've made me suspicious. I mean, why all of a sudden would they want to remain on the ground while "someone" launched off the safety of the tree house? That someone being me, I must confess to feeling a twinge of trepidation, but someone had to be first. Oh well, grasping the pipe, I took a breath and leaped into space!

I still remember a few things about that initial leap. My first thought was, "Wow! I'm traveling really fast"! My second thought was "Uh-oh! The pipe is getting warm"! Then, while my brain was digesting those two small bits of data, I noticed how quickly the distance between me and the anchor tree was narrowing! Oh man! I was moving fast! All of a sudden, I had a choice to make! Either let go of the cable, or smack into the tree! I let go of the cable just before impact and it was not a pretty scene! I was traveling so fast that, when I hit the ground, I narrowly missed the tree and flipped end over end for another 25 or 30 feet! I stopped when I either hit a tree or ended up in a ditch, or maybe both; I'm not sure! I was so dazed I just sat up and tried to get my gyroscope stabilized. My neighborhood gang wasn't much help either; they were all laughing too hard! After catching my breath, we went back to the drawing board to make a few tweaks.

The first thing was to raise the lower end of the cable up about 4 or 5 feet to decrease the angle, thus slowing it down; at least, we hoped so! Then we draped an old cotton sack full of cotton around the base of the anchor tree, just in case someone held on too long. After I made another test run, our new project was signed off and ready for use. All of us were soon experts and had developed the technique of letting go of the pipe about a half-second before tree impact, landing on our feet, then running safely by the tree. It worked great! So great, we decided it was time let our younger gang members in on all the fun we were having. My youngest brother, Larry Wayne, was about 4 years old, and a real trooper. He was sort of the quiet type, but fearless when trying out new things, and he was anxious to use our new toy!

Roy Rogers, Gene Autry, Lash Larue and Red Ryder were just a few of the many cowboy heroes we admired back then, and Larry wouldn't be caught dead outside our house without his beloved red Roy Rogers straw cowboy hat and boots. The rest of his attire was shorts and maybe a ragged, torn T-shirt from all the rough housing we did. He looked quite appropriate, probably not unlike the rest of us. And he was so eager to soar down that cable! He climbed that tree like a monkey and was soon perched on the "diving board", ready to leap into space!

Think about that! Today, if anyone let a 4-year-old boy, wearing boots, straw hat and shorts leap from a perch about 25-feet in the air, then slide down a cable at warp speed, someone would call 911! Back then, at least to our neighborhood gang, it was completely normal; just another day. Once in place, without the slightest bit of hesitation, he leaped! Screaming down that cable, he looked like a pro! He even let go of the cable, hit the ground, rolled once and was quickly back on his feet, ready to do it again! He was always a gamer!

Many years later, when Larry was hospitalized for what would be the final time before succumbing to cancer, several family members were in the room visiting and laughing about the great times we had as kids! We were all telling stories when Larry spoke up and said: "Do you remember the cable slide we had as kids?" Are you kidding? Of course, I remember! "Well," he said, "do you remember how you told me everything about jumping, but nothing about what I was supposed to do at the bottom? I remember screaming down the cable and knew I had two choices. Hit the tree or let go of the pipe! I let go of the pipe!"

He told how just the previous year, all of his company managers had been directed to attend a company-sponsored seminar. After enduring a full day of "guidance", all attendees were divided into small groups, then sent to various locations to prepare presentations of the seminar's topic, "Decision Making Under Stress". After splitting attendees into groups, everyone was deciding how to proceed when Larry began telling the team his story about the tree house and cable. Everyone broke up laughing, then elected him as their team leader. He was also designated to give the presentation. Using his story as the centerpiece of the presentation, his team got a standing ovation! They also won first place! I think he was very proud of that! He was a good man, a special brother and a wonderful storyteller! I still miss him. I guess I always will.

Another Lesson Learned

After the zip line success, I didn't think anything could scare me, but I was wrong. It happened one day when one of my buddies and me were playing catch with a baseball in the front yard. Leonard Smith, the town drunk and all-around handyman, was a rough-looking character who drove all over town on his Ford tractor, looking for odd jobs to do. He had found one on that particular day and he needed two young, dumb teenagers to help him make it happen. So, when he pulled up, stopped and asked if we could use five dollars apiece, we both said yes! "Well, then, climb up on my tractor and let's go make it happen!" he said, and off we went!

Unbeknownst to us, we were heading straight toward the cemetery, located way out on the outskirts of town. We bounced and clanged right up to where there was a rectangular outline of a three by six-foot grave. Definitely not what we wanted to do! But five dollars was five dollars, so I began wielding a shovel while my buddy heaved the pickaxe to break up the hard ground. Satisfied all was going well, our newfound boss decided to drive back to town and find more "stuff" to do, leaving us at the cemetery; just the two of us! Digging a grave is hard work, not to mention a bit scary. At least, it is if you're two young, impressionable thirteen-year-old boys.

When the sun finally settled beneath the horizon, that cemetery was no place for two impressionable teenagers to be! It was very dark and there were no lights anywhere! The place got creepier by the minute, so we threw the tools in the open grave for safekeeping, then hightailed it out of there! In the darkness, we made

a couple of wrong turns before making our way back to the nearest road, hoping to hitch-hike to town. Unfortunately, the few cars that passed by wanted no part of two dirt-covered young boys so close to a cemetery! We had to walk all the way home and never did get our five dollars. We just chalked it up to one of those lessons learned, because neither of us wanted to go back to the cemetery to finish digging that grave.

Moving On

It's probably a good thing we soon outgrew such adventures, or who knows what would've happened. After "graduating" from the cotton fields, tunnels and tree houses, I began working for my Dad, who had gone into business for himself drilling water wells. I did that for several years and ended up operating one of the rigs when I was just 16 years old. "Dragging main" on Littlefield's main street provided excitement during weeknights and weekends. Life was good, but during mid-term of my junior year of high school, my Dad made the decision to relocate from Littlefield to another small West Texas town about 90 miles south; a town named Seminole. Unless it was named for the tribe of Indians in Florida, I have no idea how the name was chosen. Their sports teams were nicknamed the Indians, and the high school yearbook was the Tepee.

There were also a couple of well-known country-western music entertainers born there: Larry Gatlin and Tanya Tucker. Also, if you're a fan of the Elmer Kelton book series, you might also be interested to know the author of those books was Paul Patterson, who also claimed Seminole as his hometown. It's a place where the scenery didn't change much from Lamb County, except farmers and ranchers were busy clearing scrub-brush from unimproved areas to plant crops. This was the main reason we moved there – to drill water wells to irrigate the new crops. It was not easy leaving life-long friends in Littlefield, but I quickly fit into the mainstream of teen-aged activity and soon found myself really enjoying it! It was a new world in which my new classmates were open, honest and friendly, and they welcomed me into their society. They added credibility to the old saying about how "God had made West Texas so unfit for man that he made up for it by populating it with some of the best people in the world!" I still believe that to be true!

My Introduction to Aviation as a Flagman

My aviation career may have sprouted roots in Seminole, Texas after I decided to ride out to the small Seminole Municipal Airport to check on a job with an aerial spray company. My high school buddy had turned in his resignation, and the company utilized an old military surplus Stearman biplane to spray local crops to eradicate insects and/or weeds. The legendary plane had been the U.S. military's primary flight trainer during World War II. Having never seen one before, I was really interested in seeing it, especially since it was all rigged up to be a spray plane.

After arriving at the airport, I was admiring the spray plane when a young man walked up and asked: "What do you think about it?" I hardly knew what to say, except to blurt out that "it was an awesome airplane and I was hoping to replace my friend who was terminating."

"Well," he said, "we'll sure need a new flag man; can you do that?"

"I could certainly learn", I responded.

"OK, you're hired. Come on out tomorrow morning and Dave can teach you what to do."

Just like that, I had a job as a flagman, whatever that was.

One of the first things I learned about the spray business was that it starts early, like about 4:00 AM early! As soon as the plane was loaded with insecticide, my pal and I took off in the pickup so we could arrive at the edge of the field about sunup to be in position to "flag" for the pilot when he arrived. My buddy grabbed the vinyl flag attached to about a 5-foot pole, and we trudged out to the edge of the field to wait for the spray plane to show up.

Just as the sun began creeping over the horizon, we could hear the distinctive sound of the Stearman's radial engine as it flew toward us. It was just another day at work for my buddy, as he started waving the black-and-white checkered flag back-and-forth in front of him, just as they do at NASCAR races. The pilot lined up on the flag at the far end of the field, then dropped down very close to the ground to begin his spray run. As soon as he flew over the far edge of the field, he unleashed a wide swath of insecticide spray that followed him like a faithful cloud as he roared directly toward us!

I had a very strong urge to run like hell, but my buddy just stood there and kept on waving that flag back and forth! Then, just before I was certain the propeller was going to make a minced-meat slurry out of us, he began running across the plowed field with me matching him, stride for stride! We probably only ran about 30 paces before stopping, just in time for the plane to roar by so close I could see the pilot's teeth! He was close, but that wide swath of insecticide vapor was even closer!

Totally mesmerized by the spray-plane roaring by, I barely had time to turn my back before we were totally enveloped by the insecticide vapor. I had no protection, but I noticed my good buddy had completely wrapped himself up in the checkered vinyl flag to protect himself from the evil smelling spray that drenched both of us! I just knew I was going to die from insecticide poisoning, but I managed to make it through the rest of the morning before we called it quits. I also bid my pal "adios" when we finished because I wanted that damned flag all to myself so I could wrap up in it when I went back out!

That was the beginning of a long, hot summer of early get-ups and never-ending trips to various fields all over Gaines County. My job was to help refill the plane's hoppers with spray, then proceed to the fields to direct the pilot by waving that checkered flag! I also learned how to wrap up in it as he roared by, spewing a white fog of poison! I cannot remember what type of insecticide or weed poison it was, but I do recall smelling like that God-awful spray all summer long, even after many baths!

That was the norm until I decided to move on, and they hired a young man to replace me. I quickly checked him out on all the intricacies of being a flagman, including how to wrap up in it! Aside from working with all that insecticide and staying out of the path of that low-flying spray-plane, I don't think either of us ever thought of that job as being dangerous. As long as you wrapped yourself up in the flag and stayed nimble enough to sprint out of the way of that whirling propeller, you were okay. But danger often lurks in the strangest places, and death can occur when you least expect it, just as it did in the early morning hours of September 10, 1964.

On that particular morning, one of the owners of the spray company had driven to Seminole just to see how things were going, then volunteered to drive the young flagman out to the field located approximately 10 miles west of Seminole. As luck would have it, the young flagman also brought along one of his buddies to help him since it was going to be a busier day than usual. After loading the spray plane, all three of them jumped into the owner's pickup and headed to the job site.

After they arrived at the field, they pulled off the narrow farm-to-market road and parked. After exiting the pickup, they picked up the flag and began walking toward a small ravine that lay between them and the field. They probably saw what appeared to be a thick layer of early morning fog lying ominously in the ravine, but thought nothing of it. Unfortunately, the fog was a deadly form of hydrogen sulfide gas capable of killing any living creature in its path! The unsuspecting trio made it just a short distance before collapsing. A farmer passing by saw the pickup with several bodies sprawled near it. Being a Good Samaritan, he parked his pickup, then stepped out to assist the victims; in a matter of seconds, he became the fourth, and final victim.

I had been to that same field many times before, walked into the same ravine, and could have just as easily been one of those victims. That's the thing about aviation; in spite of having all the experience and know-how in the world, every pilot knows luck is a major factor in how well his day goes, be it good, or bad. You just never know, because luck doesn't play favorites. It just seems to depend on a toss of the dice, and when a bad number is rolled, someone has a bad day, and all too often, pays the price! Just as it did so long ago when death came calling in the guise of an innocent-looking fog, slithering down a ravine.

My Dad was a share cropper in east Texas and these were his horses used to pull wagon to/from town, plus plow the fields in 1938 just before getting married.

Family photo of my mom (Dovie), dad (Lige), sister (Norma Gail) and me just before Dad was sworn into U.S. Navy in April, 1944.

((Left to Right) My brother Larry Wayne, sister Norma Gail, brother Freddie James and me in Littlefield, Texas (1952).

My Mom with brothers Freddie and Larry with Scotty, the family pet and my best forever friend of my childhood.

I'm wearing my shirt with spaceship & plane design my mom made from flour sack.

CHAPTER 2

The Military Comes Calling

Preparing for the Draft

The draft has been with us off-and-on dating back to the Civil War and up into the 1960's. When a young man turned 18, it was mandatory that he registered for the draft. There were no options, no extensions and no waivers! If you were a male eighteen years of age, you registered. That didn't mean the government was going to draft you right away, but if the military didn't meet their quota of volunteers to maintain a certain manpower level, you were a prime candidate to be drafted. The first step to decide one's eligibility for the draft was a pre-induction physical to ensure you were healthy enough, and there were many who prayed fervently for flat feet, warped spines, shrunken brains or oversized teeth; anything! Anything at all that would keep them from receiving the dreaded 1A rating for physical fitness! Being declared 1A, coupled with a low draft number, was the kiss-of-death, since it essentially declared you were a prime candidate for the draft!

The next step was for the government to conduct a national draft lottery, which, during my time, involved two drums stuffed with air balls. One drum consisted of 365 balls, each marked with a birthdate for every day of the year, followed by the second drum spitting out an air ball with a number from 1 through 365. As the air balls were expelled into their prospective drums, they were paired together. A birthdate coupled with a lower number meant you had a very good chance of being drafted, especially if you were in good physical condition. In the final analysis, your fate was decided by luck. Those who got the lower numbers had a good chance of dying in combat, while life didn't change much for the lucky ones who ended up with a high draft number. Fortunately, in early 1973,

Secretary of Defense Melvin Laird announced no further draft orders would be issued. The last drafted enlisted rank soldier retired from active duty in 2011; the last Vietnam-era drafted soldier of Warrant Officer rank retired on November 10, 2014. It was one heck of a career!

I don't recall being overly upset at having to register, nor did I necessarily think being drafted was the worst thing that would ever happen to me. Everyone had to register, and consequently everyone was subject to being drafted unless they were physically unfit, married, had children, were a full-time college student, went into hiding, or jumped the border into Canada. Since it was a numbers game, all the young men who claimed waivers or exemptions certainly increased the chances for the rest of us. I was in college, and had been for some time, but unfortunately, I had to work to attend college, so I could only afford 10 hours per semester instead of the 12 hours that would have qualified me for a student draft exemption.

I think the catalyst that catapulted me into the draft occurred on the afternoon of August 2, 1964. That was the day the destroyer U.S.S. Maddox fired warning shots at three North Vietnamese patrol torpedo boats that approached the ship at high speed. The combatants engaged in a series of firing exchanges, but in the end, only slight damage was done on either side. It was little more than a skirmish, but that attack was a major factor in passing the August 7th Gulf of Tonkin Resolution in Congress, thus enabling President Lyndon B. Johnson to take military action against North Vietnam.

A few months later, in the spring of 1965, I was summoned to report to my draft board in Lamesa, Texas. From there, I was transported via bus to the Abilene, Texas pre-induction center for physicals. There seemed to be hundreds of us from all over Texas, so it took a lot of doctors and assistants to indoctrinate, poke and prod us in much the same manner a rancher would manage his cattle! We had to fill out a ton of forms, some of which prompted one country boy to ask, "Sergeant, what's a spouse?" and "Does sex mean how many times I done it?" I distinctly remember the officers and sergeants discussing his idiotic responses in their office before unanimously declaring him as totally un-fit for service. Thinking back, anyone who could talk himself into a 4F rating wasn't dumb! As far as the draft was concerned, he was "4ever free", while I got tagged 1A for "prime pickings!"

On the bus ride to Abilene, I sat beside a young man I had never met before, but we hit it off right away and before long, we were talking like old friends. His name was John Lynch, and it wasn't long before our conversation was focused on the U.S. Army's helicopter flight school. It seemed we both had an interest in aviation, although he was much more knowledgeable about the Army's flight school. Before parting, we both agreed to meet up and visit an Army recruitment center to see if they could help us get into flight school. What made the Army's flight school so appealing was the fact it required no college degree for admission, which was a big plus, since neither of us had graduated from college. Heck, we felt fortunate just to have graduated from high school! Anyway, within a day or two, we met up and headed straight for the U.S. Army Recruiting Center in Lubbock, Texas.

The young sergeant who welcomed us into the center was a pleasant fellow by the name of Sergeant Larry Dooley, who quickly informed us "he was there to help." Surprisingly, he really was. In response to such a warm welcome, John and I advised him we were particularly interested in the Army's helicopter flight school and were hopeful he could help us. Unfortunately, he frowned at us while patiently explaining that enrolling for flight school was basically out of his hands, or as he put it, "above my pay grade." It seemed the U.S. Army's helicopter flight school cadet ranks were filled primarily by those who had prior service or were currently serving on active duty; not by individuals like us, trying to enlist into those hallowed ranks! However, true to his word of being there to help, he did agree to find out all he could about the program and even arranged for us to take a series of tests to see if we could even qualify for flight school.

The recruitment center for Army indoctrination was in Amarillo, Texas, so off we went to "make it happen" one Friday afternoon, via a long bus ride, since the tests were always scheduled on Saturday mornings. I don't remember much about the tests because John and me, along with several other young Army recruits we met, went to a local bar the previous evening to have some drinks. Although we enjoyed the time and good conversation, we eventually began to realize it was very, very late and we were seriously drunk! In fact, I don't remember much at all about that evening, or even going to bed! What I do remember is the tests all seemed to be an endless series of page after page of nothing but squiggly lines and unknown hieroglyphics we all desperately tried to decipher. How I made it through, God only knows! I was extremely fortunate, in that I passed with flying colors, but my new friend and pilot-to-be, John, did not. He was totally devastated, and I'm sure I represented all the dreams he had so desperately wanted for himself, but lost. I've never seen him since, and I've often thought how easily those roles could have been reversed.

Alone in my quest to become an Army helicopter pilot, faithful Sergeant Dooley came through once again by arranging for me to take a flight physical, another step forward in my quest to become an Army helicopter pilot! The next big step was a trip to the U.S. Air Force Air Training Command (ATC) base at Amarillo, Texas. It was a small base, but it had once been a home for B-52 bombers ready to launch a nuclear counter-attack anywhere in the world in response to an attack on the United States. It consisted of one runway stretching almost two miles across incredibly flat terrain, and it was a mass of concrete and steel almost five feet thick! Those Air Force guys took their business seriously, and it was certainly very interesting for a young man from the cotton fields, trying to break into the U.S. Army aviation business.

The entrance swarmed with MPs and it took me a lot of talking, not to mention a phone call from Sergeant Dooley, to convince them

I was there to utilize their medical facilities for a flight physical for the U.S. Army flight school. Yes, I explained, I realized I was at an Air Force Base, but in that instance, Uncle Sam wanted them to share their facilities with me, a young cotton-chopper from Littlefield! Finally satisfied I wasn't a spy, with orders in hand, I boldly marched into the assigned building where I was met by a very polite, courteous young Captain who was probably a chick-magnet due to his good looks. He was one handsome dude, but also pleasant and helpful. He was a doctor and soon had me going in and out of a myriad of offices where every doctor, nurse or assistant could push, pull, prod or poke every square inch of my body. It was a seemingly endless day of such torture, but everything seemed to be going well, with only one doctor left to visit. He was a dentist who almost ended my career before it began.

The dentist was a much older man, and quite tall. He was far from handsome and his wrinkled skin, silver hair and matching mustache removed any vestiges of youth. He rudely grabbed the paperwork from my hand while directing me to sit down in an old, dilapidated looking dental chair. After commanding me to "open wide", he immediately began to berate me and demanded to know "who the hell" was the idiot who sent me in for a dental examination! Hell, I had several fillings in my teeth, and everyone knew an Air Force pilot had perfect teeth, not to mention a killer smile. I tried arguing with him, but since he was holding my mouth open, all I could muster was "I'm not joining the (drool, gurgle) Air Force to be a pilot! I want to be a (slurp, drool) Army helicopter pilot!"

But it was like arguing with a pig in mud immensely enjoying himself. Ordering me out of the chair, he wrote in bold red letters NOT QUALIFIED TO BE A PILOT across the entire length of the paper, threw it back to me and told me to get the hell out of his office! I was so stunned by his actions; I was at a loss for words! Holding the crumpled piece of paper, I struggled to hold my composure as I slowly walked out of his office into the hallway. With one bold stroke

of his pen, that man had single-handedly destroyed my dreams of becoming a helicopter pilot! There were no words to describe my feelings, as I hung my head and slowly shuffled down the long corridor toward the door that had suddenly become the entrance into the rest of my life, whatever that might be!

The world had become a blur and all I could see were those bold red letters that ended my life-long dream of becoming a pilot. In fact, I was so engrossed in that sudden turn of events I didn't pay much attention to the sharp looking young officer who opened the door just as I was reaching for the door knob to exit the building. His query of "Hey, how's your physical going?" shook me out my suspended animation, and I recognized him as the young Captain who had been so helpful when I first arrived. My response was: "Not very damned good!" I then explained my dilemma with the dentist. The captain's response was to grab the papers from my hand and swiftly head back down that long corridor to the dentist's office, taking me along with him. Barging in, he barked at the dentist, sternly telling him he had no authority to disqualify me! I was taking a flight physical for the U.S. Army, not the Air Force! Surprisingly, the dentist quietly motioned me back into the chair, took a new sheet of paper and complied with the captain's orders. When it was over, I grasped the captain's hand and thanked him profusely! I've often thought of him and I wish I had kept his name. He saved my career and I still owe him a tremendous debt!

Receiving My Draft Notice from Uncle Sam

Having gone through the ordeal to take my flight physical, I was feeling pretty good about my chances of getting into flight school. But my feeling of euphoria was short-lived. Shortly thereafter, the U.S. Postal Service delivered a standard, non-descript letter from the draft board. It certainly didn't look like the kiss-of-death, but it was! Damn! Everything had been trucking along just fine, then came the official invitation from Uncle Sam himself! And he wanted me! Letter in hand, I made a very quick trip to the recruitment center hoping Sergeant Dooley could help me out.

The minute I walked through the door, he knew what the score was! I guess my stunned look of depression gave it away! Eyeing the letter, he asked me if I had opened it. My response was "No". He visibly relaxed a bit, then simply said: "Good; had you opened it, you would already be the property of the U.S. Army." I still don't know why I didn't open it, but thank goodness I didn't, because I would've gone directly to Vietnam as an infantryman. Not a good thought! As it turned out, I enlisted for three years to get into helicopter maintenance training. It wasn't what I wanted, but Sergeant Dooley explained at least I would be at Fort Rucker, Alabama, the home of Army Aviation and maybe they could help with my quest of becoming a helicopter pilot. Thinking back on it, Sergeant Dooley was a good man who did his best to help me out when I needed it. He's another man I owe!

Hello Uncle Sam

June 7, 1965 was the day I was inducted into the U.S. Army in Amarillo, Texas. I think there were about 50 of us who boarded that shiny, four-engine Lockheed Electra that afternoon for Love Field in Dallas, Texas on our first leg toward Fort Polk, Louisiana. It was actually an exciting trip for me, since that was the first time I had ever ridden in an honest-to-gosh airliner. After spending several hours at Love Field, we finally took off in a Trans Texas Airways DC-3 (commonly called "Tree Top Airlines") for Fort Polk and arrived there about midnight. Stepping off that old DC-3 was like stepping off into another world. A world of chirping crickets and cicadas in stifling heat matched only by the humidity!

Everyone was dripping with sweat in a matter of minutes, but the best was yet to come, and it arrived in the form of a snarling, screaming Army sergeant who unloaded on us for daring to show up at such an un-Godly hour! Plus, he was going to teach us all about "Colonel Peter's Pivot Points", little dots painted on the sidewalk where we would pivot on one foot! And how to make a bed so tight you could bounce a quarter off it! Everyone was suddenly thinking what in the hell have we gotten ourselves into? But we learned how to march along the sidewalk, pivot on "Colonel Peter's Pivot Points" and how to make a bed before being allowed to go to bed at about 4:00 AM. We then rolled out of bed at 5:00 AM, so we could stand in line for breakfast. We learned quickly you do a lot of standing in line in the Army, and we had eight more weeks of that regimen.

The first few days were spent getting boots, uniforms and SHOTS! God almighty! There were multiple kinds of shots the military deemed necessary, and all were administered in typical military fashion. A long line was formed just outside the medical barracks and all recruits were ordered to remove their "blouse" (shirt). The line was long and it was HOT! But we were the property of Uncle Sam, and the task at hand was to receive every shot known to man, all of them waiting just beyond those swinging doors. We all had the feeling we were being taken into a slaughterhouse, as we slowly inched our way closer and closer to those doors; those dreaded doors behind which were doctors and nurses eagerly awaiting our arrival! Armed with a vast assortment of needles, some of which resembled a gun, they were the enemy!

I managed to run that gauntlet without too much damage, but others weren't so lucky! Some of them staggered out the door and collapsed! The pitiless sergeants screamed and yelled at them, but it was no use. Then they directed their screaming and yelling toward us, telling us to get those "piss-ants" out of there! Thank God, they gave us shots on a Friday! The following Saturday and Sunday were the only times they ever allowed us both days off. A good thing, too, because everyone was either deathly sick, or felt too bad to even get out of bed!

I still recall the next eight weeks of basic training were some of the most miserable days of my life. Not because of the physicality of it; what I hated most was having to endure it all in one of the

most miserably hot and humid places I'd ever been, even to this day! Anyone who's ever been to Fort Polk, Louisiana will probably tell you the same thing.

It was non-stop training, classrooms, weapons firing and physical fitness to turn "momma's boys" into trained killers. After all that training, we were required to complete a mandatory multi-part Physical Training (PT) test just before the final week. Everyone had to pass it. The maximum score was 500 points and I almost maxed it! My reward was being promoted from Private E-1 to Private First-Class E-2! My pay jumped from $64.00 a month to $70.00. It was a good beginning to my military career because as soon as I completed basic training, I went home and married my young wife on August 12, 1965.

Although I had passed the written examination, the draft notice I received forced me to join the Army before I could accomplish all requirements to attend flight school. Consequently, after completing basic training, I was assigned to Fort Rucker in August, 1965 to attend aircraft maintenance training. While stationed there, I happened to meet a young Specialist 4th Class who was the training company's administration clerk, and he knew all the administrative prerequisites for flight school. He was a draftee and only had a few months left in the military, but after I explained my dilemma, he quickly offered to assist me in my efforts to get into flight school. His assistance was incredible because he knew what type of paperwork was required, who needed to sign off on the paperwork, and how to funnel all that into the proper channels! He was successful in his efforts because I received orders to report to Fort Wolters, Texas to attend flight school in December, 1965. When I rushed over to thank him for his efforts, I was told he had already been discharged from the military. I still regret that I can't recall his name, because he played a huge part in my efforts to get into flight school!

CHAPTER 3

Attending Flight School as a Warrant Officer Candidate (WOC)

Fort Wolters, Texas was located just east of Mineral Wells, a quaint western town that claimed an old west ranching history dating back to the late 1880's. The surrounding area of canyons and mesas blended together to form a rugged, beautiful countryside through which the famous Brazos River looped and twisted in serpentine fashion amid water-carved canyons. It was a place where the fierce Comanche Indians once pitched their tepees and hunted the rich bounty of deer, turkey and antelope. One of the branches of the famed Chisolm Trail also pierced the heart of that area in the late 1800's, when cowboys drove vast herds of cattle from south Texas all the way up to the railheads in Kansas.

It was also the same area that General Jacob F. Wolters, then commanding general of the Texas National Guard 56th Cavalry Brigade, deemed to be suitable for cavalry training after being mandated to do so by the state of Texas in the early 1920's. Officially designated as Camp Wolters in 1925, it was still a wilderness of sagebrush, mesquite trees and rugged terrain when it was considered useful for training mounted cavalry personnel in Texas. I doubt whether he, or anyone else for that matter, had the slightest idea it would not only become a major basic training camp during World War II, but also home to the U.S. Army's flight school for both helicopters and small airplanes on July 1, 1956. It would remain so until 1973.

Thousands of young men who attended helicopter flight school will never forget the place where so many memories of their youth still exist. Officially referred to as the United States Army Primary Helicopter School (USAPHS), Fort Wolters became the temporary home for every young man who attempted to master the art of flying helicopters during the Vietnam war. That war, often referred to as a "helicopter war", had an insatiable appetite for helicopter pilots to fill the cockpits of thousands of helicopters that steadily funneled into Vietnam from Bell Helicopter's busy plant in Fort Worth, Texas. It still remains hallowed ground that honors the memory of so many young men who didn't survive the battles fought in Vietnam.

My own arrival at Fort Wolters on December 10, 1965, was unheralded, since I was just one of many such young men arriving during that time. My beautiful young wife of barely four months drove the car through the main gate on that chilly morning, and I couldn't help but notice two helicopters standing like sentinels on either side of the entrance into the base. Their large bulbous plastic windshields resembled the eyes of a dragonfly, and I almost felt like they were eyeing me curiously as we passed silently beneath them. At the time, I couldn't decide if their stares were a welcome or challenge but, full of excitement and eager to get on with my latest adventure, I didn't give it much thought, at least until my first attempt to hover. That's when I realized that look had most definitely been a challenge. Maybe even a sneer!

But on that day, we blissfully continued toward our destination, guided by signs directing us to the Warrant Officer Candidate (WOC) billets, even while taking note of various buildings that seemed so typical of military bases. Everything visible was World War II vintage with little else to show progress. However, as we topped a small hill, I was pleasantly surprised to see several all-brick barracks sitting high on a distant hill, all looking like new. That was WOC territory! It was the place that would become my home for the next four or five months while I underwent the trials and tribulations of helicopter flight training. It became the place we would commonly refer to as "the hill".

After parking in front of a large building, I exited the car, then reached inside for my military-issued duffel bag. Looking around the area, I couldn't help but notice many fellow candidates in various stages of standing, marching or double-timing in all directions. As I hugged my wife goodbye, I happened to notice a formation of candidates marching by while counting cadence to keep in step. They were very sharp looking in their red baseball caps, plus their manner of marching seemed more orderly. Wow, senior candidates, I thought! Throwing the duffel bag over my shoulder, I hesitated for a final goodbye to my wife, then turned and began walking up the sidewalk toward the large, imposing building with a sign that read "3rd Warrant Officer Candidate Company". The false bravado I exhibited after bidding my wife goodbye seemed to evaporate with each step toward my new home. In fact, by the time I reached the entrance, it had totally transformed into a feeling of dread and anxiety as I proceeded into the orderly room to report in. Expecting the worst, I was pleasantly surprised everyone seemed to be both helpful and friendly during the sign-in process. Hey, this might not be a bad gig after all!

It was a false sense of peace and tranquility that ended somewhat abruptly in the wee hours of the next morning. Lights were flipped on and every candidate rousted from bed while the environment around us rapidly deteriorated into a chaotic din of shouts, snarls, grunts and curses, all of which were being expelled by men dressed immaculately in military uniforms and glistening chrome helmets! They were also carrying swagger sticks being expertly wielded to poke, prod or punch the closest available candidate whenever they felt the need to emphasize whatever they were shouting, snarling or grunting about! I was impressed! It was a very effective method of getting their points across, plus turning a barrack full of sleepy, confused candidates into speed demons. Suddenly, we were enlightened! If there wasn't a door to run through, you made one! They no sooner got the word "jump" out of their mouths and every candidate within ear shot was already leaping toward the heavens!

We were also made acutely aware that all candidates were unworthy of breathing air in the same domain of those immaculately-dressed dudes in their glistening chrome helmets!

It didn't take long to learn those immaculately dressed gentlemen were Training, Advising and Counseling (TAC) Officers who were responsible for training, mentoring and coaching Warrant Officer Candidates. They replaced mothers, fathers, wives, pets and any other vestige of civilian life! All were combat veterans selected to guide us through every aspect of becoming Warrant Officers. Much like shepherds guarding a flock of sheep such as us, they monitored our lives on a 24/7 basis and had but one goal – teach us how to become officers, gentlemen and helicopter pilots in the U.S. Army! Or not! They were not shy about telling us that not being up to the task at hand might well result in immediate removal from the program, followed quickly by an official invitation to Vietnam as an infantryman.

True or not, that threat remained a common theme throughout our time at Fort Wolters. No one took it lightly, because everyone knew as badly as the Army needed helicopter pilots, there was an even greater need for the U.S. Infantry branch, often referred to as the "Queen of Battle". Infantrymen were the guys who met the foe face-to-face in terrain the enemy knew like the back of their hand! Commonly referred to as "grunts", those brave men bore the brunt of battle in Vietnam and suffered the highest number of casualties. They, and they alone, met the enemy on its terms on the ground! They fought for, bled and died just so they could occupy the highest terrain to become "king of the hill". We took those veiled threats seriously, and everyone picked up on the fact that no one individual was singled out for the TAC Officer's abuse. We were all fair game and if one of us screwed up, we all paid the price! Their tactics were brilliant in their simplicity because they taught us the importance of teamwork. That teamwork drilled into us at Fort Wolters saved many lives in Vietnam.

The Third WOC candidates were identifiable by blue caps, as in baseball style caps. Other companies had brown hats, green hats, yellow hats and red hats, such as I had seen during my arrival. But I was thrilled with my blue cap, and it wasn't long before we developed a friendly rivalry with all the other WOC companies. Each company was defined by the color of their cap and our blue caps were always worn with a great deal of pride.

All WOC Companies were made up of three platoons, and I was assigned to the Third Platoon along with approximately 40 other candidates. Since our company was one of the first expanded classes due to the accelerated military build-up in Vietnam, The Third WOC was too big for one building, so the only option was for us to relocate to one of the other buildings. We were also divided up among the TAC Officers. I can still remember there was one TAC Officer that looked exactly like the sadistic guard in the movie "Cool Hand Luke". He was the one who was never seen without his mirrored glasses and the inmates all called him "No Eyes"! They could never see his eyes and that projected fear since they never knew where he was looking. Despite our prayers pleading that he not be assigned to our group, he was the man who became our TAC Officer! So much for that prayer; it seemed God had far too many prayers to contend with in Vietnam rather than deal with our pipsqueak plea for help. "No-Eyes" became our mother, father and most despised uncle! Making matters worse, I was in a group of free-spirited characters that soon gained the Third Platoon the dubious nickname of "Third Herd".

"No-Eyes" wasn't overly tall, but he had that ramrod-straight military bearing of a fit body that perfectly filled his immaculately starched and pressed fatigues, with no hint of a beer-belly! Wearing mirrored sunglasses accented by a glistening chrome helmet and armed with a swagger stick made from a 50-caliber slug, our TAC Officer looked formidable, fearfully so! That fear turned to respect on a cold, miserable day in February, 1966, just about the time all candidates were involved in mastering the fine art of hovering a helicopter. The stress was intense, and sensing that, "No-Eyes"

ordered all of us outside to hold a formation on that cold, snowy Saturday morning before marching us up onto a small bleacher perched alongside our drilling field.

Ordering us to take a seat, he slowly laid his swagger stick on the bottom bench, then proceeded to pull his gloves off before carefully placing them atop it. Then, in slow-motion, he reached up, pulled his glasses off and laid them on his gloves. My God! An audible gasp emerged from our assembled group! He had eyes! Eyes that seemed to look directly at each of us before he began to address us in a calm voice: "Candidates! I know flight school is tough. It's meant to be tough, because it's important to know you're up to the task now, here at Fort Wolters. That will save your life in Vietnam." He then proceeded to talk to us for another hour or so about anything and everything we wanted to talk about. No one seemed to mind the frigid temperature! We were too excited to learn he was human after all. When we had no more questions, he put on his glasses, picked up his gloves and swagger stick and returned to the tough-assed TAC Officer he was. But our fear was now one of respect. He was in charge, but he understood what we were going through.

We needed that pep-talk, because we had already completed the preflight portion of flight school that had begun shortly after we reported in on December 10. It was a four-week course dedicated to teaching us how to become Warrant Officers; in fact, it was a prerequisite before we could begin flight training. As for the flight training, I don't think I expected it to be easy, but I sure didn't think it was going to be hard as it was, especially the hovering part! That rather simple task can be equated to a person jumping up and down while patting the top of their head with the left hand, and rubbing their stomach with their right hand – simultaneously! I say that because there is a control for the right hand, one for the left hand, plus one for each foot. Unfortunately, the slightest movement of any control dictated a similar movement on the other controls. Everyone's first attempt to hover resulted in an indescribable

series of aerobatics performed dangerously close to the ground, and it was horrifying!

During one's first attempt to hover, the helicopter took on the resemblance of an out-of-control beast determined to wreak as much havoc upon anyone foolish enough to even attempt hovering it. It was a dance of death! A delicate balance between man and machine, whereby the student pilot desperately tried to keep sufficient space between himself and the unforgiving ground, while the helicopter was equally determined to make the student sweat and scream piteously before delivering a final death blow! Even if that final act entailed turning itself into shredded, metallic confetti! The only human being on earth that could prevent such an occurrence sat silently beside you. He was a brave man and yes, sometimes he screamed at you when you made a mistake. Maybe a lot! He was the flight instructor, a man second only to God when learning to hover!

He was there to assist you on your first flight and, if you were good enough, he would be there at the end. He, and he alone, controlled your destiny as to whether or not you would one day wear those coveted silver wings. He was a brave soul who was challenged daily by all of us who tried so hard to master the simple task of hovering, in hopes of moving on to other challenges in our mission of becoming Army aviators. The ease with which the instructors flew that beast elevated them to God-like status, at least in our eyes! Fortunately, I was up to the challenge because on February 16, 1966, I accomplished my first supervised solo and was elated when my classmates threw me in into an old cattle water tank just south of stage field No. 3 to celebrate the occasion! That was the tradition, and I was absolutely thrilled to be a part of it. Being winter, the water was icy cold, but I scarcely felt it! The hard part was trying to find a "solo stone" that would become a prized memento after it had been scrubbed clean, painted and had our first solo flight date written on it. Those rocks were scarce after so many students before us had undergone the same treatment.

To my great sorrow, my treasured solo stone has long since disappeared, but I still have my Hoverbug card presented to me when I soloed. It states: "Be it known by these contents that E.D. Williams, having remained motionless in space, flown backward, forward, sideways and vertically in U.S. Army Helicopters, is hereby designated a genuine U.S. Army Hoverbug."

February 26, 1966, turned out to be another memorable day for me. About nine days after the excitement of soloing in a helicopter, I was in class when I was told to report to our TAC Officer's office, immediately! Being told "No-Eyes" wanted me in his office felt much like being named to the FBI's most wanted list! I was stunned and my mind was racing, trying to understand why I had been summoned. I had good grades, flight training was going well, my boots were spit-shined! What did I do? Was anything wrong? Finally, I was standing at the TAC Officer's door and pounded three times on the pounding board mounted alongside the door-facing. In a loud voice, I announced: "Sir! Candidate Williams requests permission to enter!" Looking up, "No-Eyes" told me to enter and stand at attention! That was most unfortunate because my quivering legs didn't want to hold me up. His glasses were laying on his desk, but his eyes drilled through me before he asked: "What are you doing standing there?" He slowly arose from behind the desk, picked up a cigar and walked around his desk and stood before me. Again, he said: "What are you doing here when you have a brand-new baby boy waiting for you at home?" Smiling, he handed me a cigar, shook my hand and said: "Congratulations! Now get the hell out of here and I don't want to see you again until next Monday when your three-day pass expires!" Thanking him profusely in a proper military manner, I left quickly as I could! I still hold Mr. "No-Eyes" in the highest regard. Mr. Bennett J. Locke was a fine man! A damned fine man whom I still respectfully remember.

In the meantime, life in the barracks had become rather humdrum since all of us in the "Third Herd" worked together as a team, akin to a well-oiled machine! Even though we had started out with

40 candidates in the platoon, we lost about 10 for various reasons, so only about 30 candidates occupied a large barrack that had three floors. We basically occupied only the bottom floor, but were responsible for keeping the entire building clean! As in spotlessly clean! To make that happen, every candidate had a specific task, sometimes even two. One team kept the latrines clean, while another team waxed the floor, and so on. Since I was the only one who had previously operated one of those big 36-inch electric-powered buffers, my job was to buff that wax to a high gloss, and I did! I mean, I kept that floor spit-shined to a high gloss all the time! We also made sure it stayed that way because everyone had an extra-large pair of military socks they slipped over their boots every time they walked into the barracks. Each room housed two candidates and it was everyone's responsibility to keep their room not just clean, but immaculate! That meant everything! Anyone's indiscretion meant we all paid the price. It was all about teamwork, and we excelled at that!

There had also been one instance where a TAC Officer had – gasp – found a dead bug in the overhead light! To ensure we could find the dead bug, he had taken a Marks-A-Lot pen and drawn an arrow stretching from the door, across the floor, up the wall, across the ceiling before ending up with an arrow pointing to the DEAD BUG, written in bold letters on the light globe. We couldn't just wipe the critter into the trash and be done with it. Oh, no! Someone had blatantly killed one of God's creatures on military property, so we had to have a military funeral for it! It would be held the next Saturday, during our free time! We took a small matchbox, dressed it up nicely, then put cotton in it so the bug could rest comfortably for eternity. We had a military funeral, as ordered, and he had plenty of company since he was buried alongside some of his buddies in a small cemetery who had suffered the same fate. Some of them even had little tombstones! It was a sad affair, but not necessarily because of the dead bug. We were late getting to our favorite pub!

Soon enough, we all graduated from Fort Wolters' Primary Helicopter Training School and departed for Fort Rucker, Alabama in May 1966. My wife, our new son and I felt on top of the world when we set off for Fort Rucker, home of U.S. Army Aviation! Although it was a new adventure, the trip was uneventful, at least until we crossed the state line into Alabama. It became even more unsettling when we passed the city limit sign into Selma. I can't say I saw anything really bad; it was just a feeling. We had watched the news enough to know Selma was the epicenter for unrest in the South. Changes were happening there in the deep south, and it didn't matter which side you were on.

One thing that hadn't changed was the famous southern hospitality of southern Alabama. When we pulled into Ozark, Alabama, we both knew we were at our home away-from-home. Being a Warrant Officer Candidate, we struggled to get by on our once-a-month paycheck and lived on a very tight budget. Fortunately, we managed to find a small garage-type apartment behind one of those enormous southern homes in Ozark, Alabama. At $100 dollars a month, the rent was hard on our meager pocketbook, but we were glad to be living together again after such a long time apart while I was a student at Fort Wolters. I even learned to change diapers, warm bottles and feed my young son! Life was good and Vietnam was far off in the future! Signing in at Fort Rucker was a happy time; not only did I have my small family with me, it was also where I would be awarded those coveted silver wings. Of that, I was certain!

Fort Rucker, Alabama

Fort Rucker still remains a U.S. Army post located in the southeastern corner of Alabama, lying about half-way between Ozark and Daleville. Both were quaint little southern towns, as was the larger town of Dothan; at least back then. It was an area commonly referred to as the "wiregrass area", so named in homage to a bunch-type of grass native to that area, along with all the pine trees. It

was a hearty grass that survived multiple wildfires and other catastrophes throughout hundreds of years, but it couldn't survive the worst enemy of all – early settlers! Cutting down so many trees and cultivating the land soon removed its favorite habitat, so the wire grass is long gone. Only the name remains.

The original name for the post was Ozark Triangular Division Camp, but before it was officially opened on May 1, 1942 it was renamed Camp Rucker in honor of a Civil War Confederate Colonel by the name of Edmund Rucker, who was given the honorary title of "General". The Army's 81st Infantry Division was the first to train at Camp Rucker before departing for the Pacific Theater in 1943. Other Divisions trained at Camp Rucker prior to the end of World War II, and it even had German and Italian prisoners-of-war in stockades at the southern end of the post. It was inactive from March 1946 until August 1950, when the Korean War began. Closing down shortly after the Korean War ended, it was again reopened and expanded when it became a helicopter training base and officially became Fort Rucker in October 1955. It quickly became referred to as "Mother Rucker" since everyone in Army aviation returns there for continued training throughout their military career.

Fort Rucker also had a bit of history in the film industry as well. The opening scenes in one of my favorite movies, "Twelve O'clock High", was filmed at the old Ozark Army Airfield (now Cairns Field) in 1948, as was one of the most famous flying stunts in film history. Old battle-weary B-17 "Flying Fortresses" making takeoffs and landings had literally turned the airfield into an old English aerodrome for the movie, and Paul Mantz, Hollywood's most famous stunt pilot, crash-landed a B-17 "Flying Fortress" there. It may be the only time a B-17 was ever flown solo.

Upon our arrival in May 1966, Fort Rucker still resembled a typical World War II military post. Most of the buildings were made of wood, save for a few newer brick buildings. There were no guard gates at any of the entrances and everyone, locals and visitors included, were able to traverse throughout the entire post. We all thought it was rather strange, especially since it was a very important training base for pilots and maintenance training for all U.S. Army helicopter training, plus even some fixed wing training. But we weren't in charge, and far as I know, there was never any clandestine activity on the base.

Our first training was learning how to fly in inclement weather regulated by Instrument Flight Rules (IFR) in the Bell TH-13 helicopter. I still recall how much I enjoyed flying the venerable old Bell, but I cannot recall my Instructor Pilot's (IP) name even though he was a very pleasant individual to fly with, and a good instructor. The only time he ever raised his voice was right after my end-of-course check ride. The next morning, my IP sat down at the table and said, "You really made me look bad!" I was stunned because the standardization pilot had said very little except that I had done a good job which, in my mind, meant I had passed the check ride! He then smiled broadly and proceeded to tell me I had made a 96 on my check ride, a wonderful surprise! The next phase of training was transitioning into the Army's new UH-1 "Huey" helicopter, followed by tactical training and minimal gunnery training before graduating on October 11, 1966. The easy part was over; my next stop was Vietnam.

I'm a new Warrant Officer with wings after completing U.S. Army flight school at Fort Rucker, Al. (10/11/1966).

My HOVERBUG card presented to me by my flight instructor after making my first solo flight in a helicopter at Fort Wolters, Texas (02/16/1966).

CHAPTER 4

Welcome to Vietnam

Vietnam has a long history of fighting. Dating back to 1858, the Vietnamese had chafed under French rule after the French occupied Vietnamese cities, then along came the Japanese in 1937. The U.S. military actually supported the Viet Minh, a Communist guerilla force formed by Ho Chi Minh in 1941, in clandestine operations against the Japanese, but at the end of World War II, the French returned. They remained there until France surrendered to the Viet Minh following the decisive and bloody battle of Dien Bien Phu on May 7, 1954. That same year, Vietnam was divided along the 17th parallel, and that created internal friction between what had become North and South Vietnam.

In 1955, the United States began sending military advisors to train the South Vietnam military, and things began escalating from that time forward as more and more American military advisors entered the fray. Then, on May 8, 1959, Major Dale R. Buis and Master Sgt. Chester M. Ovnand became the first Americans killed in combat during the Vietnam War. On February 24, 1963, two U.S. H-21 helicopters were shot down and from that point forward, helicopter gunners were ordered to shoot enemy soldiers on sight. A storm was brewing, and the winds of war had begun increasing in intensity.

1964

On August 2, 1964, everything changed dramatically. That's when the USS Maddox reported being attacked by North Vietnamese gunboats while sailing in the Gulf of Tonkin. There are some critics who say the reports were contrived, after North Vietnamese fishing boats had accidentally bumped into the destroyers. But who did what to whom really didn't matter; the U.S. quickly retaliated

by bombing naval bases and oil depots along the Tonkin Gulf Coast. More importantly, on August 7, 1964, Congress passed what would become known as the Tonkin Gulf Resolution. That gave the sitting president, Lyndon Baines Johnson, "all necessary measures" to deter future attacks on U.S. forces. On November 3, President Johnson won the presidency in a landslide election and by year's end, there were 17,280 U.S. military servicemen in South Vietnam. Just to make things more interesting, on September 30, 1964, the first large-scale antiwar protest of the Vietnamese war was held at the University of California-Berkeley. It would be the first of many. The death toll in Vietnam was 216.

1965

Everything began escalating rapidly at the beginning of 1965. More American servicemen were being funneled into South Vietnam, and a car bomb destroyed the U.S. Embassy in Saigon, an act that was considered a slap in the face of Uncle Sam. On July 28, the U.S. monthly draft call was almost doubled to meet the increasing need for U.S. military servicemen in Vietnam. That sent ripples of fear down the spine of every able-bodied young man between the ages of 18 to 26! But the most important dates for me were June 7, and August 12, 1965. On June 7, I was sworn into the U.S. Army; on August 12, I married the love of my life, Lynnette Hall. In December, my new wife dropped me off at the gates of Fort Wolters, Texas, where I would begin helicopter flight training to become a U.S. Army aviator. There were 129,611 American servicemen in South Vietnam by the end of 1965. The death toll was 1,928.

1966

I was too busy trying to learn how to fly a helicopter in 1966 to pay too much attention to world events, aside from the news from South Vietnam. I think most of us were hoping things would begin to wind down by the time we graduated later in the year. Had we known the long war-like history of Vietnam, I think we would have realized the futility of such thinking! The North Vietnamese were determined to release themselves from the yoke of so many masters, and were more than willing to pay the price! The swooshing sound Americans heard at the end of that year was all the young American servicemen being funneled into South Vietnam. At the end of 1966, there were 317,000 U.S. servicemen in Vietnam, and I was one of them, having arrived just prior to Thanksgiving. The death toll for the year was 6,350.

My graduation from flight school had been the culmination of almost a year's worth of hard work, and the most memorable part was my young wife pinning the silver U.S. Army flight wings on my chest! It had been a long struggle, so I was over-the-top excited, and I think she was excited for me as well. In fact, I will tell you right now, those wings were as much for her as they were for me. She was with me every step of the way, from dropping me off at Fort Wolters, all the way to graduation at Fort Rucker. Whenever we were able to live together, she kept my army fatigues starched and pressed, my flight suits clean, and she could spit-shine my boots better than I could! While at Fort Rucker, she knew the Huey checklist as well as I did. And when I arrived home one morning about 6:00 AM after 72 hours of starving while undergoing the rigors of escape and evasion training, my marriage was sorely tested.

Being a Texan, at some point during that long stretch of starving, I became fixated on tacos. Nothing else would do. Consequently, when I shook her awake and told her I was starving, she wrinkled her nose and yawned. Being every bit the trooper, she said: "OK, I'll fix you some bacon and eggs," but my reply shocked her. "No", I said, "I want some TACOS!" Then she really wrinkled her nose! "TACOS? At 6:00 o'clock in the morning!? Are you kidding me?" But she fixed me six tacos that I wolfed down with great gusto! To this very day, I will tell you those were the best tacos I have ever eaten! She was my biggest supporter, and I could not have made it through flight school without her.

Following our graduation at Fort Rucker, Alabama, all of us were

given a set of orders. They were orders to Vietnam, but strangely, we were told they weren't official orders. "Go home", they said, "and official orders will arrive at your respective homes of record via the U.S. mail". "Trust me", they said. "In the meantime, take advantage of your leave, go home to your families and enjoy what might be the rest of your young, short lives". Prior to departing for home, most graduates tried to identify other new grads whose names were on the same set of orders as yours. In my case, a young man by the name of Lloyd William (Bill) Whitlow, Jr. happened to be on the same set of orders and, even though we had barely known each other in flight school, we sought each other out so we could coordinate our trip to Vietnam. After exchanging phone numbers and addresses, we headed out to our respective homes. Bill's home was in Chicago; mine was in Littlefield, Texas. As the crow flies, that was a pretty far distance from each other.

It was November. Leaves were changing colors and there was so much excitement at the football games. It was the time for turkey, dressing and all the trimmings to be shared with family; a day of thanksgiving. But I wouldn't be celebrating that, at least not in the U.S.! It was hard for me to even think about such things. After all, I was going to Vietnam and everything had taken a back seat to that. In true military fashion, my official orders never arrived, nor had Bill Whitlow's. Eventually, we both began to get a bit anxious, so just before our 30-day leave was over, we both decided it might be best to meet up in San Francisco and try to get orders at Presidio, one of the U.S. military's most beautiful Army posts.

Fort Presidio occupied prime real estate in San Francisco that was absolutely spectacular! It also offered a picturesque view of the Golden Gate Bridge. Finding it was certainly easy enough, but convincing gate guards we were really two forlorn, newly minted Warrant Officers desperately trying to get to Vietnam was not! The suspicious attitude was: "why would anyone be trying to get to Vietnam?!" After all, thousands of guys were using any excuse they could think of to stay as far away as possible. Some even fled to Canada! Finally, we were able to convince them of our sincerity, and they were most helpful in providing directions to the post Personnel Center where we hoped to get some assistance in finding transportation to Vietnam.

Fortunately, we finally found ourselves in the presence of a very helpful lady in finance who listened to our sad story, then promptly purchased each of us a ticket on PAN AM Airlines that would take us to Vietnam. On November 17, 1966, Bill and I were booked on PAN AM Flight 841 that would take us first to Honolulu, Hawaii, then on to Saigon, arriving there at 09:55 AM on November 19. It cost the U.S. Army a total of $569.90 to fly me to Saigon; $295.20 for me, $274.70 for my duffel bag. My duffel bag was nearly worth as much as I was!

Hello Vietnam!

After having a pleasant meal during our stopover in Honolulu, I couldn't help but feel much like a condemned man eating his last meal. Our last link to U.S. soil was severed the moment our flight left the runway on the final leg of our journey. We arrived at Saigon's Than Son Nhut International airport just before lunch-time on November 17, 1966. After what seemed like an eternity sitting on the tarmac, stepping out of that aircraft into South Vietnam for the first time was like stepping into the fringes of Hell! In fact, you could actually feel the heat rushing to embrace you even before you stepped out the door! Beads of sweat immediately erupted out of every pore of my body, while my uniform sucked in the dank, humid air like a thick, thirsty towel! Everyone was immediately soaked, and my immediate thought was: "it can't get much worse than this", but I was wrong. It could. It could get much worse. And it did!

Just as you're trying to get past the intense heat and humidity, the smell hit you right in the face! Oh my God! It was a smell that seemed like a living creature; it felt like it wrapped itself around your sweaty body, then poked its ugly snout up your nose, causing

your head to snap backward in revulsion! It was the smell of a thousand years of poop, piss, sweat, body odor and the decay of those who were sick, had died or were about to die! Oh man! All of a sudden, I really needed a shower!

Naturally, after deplaning, the first order of business was to organize the few enlisted men into a small formation. That's just the military way. The enlisted men were told to form up, but all officers were exempt. The enlisted men were also ordered to pick up the officer's duffel bags, but we all declined that generous offer and picked up our own before making our way into the terminal. Of course, after such a long trip we all needed to use the nearest toilet, and fast! We finally located one, and just as we were approaching the entrance, the first wave of nauseating stench hit us! The foul odor we encountered when we exited the aircraft was bad enough, but that new "aroma" was elevated to even greater heights! Stopping at the entrance, we were all dumbfounded, because we could now see the entire toilet area was awash in urine! There was at least an inch or two of urine, plus whatever else had seeped into it!

In the middle of all that swill was a smiling, toothless old Vietnamese gentleman who was stooped over in the classic oriental squat with waves of urine slapping against his butt. He was also barefoot! In his hand was an elongated brush about two feet in length and he was happily swishing the greenish, putrid looking mess from one side of the room towards a miniscule drain hole chopped into the concrete of the far wall. A quick glance toward the urinals provided the first clue where all the urine was coming from. Since it was a large international airport, there were about eight urinals mounted on the entire length of one of the walls and out the bottom of each one protruded a drainpipe, perhaps a foot in length. In the U.S., that drainpipe would have continued to the floor, or into the wall, for proper drainage into a sewage outlet. But not so there! There was a concrete trench directly below the urinals that stretched the length of the wall and everyone's urine simply emerged from the pipe, then dropped another foot into the trench!

I'm sure there was supposed to be a drain at one end of the trench but unfortunately, it was totally plugged. As a result, the urine could only overflow across the floor. And it had, apparently for some time! Truth be known, the toothless old man was probably smiling because he knew he had a lifetime job, or at least until someone unplugged the trench drain, assuming they could find it. I guess we all could have just unzipped where we stood and emptied our bladders onto the floor, but officers and gentlemen that we were, we held our noses and slowly waded to the urinals! It was quite a welcome to Vietnam!

After finally taking care of business, we made our way outside to board buses outfitted with what looked like cyclone fencing mounted on every window to prevent grenades from being tossed into the bus! That may have been our first inkling that war is a dangerous business! Fortunately, we made it to the 90th Replacement Battalion and were directed to a row of tent-like "billets" (structures surrounded by stacks of sandbags) that would be our home for the next couple of days while they figured out what to do with us. While Bill and I were carrying our duffel bags across a big open field to the billets, we met 5 or 6 Warrant Officers headed toward the mess hall. Quickly recognizing them as our classmates from flight school, we were quite excited to meet up with them. They had all arrived just a day or so before us, and the first thing one of them said was: "Hey, did you hear about Bob Pruhs?"

Well, no, we hadn't; but we knew he had arrived just a few weeks before us. "He's dead!", they said. What do you mean dead? He just got here! "Yeah! He was in country about a week and was killed on his first day of flying; on his orientation flight, for God's sake!" Nobody had any details of his death, but all of a sudden, I felt old and defeated! Damn! In-country a week and killed on his first flight! Now he's dead! Everything inside my brain seemed to evaporate, and I was stuck in a gigantic void where nothing made sense.

Bob Pruhs, our classmate who had completed flight school with us was already dead, and I'm looking at 365 days! Suddenly, 365 days seemed like an eternity! It was going to be a very long year!

It was two chastened Warrant Officers who strolled into the Replacement Center early the next morning to learn our fate. Pruh's death had hit us hard and we had scarcely slept just thinking about it! Now, we were standing in the exact spot where he had probably stood, scarcely a couple of weeks ago. Now, our time had come to learn where our new home for the next year would be. We soon found ourselves standing in front of a young Captain sitting behind a desk, shuffling paper.

He seemed totally unconcerned about the momentous decision he was going to make in the next few minutes. Most likely, he would probably be sitting in the safety of that same chair until receiving his orders back to the U.S., but he was very nonchalant about our placement. That officer, who neither knew, nor cared, about either of us was the master of our fate. Picking up a sheet of paper and taking a cursory look at it, our fate was sealed when the captain announced: "Gentlemen, I have two vacancies to be filled down in the Mekong Delta with the 13th Aviation Battalion in Can Tho. A plane will take you both down tomorrow evening. Any questions?" Well, no, we didn't know enough to ask questions! Simple as that, our destination was no longer in doubt. We were going to Can Tho, down in IV Corp! We sure wished we knew where Can Tho was.

During the Vietnam war, South Vietnam was divided into four Corp tactical zones for purposes of administrative and command area for military operations. I (pronounced "eye") Corp was the most northern since it bordered the DMZ with North Vietnam; the others extended all the way down to IV Corp, which included the entire Mekong Delta, often referred to as the "rice bowl of Asia". II and III Corps were primarily the U.S. Army's domain, while our primary duty in IV (pronounced "four") Corp was to support various elements of the South Vietnamese military.

The ride from Saigon to Can Tho was provided courtesy of the U.S. Army in a large, twin-engine airplane known as a "Caribou". We departed late in the afternoon and the sun was hovering just above the horizon when we finally landed at the Can Tho airport. After deplaning and grabbing our duffel bags, it wasn't long before we were hugging onto a jeep driving through the fringes of town, heading toward the Mekong Delta Hotel that had been militarized for military personnel assigned to that area. Our driver was a U.S. enlisted man who didn't look very old, but he certainly knew how to navigate expertly through a wide variety of cars, jeeps, bicycles, buses and mopeds, all of which noisily announced their presence by an unending mixture of toots, tweets, whistles and other assorted forms of clamor! He also knew his way around, because it didn't take long before we arrived at the hotel unscathed. At the entrance to the hotel was a heavily guarded gate manned by armed MP's, a very ominous sign! Even though the gate had already gotten our attention, the sight of barbed wire and sandbags stacked around the building ratcheted up our concerns even more. However, no one else seemed the least bit concerned about that strange new world we had just found ourselves in; consequently, we managed to relax, at least for the moment.

As usual, we learned the best place to relax was at the hotel bar. Not wanting to miss out on an opportunity to have a cold beer after such a long, exhausting trip, we soon found ourselves sitting in a neat little bar. From that vantage point, we could not only drink, but could also forget our homesickness by listening to non-stop war stories, courtesy of the local veterans situated all around the bar. After a while, someone suggested we go up on the roof where we could get a much better view of the city and surrounding countryside. So, with a couple of beers in hand, we went topside and managed to squeeze into a couple of chairs that overlooked the lights of Can Tho. The lights of the city were a beautiful sight; beyond that, the vast, mysterious Mekong Delta was bathed in an inky darkness.

Sitting there enjoying our beer, it wasn't long before we saw

what appeared to be lightning flashes far out on the horizon, followed shortly afterward by faint booms that didn't sound like thunder, at all! Suddenly, a few red tracer rounds spewed upward from that same area; arching upward, they disappeared into oblivion as the tracers burned out. Then, seemingly out of nowhere, a firehose of red tracer rounds appeared in the middle of the sky! Weaving back and forth, they were directed downward toward the area where the first set of tracer rounds had erupted from. From far in the distance, we could hear a very distinctive "brrrt" sound that seemed to be out of synch with the firehose of tracers. Then a few more sporadic tracer rounds arched upward, seemingly directed toward the source of that firehose of tracers! Immediately, another dose of red tracer rounds fire hosed downward! We were not only mesmerized by what was going on, but also curious. Straining to see what type of invisible aircraft was spewing all that fire, we got our answer when someone nearby announced: "Spooky is giving Charlie hell tonight!"

That was our introduction not only to the war in Vietnam, but also the legendary DC-3 that had been dusted off and resurrected for duty in Vietnam as a heavily armed USAF gunship! What we had witnessed from the safety of our hotel was a pretty intense fire fight between unknown combatants and a Spooky Bird on our first night in the Delta; an awesome introduction! I think it was yet another inkling that Bill and I were looking at a long, deadly year in Vietnam; more specifically, in the Mekong Delta!

The next day, Bill and I reported into the 13th Battalion headquarters for our assignments. It was strictly luck-of-the-draw, but Bill was assigned to the 114th Aviation Helicopter Company, and I was assigned to the 175th Aviation Helicopter Company, both of which were located at Vinh Long Army Airfield. It was a small U.S. Army airfield that lay deep in the Mekong Delta, a huge, lush area of rice paddies so fertile and productive it was considered to be the rice bowl of Southeast Asia. Simple as that, Bill became a Red Knight, and I was an Outlaw!

Vinh Long itself was an old, moderate-sized city, comfortably nestled alongside the famous Mekong River which was utilized as the primary mode of transportation throughout the Delta. In fact, if it were in the Unites States, it would be called an interstate waterway rather than a highway. The area itself was a vast myriad of rice paddies, rivers, tributaries, streams and canals, all filled with boats. There were big boats, little boats, sampans, canoes, etc.; in fact, just about every type of boat imaginable was a common sight in the Mekong Delta.

Farmers utilized these vessels to carry rice and other goods to markets in the larger cities, while locals used them as their primary means of transportation throughout the Delta. Then there was that well-known group of individuals that utilized them at night for more clandestine operations. They're the ones who bothered us the most, because they were always up to no good. They were always shooting at us, blowing up buildings or dropping mortar rounds into our company area. In a nutshell, they were trying to kill us. That's what concerned us the most!

Those people were known as the Viet Cong, or VC for short, and it seemed like they were everywhere at night, yet nowhere to be seen by day. They were also smart, elusive, disciplined and well-armed! Their armament could range anywhere from an ancient one-shot blunderbuss or crossbow, right up to the most modern weapons the Chinese, Russians, or other supportive countries had in their stockpile of weapons to offer them. Any of them could be deadly, especially in the arms of someone who knew how to use them; but it was the big, heavy automatic weapons we hated the most! That was because they could blast our helicopters out of the sky! They were particularly deadly when the helicopter was most vulnerable, such as when it was landing in a rice paddy. Sort of like a duck in a pond.

If safely parked in revetments (parking spaces surrounded by 55-gal drums filled with water) located alongside the airfield's runways, there wasn't too much for flight crews to worry about except for random mortar attacks in the wee hours of the morning. And we did lose several helicopters in mortar attacks during my year in Vietnam. It was those critical 30 to 60 seconds when helicopters were either hovering, or sitting in a Landing Zone (LZ), that flight crews felt so helpless! Strapped in armored seats, pilots endured a cruel form of torture while constantly scanning potential locations for deadly weapons almost always hidden in tree lines.

One pilot, a very brave and respected individual, once confessed to being so scared he flexed his chest muscles so tight he suffered a severe muscle strain! It had happened around dusk one evening when they were going into a small, hot LZ and the captain on the radio had specifically told the pilots that all wounded soldiers would be on their right side when they landed. What my buddy thought he heard was *all the soldiers* would be on the right side, instead of *all the wounded*! Consequently, after they had settled into the LZ, the pilot looked to his left and could see the silhouette of three or four soldiers with weapons in their hands standing just outside the rotor downwash! He was sitting in the LZ and felt so totally helpless and scared, he tensed his chest muscles knowing he was about to shot by the VC! As it turned out, the soldiers had all been U.S. Army soldiers. He endured his pain for several days, but never told anyone because he was too ashamed. But he never made the same mistake again. He always made sure there was no miscommunication!

Dense, jungle-like trees hugged virtually every river, stream or canal throughout the vast rice bowl regions, all of which provided excellent concealment for roving bands of VC combatants. Those same waterways also allowed enemy forces located all over the Delta to navigate from one well-concealed camp to another, conduct an attack, inflict maximum damage, then withdraw back into the tree lines to literally disappear during the day. They would then re-emerge in the dark of night; the kind of dark that seemed to smother you with its inky blackness! That was the perfect time for

them to drop in a few mortar rounds, switch into stealth-mode, slip quietly into the water, then disappear phantom-like into the black abyss. They always re-appeared at another time or another place of their own choosing. That was the enemy we were always pursuing. Truth be known, we only caught up with them when we were either lucky, or they preferred to stand and fight. Those times when they chose to stand and fight were when we had our most savage battles, often with heavy casualties on both sides.

The 175th AHC was the company I served with the entire time I was in Vietnam, and their call signs were Outlaws and Mavericks. The transport helicopters, commonly referred to as "slicks", were called the Outlaws; the gunships were the Mavericks. As an Outlaw pilot, I was always focused on being the best pilot I could be, but I couldn't help but notice the Maverick pilots and crewmen who wore red bandanas around their necks. I also remember feeling quite comfortable whenever they were providing cover for us during combat assaults. After flying with the Outlaws for several months and enduring heavy machine gun fire crisscrossing Landing Zones (LZ's), often accompanied by mortars dropping in, I began to have second thoughts about flying the slicks! I think the final straw was when the helicopter I was piloting sank into the deep mud of a rice paddy and we couldn't get out, even while receiving heavy machine gun fire that shattered our windshields and wounded our gunner!

The Mavericks had come to our rescue by putting in a fierce attack on the enemy, then Maverick Lead had landed alongside us between a determined enemy and our immovable aircraft! While his gunners covered us, some of the troops leapt out of our aircraft, then sloshed through the rice paddy to his aircraft, allowing us to get our much lighter aircraft out of the mud! We were soon out of harm's way and flew back to the staging area, but that incident pretty much convinced me I wanted to be a Maverick pilot! A few weeks later, I officially transferred into the Mavericks.

All Maverick crewmen proudly wore a red bandana tied securely

around their necks. Bandanas, Outlaws and Mavericks may conjure up images of the wild and wooly West, but the bandanas we wore were the kind worn by old steam locomotive engineers and train crewmen. That's because they were larger than those worn by cowboys, plus they prevented red-hot cinders from sliding down the railroad engineers' shirts! In our case, they prevented hot brass shell casings from flying into our open collars and sliding down our shirts, resulting in severe burns in some instances. Those bandanas also created a tremendous esprit-de-corps and sense of pride for all Maverick pilots and crewmen.

The price you paid for wearing the bandana was having to fly an old Bell UH-1B or UH-1C helicopter gunship in combat, and it's difficult to adequately describe what that was like. It's excruciatingly loud because the rear cargo doors were fully open, as were the pilots side windows. As if the ship's vibration and extremely loud turbine engine were not enough, two gunners firing M-60 machine guns generated a tremendous amount of noise and hot brass, much of which pinged around the inside of the helicopter like a pin-ball machine! When you added the copilot's minigun, plus the pilot's 2.75 Folding Fin Aerial Rockets (FFAR) to that chaotic din, the noise level was off the chart! Little wonder why most of us are deaf today!

Life at Vinh Long Army Airfield

Life at Vinh Long Army Airfield was probably not all that different from other helicopter bases located throughout the length and width of Vietnam; however, it had one really neat thing. Rather than being a "tent city", like so many men in the companies based north of us had to endure, we had permanent structures! Inside our compound, we also had a movie theater where we could enjoy movies once or twice a week, plus a small library where everyone could share paperback books. I would have to say the more popular books had been pretty rough looking with a lot of dog-ears because of so many folks reading them.

The Mekong Manor was our officer's club, and it was the center of entertainment because that's where we ate, drank beer and indulged in whatever else was available in the form of entertainment. I don't remember the food as being over-the-top delicious, but it was adequate once you got used to the weevils that filled the bread like raisins. A new guy would attempt to pick weevils out of his bread; for the veterans, weevils had become a regular part of their diet. The one thing I didn't like was the limp lettuce! In fact, in Vietnam everything that made up a salad was limp, and draped lazily over your fork! The one time everyone really enjoyed eating at the Manor was on Sunday night. That's the night everyone could pick out their own steak, then grill it on the outdoors grill while drinking beer. Grilling steaks kind of took you home mentally, and it made everyone feel just a bit normal. Sundays were good nights!

There had been three platoons of pilots assigned to the 175th AHC. The First and Second Platoons were the Outlaws and they flew the transport helicopters, or slicks. The Third Platoon was the Mavericks and they flew gunships. Each platoon was made up of anywhere from 10 to 14 pilots, all of whom spent one year in country. There was always a lot of camaraderie among the company as a whole, just as it was with every other helicopter company in Vietnam. I also recall being very protective of our slick pilots during combat assaults. I would like to say the Mavericks had been the best gunship platoon in all of Vietnam, but I'm absolutely certain every gunship platoon in Vietnam felt the same way. Perhaps that's what made us all so proud of what we did!

I'm standing beside cockpit of an Outlaw helicopter (commonly referred to as a "slick") after my first combat assault (11/27/1966).

I'm standing beside a 175th Aviation Helicopter Company gunship named Satan's Playmate at Vinh Long Army Air Field (1967).

CHAPTER 5

Memorable Battles and Events

"The Helicopter War" became the most common phrase that defined the Vietnam War. There's a reason for it, too. In its entirety, in all branches of the military, there were more than 12,000 helicopters of all sizes, shapes, makes, models and descriptions. More than 5,600 were lost. Almost half! The iconic Huey and Cobra helicopters also suffered heavy losses, since more than 3,300 were lost out of 7,000 Hueys sent to Vietnam. Again, almost half were lost! Out of 1,110 Cobras sent to Vietnam, more than 300 were lost.

Based on those sobering facts, if you were a helicopter crewman in Vietnam, your chances of becoming a casualty were far greater than almost any other job in Vietnam. In fact, I read one article that listed the top three jobs in terms of being most dangerous. At the top of the list was helicopter crewman, second was Long-Range Recon Patrol (LRRP) member and last, a tunnel rat.

Elite LRRP teams were dropped off deep into enemy territory by helicopters so they could perform reconnaissance missions that sometimes lasted for days! Those same men often-times had to resort to "running & gunning" tactics just to escape angry hordes of enemy soldiers who had discovered them. Those incredibly brave men suffered heavy casualties, and one of my brother's best friends won the Congressional Medal of Honor, posthumously.

Tunnel rats were usually small-framed, incredibly brave, crazy-assed soldiers who were specially trained to descend into the depths of multiple tunnels chiseled deep underground by an elusive enemy! Typically armed with a .45 pistol and GI-issued flashlight, they would slither into the tunnels often "guarded" by poisonous snakes (usually Cobras) to ferret out enemy troops and destroy supplies! I am still in awe of LRRPS and tunnel rats, all of them brave men and awesome soldiers!

I was a helicopter gunship pilot and my younger brother was a team leader in the LRRPs during the same year, but we both agreed neither of us would have even considered being a tunnel rat! Consequently, I'm not sure about the order of listing, but suffice to say, any of those jobs were tough, with high mortality rates!

The U.S. Army and Marine infantry units took far greater losses than any of those listed above, and I would never detract from what they endured. Those guys on the ground, the "grunts", as they liked to call themselves, were indeed the folks who saw the worst in Vietnam! They were, and still are, the real heroes in a very confusing, but deadly war. But in terms of per capita statistics, helicopter crewmen suffered the greatest losses, with more than 6,000 helicopter crewmen killed in helicopters. Helicopters always seemed to be at the epicenter of the heaviest action. We never left our American brothers-in-arms on the battlefield and that made for some truly dangerous moments!

My First Combat Assault
November 27, 1966

Most aviation companies tried to be nice to newly commissioned Warrant Officers straight out of flight school, especially when they first arrived in Vietnam's war zone. They greeted you, then assigned you to a platoon of young men just like yourself, all of whom became closer than brothers! Then they proceeded to provide orientation flights in the local area As-Soon-As-Possible (ASAP). They needed your services as a pilot right then because it was rapidly turning into a "Helicopter War", and there was a critical shortage of pilots needed to fill the cockpits of the huge influx of Huey helicopters arriving in Vietnam. Consequently, the first order of business was to take care of such niceties as getting you at least five hours in the local area before declaring you ready for combat assaults. After enjoying a Thanksgiving meal on the 25th, I got all requirements completed by November 26th, then put on the assignment board as copilot for Lt. Alexander in Outlaw Trail, the last aircraft in an assault formation. I'm not sure what the appropriate number of hours would have been to prepare me for a combat assault, but I was thinking definitely more than five! However, senior officers running the show were smarter than I was; they knew no amount of flight time would prepare you for your first combat assault!

I can still recall how nervous I felt during the briefing, especially when the subject of "what to expect" in terms of weaponry were discussed! The briefing officer topics were such things as "there are reports of heavy automatic weapons fire in this area" as he pointed to some unknown point on the map, or "we have reason to believe the enemy has repositioned mortars close to the LZ"! They always seemed to "suspect", have a "reason to believe", due to some "reported" this-or-that, or some other cliché that never added much to the briefing! They seemed to thoroughly enjoy using any term that fit the occasion in describing enemy positions and weapons! Their cliches also scared the hell out of you, especially if it happened to be your first combat assault!

Warrant Officer Phil Reichard and I were the only "newbies" in the Outlaws, so that was our first combat briefing. I remember looking around at the faces of the "old" veterans, who's ages probably ranged anywhere from 19 to 30, to see if they were exhibiting any signs of fear that might give me a clue about what to expect; but there were none - just blank faces! Hmmm, maybe it wouldn't be so bad after all, I thought. But I sure was glad to be sitting down, because the longer the briefer talked, the more my legs assumed the viscosity of jelly! As soon as the briefing ended, Outlaw Lead ordered all pilots to "man your helicopters and prepare for start-up at so-and-so time!" I recall thinking I had been training for more than a year for that exact moment! It was time to test my nerves, skills and "manhood", but I had a problem. I was scared as hell!

Seated in the cockpit, I watched five Maverick gunships depart Vinh Long, headed toward the LZ located about 30 kilometers away so they could recon the LZ and surrounding area for any signs of an elusive enemy. Soon enough, the Outlaws were ordered to start

up and, before long, all helicopters had rotors turning at 100%. All Outlaws checked-in ready to go, so Outlaw Lead made a call to Vinh Long tower for repositioning to the runway to pick up the "sticks" of troops. "What are sticks?", I asked. The aircraft commander explained a stick was a group of 10 men for each helicopter. Then we were cleared to reposition with each helicopter landing near a stick of troops. The pilot of our helicopter called out, "Lead, this is Trail. All troops are loaded". Outlaw Lead responded with a "Roger that", then called Vinh Long tower for takeoff clearance. We were off and running on my first combat assault! I had endured a year of flight school for that exact moment! Unfortunately, I would've preferred to be just about anywhere rather than sitting in that Huey cockpit!

Shortly after takeoff, Outlaw Lead called Maverick Lead, "Maverick Lead, Outlaw Lead. We're off Vinh Long for RP (Reporting Point) 1". Maverick Lead responded with "Roger that. Mavs will pick you up at the RP, then escort you into the Landing Zone (LZ). Saw some activity so the LZ might be hot". "Roger that". Crap! I was trying to process that! Did Maverick Lead just say the LZ might be hot?! I glance over to the pilot and see no concern whatsoever. In fact, everybody's so damned calm! Am I the only one who gets it? We could be shot at! The flight of 10 Outlaw slicks droned steadily forward. All too soon, Outlaw Lead reported in at the RP.

Very calmly, Outlaw Lead said, "Outlaw flight is at RP 1, turning to heading of 0-7-0 inbound to the LZ. Starting descent in five minutes". Maverick Lead calmly acknowledged him with a simple "Roger". Soon, we're starting our descent into the LZ, an insignificant looking rice paddy that seemed pretty close to a lot of tree lines and thickets. Just then, Maverick Lead announced his fireteam would commence firing into the tree lines just to "keep the enemy's heads down"! An overhead Forward Aircraft Control (FAC), a small Army airplane, chimes in with "I'll fire some Willie-Pete" (white phosphorous rockets) into the area to "mark the area in case we have to call in the fast-movers (jets)".

On final approach into the LZ, I had a panoramic view of the LZ where gunships were swarming like angry bees, firing rockets into tree lines where they exploded into fireballs and white smoke, then more radio calls alerting gunners "to stay alert but don't fire unless we take fire" and "throw smoke if receiving fire"! My job as copilot was to monitor the instruments and be prepared to grab the controls if the flying pilot is hit; aside from that, I had very little to do unless the pilot got hit! Sitting much like a bump on a log, I remember thinking "the only time I have left to live is however much time there is between where I was sitting, and the LZ!" I wasn't scared so much as being resigned to that simple fact. There wasn't anything I could do, yet everything seemed so calm; and so surreal! But everything was about to change.

The Outlaw helicopters in a two-vees-of-five formation began settling into the muddy rice paddies to disgorge their "sticks" of troops. That's when I saw little splatters of water popping up just ahead of us, then between us and the next helicopter. The door gunner on the left side, my side, opened up with his M60! The radio was filled with shouts of pilots in other aircraft screaming "receiving fire from 9:00 o'clock, 10:00 o'clock, now 3:00 o'clock!" Damn! We were taking fire from everywhere! It was difficult to understand the calls, since the background noise of M-60 machine guns was so thunderous! Maverick Lead told the Outlaws to "get the hell out of there!" I agreed; I wanted to get the hell out of that LZ, too! Then "thunk, thunk, thunk!" Tinny sounds of bullets slammed into our tail boom! Our gunner seemed to be firing with even greater fury! Finally, Lt. Alexander announced to Outlaw Lead that "all troops are out of the aircraft", and Outlaw Lead responded with "Outlaws coming out of the LZ"! Maverick Lead's response was a double-click with his transmit button, and we were soon out of the LZ. Everything became quiet again. Amazing!

The flight back to Vinh Long was uneventful, and the short lifespan I had been contemplating just a few short moments ago had passed. God had granted me an extension for the day, or until our next flight into that LZ. After we had landed and shut down, we counted nine bullet holes in the tail boom alone! Other helicopters had similar battle damage. The mechanics were very philosophical about it; kind of like "been there, done that". They nonchalantly began patching each one of them, then spraying the patches with lime-green primer to keep them from rusting. They would all be properly patched and painted when the helicopters went in for major maintenance. In the meantime, we had a war to fight and could be called back into action at any time!

Stuck in the Rice Paddy
December 5, 1966

Not long after arriving in Vietnam, I flew what was probably the scariest mission I was involved in as a slick pilot. On that particular day I was flying as copilot with Captain Ray Leuty, our 1st platoon leader and experienced Outlaw pilot on yet another combat assault that seemed to happen with great regularity in the Mekong Delta. It was a rather small operation that required only ten Outlaw slicks, and we were flying as Outlaw Trail, the last helicopter in the formation. Outlaw Lead always led the helicopter formation, but it was Outlaw Trail's job to keep him informed of how the flight was progressing, when troops were loaded, and when all troops were off the helicopters in the LZ. It was a very important position, because Outlaw Lead made his decisions based on information passed along to him from Outlaw Trail.

We made two initial lifts into the LZ earlier in the morning, and after that was accomplished, we flew back to the staging area, shut down and began standing by until our services would be required later in the day after the troops had completed their sweep through the Area of Operations (AO). The gunships flew missions all day long, but only one or two slicks might be called upon to pick up wounded troops, or carry in supplies. It was always the "luck of the draw" which aircraft & crew went on such missions; the rest of us simply continued to wait in the hot, humid weather environment of the Delta. Finally, late in the afternoon, we were given the

signal to start engines and prepare for the flights into the PZ. What had been the Landing Zone, or LZ, was now the Pickup Zone, or PZ. Sometimes, the PZ was just as deadly as initial landings in a hot LZ.

The sun was getting low in the western sky when we finally took off to pick up the remaining troops who had slogged through mud throughout the day. I really didn't know what they accomplished, but I did know they would be tired. The first flight into the PZ had gone as well as could be expected; it was a quick in and out while the Mavericks covered us. The troops boarded our helicopters so quickly, we were out of the PZ within a minute or so. In fact, it had gone so well, it probably gave all of us a false sense of security before the last lift was accomplished. Unfortunately, things can change in a heartbeat, especially in combat situations!

Since Captain Leuty and I alternated flying duties, I was at the controls when we were on a final approach for our final trip into the PZ and I noticed most of the remaining troops had already begun moving out of the tree line, heading directly for their designated pickup sites. Unfortunately, when it came their turn to be extracted, there were no troops left to cover the last pickup in the PZ. The enemy had patiently waited for the last group of the Army of Republic of Vietnam (ARVN, pronounced "Arvin") troops to leave the protective cover of the tree line where they had provided cover for the first lift. They had also waited until the Outlaw slicks were slow enough to settle vertically into the rice paddies to gather the remaining troops. That was always our most vulnerable position, and that's the exact moment they chose to open fire on us!

The closest tree line lay to the right of the Outlaw formation, at our 3:00 o'clock position. Unfortunately, all the Outlaw slicks had to hover around just for a second or two before landing to ensure they wouldn't sink into the rice paddies, thus forcing us in Outlaw Trail to take what was left. Not only did we have to land much closer to the tree line than we would have liked, but our helicopter also sank much deeper into the mud. That's just about the time all hell broke loose!

What had been a tranquil PZ on the first lift had suddenly erupted in heavy automatic weapons fire coming from the tree line at our 3:00 o'clock position! Other pilots began screaming "receiving fire" over the radio and the troops clamoring aboard our helicopters were being hit! Just as our gunner opened up with his M-60, some gunners in other helicopters also opened up. You could even see enemy soldiers flitting about just inside the tree line, trying to position themselves closer to our helicopter to improve their accuracy! After what seemed like an eternity, all troops still alive were on board so we made a call over the radio informing Outlaw Lead that "all troops are loaded! Let's get the hell out of here!" Outlaw Lead immediately responded with "Mavericks, Outlaws departing the PZ! Hit that tree line at 3:00 o'clock!" Maverick Lead responded with "Will do!"

Responding to "getting the hell out of here", I began pulling the collective lever upward to increase power into the blades, thus enabling us to lift off! But nothing happened, except the rotor RPM drooped! A helicopter's main rotors must maintain 100% operating efficiency to fly. If the operating rpm should "bleed off", or decrease, the helicopter will not fly! The other nine slicks flew out of the PZ, leaving us all alone in the rice paddy. We had settled deep into the mud, and water from the rice paddy had filled the empty belly of the helicopter, rendering us too heavy to fly! We had suddenly become a single duck, sitting in a pond surrounded by God only knows how many hunters, all with one single goal: destroy our helicopter and kill everyone in it! Or worse, take all of us as prisoners!

Captain Leuty always preferred to fly in the left seat as aircraft commander; consequently, I was sitting in the right seat, and that just happened to be the side the enemy was firing at. Our gunner on the right side had been steadily firing, and was joined by ARVN troops in the back of our slick who began firing back with their own weapons! We were in a bad situation! I could even see the enemy darting and maneuvering, trying to get closer to us! Sensing our dilemma, Captain Leuty took over the controls and kept trying to maneuver the helicopter out of the mud, but it was no use! We weren't going anywhere, and the enemy was determined to kill a helicopter crew! I even grabbed my personal .45 pistol and began firing, but they were already smelling victory! Fortunately, the ebb and flow of combat can shift very quickly, even when things look the bleakest.

The pilot flying as Maverick Lead that day was a man by the name of Major Jerry Hileman, a real combat leader; he was the kind of man who made a difference! He did so that day by leading his Maverick fire team on a low-level rocket attack and strafing run that literally took out the entire leading element of enemy troops! I think the enemy had been so focused on killing us, they forgot to keep alert for the Maverick gunships! On the second firing run, he fired all his 2.75 FFAR rockets, then circled to land between our helicopter and the tree line! While both of his gunners began unleashing torrents of withering machine-gun fire into the enemy, Major Hileman made a call for us to send some of the troops from our helicopter over to his gunship. Reducing weight from our helicopter might enable us to break free from the mud!

Time seems to compress when you're in combat, yet everything I just described was happening at high speed! All of the talking, shooting, troops slogging over to the gunship, trying to get the aircraft free from the clutches of the mud, everything was happening fast, yet my thought process was able to keep up with it. It was all happening at warp speed, yet I was watching in SLO-MO. Finally, Captain Leuty began to work the helicopter free from the mud! Slowly, ever so slowly, the helicopter began lifting free of the mud while water poured from the helicopter's belly! We were going to make it! By God, we were going to live another day!

When we were finally airborne, there was a lot of yelling and screaming just for the pure joy of survival! We would live another day! After we returned to the staging area and shut down to inspect the helicopter, we all made it a point to personally thank the

Mavericks for saving our lives! That was also the day I decided I wanted to become a Maverick pilot. Instead of having to sit in the LZ and pray for divine intervention to save me, I wanted to be in the position to make a difference. Maybe even save lives.

Receiving Fire on the Flight Line
December 17, 1966

Everyone was up early and preparing for another pre-dawn takeoff by performing preflight inspections on our assigned helicopters when a bright flash in conjunction with two very loud gunshots echoed across the ramp, followed by an un-earthly scream! I was on the roof of my helicopter when it occurred, so being very new in country and fearing it was a sniper attack, I took the quickest route down by jumping into a black void toward the ground! It was so dark I couldn't see the ground, but managed to land on my feet in an upright position, even though I twisted my knee and jammed my back in the process. Neither injury slowed me down in my quest to find a bunker or revetment to hide behind! It was deadly quiet for a few seconds until we heard loud moans coming from the helicopter directly across the flight line from my assigned helicopter, but still no one stirred. Only when we heard a voice pleading for help did we dare venture out of our safe areas to rush over to where the shots had come from.

We found one of the Outlaw gunners writhing on the ground in pain! Somehow, he had managed to shoot himself in the groin while mounting his M60 machine gun onto the helicopter's tripod mount. He was in a lot of pain, but also pretty stoic about it; he even apologized for his stupidity. We helped stem the bleeding from two bullet holes in his groin before the ambulance arrived. It wasn't long before he was whisked off to our small hospital, then prepped for medevac to Bien Hoa, and we never saw him again. My logbook entry indicates that incident didn't slow us down since we launched on time and I logged a total of 8.5 hours of flying combat assaults that long day! I did go visit the Doc later that evening, and

he gave me either Anacin or Aspirin, and I recall limping around the compound for a few days, but took no time off since there was a critical shortage of pilots due to the sudden buildup of U.S. troops. We were always short of pilots, so most of us were flying every day!

Massacre at Vi Thanh
December 20, 1966

Vi Thanh. Just the name sparked fear! Located on a major canal leading out of Can Tho towards the southwest, it was of vital importance to the entire region, and both sides knew that. Consequently, there was a South Vietnamese army base with a nice runway that we could stage combat assaults out of. I hadn't been involved in a combat assault there, but it had sparked fear, and horror in me, for another reason. It happened when I was assigned as copilot on one of two Outlaw helicopters that had been assigned missions that carried us down to a small village not far from Vi Thanh. I think the news media referred to it as a VC prison camp, but the VC had come calling the previous night and it had not been a good visit! The village elders had been accused of assisting ARVN troops and American advisors, so punishment had been dealt out in the most horrible fashion. They decided that not only the accused should be punished, but their entire families as well. They then proceeded to massacre every man, woman and child in the families selected for punishment! Entire generations in those small families were totally wiped out. The accused were also beheaded, just to drive home the fact that anyone who dared to challenge them would meet the same fate! I don't know how many innocent villagers lost their lives that night but there appeared to be at least 60 or 70 in all. Even babies and toddlers!

Our mission on that day had been to fly ARVN officers and their American advisors to the village to investigate the events of that horrible night, but they could have saved themselves the time and effort. Those poor villagers were so traumatized and disillusioned, they were never going to recover. The sheer brutality of the VC was

indescribable! Even today, some 50 years later, it is still difficult to think about that horrible day. It elevated hatred to another level!

Mortar Attack on Vinh Long Army Airfield
January 1, 1967

The usual group of pilots had played a few rounds of poker before going to bed because after completing a long day, we were facing another early morning takeoff. I was sound asleep when the first mortar round hit very close to the crew quarters, resulting in a huge explosion and bright flashes of red-hot shrapnel, capable of tremendous damage to whatever it hit! Even though I had been sound asleep, I was jarred instantly awake, then lay there wondering what the hell was going on! I had been in country barely a month when that happened, so I was a bit confused, even as other pilots were charging down the hall toward the bunker, yelling "Incoming! Incoming!" I jumped up and managed to get my flight suit on, just as other rounds began dropping in all around the living quarters! Multiple explosions and flashes were silhouetting everything in an eerie fashion. A round slammed in very close to the hootch and getting my boots on was no longer my priority! Barefooted, I scrambled out behind the last pilot charging down the hall and we scrambled into the bunker in a dead heat!

Sitting inside a bunker during a mortar attack for the first time was not fun! Even though there were no lights in the bunker, we could still see each other with mortar rounds dropping around us, providing all the illumination we needed! No one was talking, and everyone's teeth were clenched. Some were mumbling prayers! We counted 40 rounds that dropped in all over the airfield! In all that noise, I heard the unmistakable sound of a Huey cranking! I was stunned! My God! What idiot would be cranking their helicopter while mortar rounds were dropping in all around you? Whoever was cranking had company, because between all the mortar rounds going off, I could tell there was more than one helicopter taking off. Someone in the darkness quietly said "The Mavericks are airborne.

The mortars should end soon". And sure enough, after what seemed like hours, it ended! The silence was deafening. The mortars had finally stopped, but our night was far from over.

Shortly after the mortars stopped dropping in, Captain Ray Leuty, the Outlaw's 1st platoon leader, ordered all pilots to get their gear and report to the flight line ASAP! He had gotten a call from flight-ops telling him the Mavericks had made contact, and were calling for troops to be flown into the area. In the wee hours of the morning, several Outlaw flight crews found themselves in the air with troops on board to punish the VC. It was so dark, I don't know how we managed to even find each other, much less a rice paddy to land in; but somehow, we did. Successful or not, I was just glad to survive that long night! I was also impressed how those Maverick flight crews had run through all those mortar shells dropping in to get to their gunships in the air! I also thought they were really crazy! Crazy, but extremely brave! Just like that, they had become my heroes.

That was my first mortar attack and it was absolutely horrifying! I had barely been in Vietnam a month and nothing can prepare you for a mortar attack! Being alone in your bunk in a dead sleep after a long, hard day of flying, then being awakened by loud explosions and veterans racing through the hootch screaming "incoming" is a rude introduction to the realities of war! Self-preservation kicks in, and even the bravest men raced for the safety of the bunkers! It was another reminder it was going to be a long year. Plus, after the adrenaline stopped flowing, my feet began to hurt after racing across all those rocks and stones!

If someone had told me I would be joining those crazy-ass Mavericks racing through all those mortar rounds just a scant two months later, I would have laughed in their face!

Operation Deckhouse V
January 5, 1967

In the early morning hours of January 5th, I was sound asleep, but something made a noise somewhere close by. Just a whisper of noise that snapped me instantly wide-awake, and fully alert! Soon, I could hear quiet footsteps coming down the hallway towards my little cubicle. In just a few seconds, the hanging strands of bamboo acting as a door were pulled back and a voice whispered, "Mr. Williams, Mr. Williams. Are you awake?" I replied "Yep; now I am!" "Sir, they need you over at flight-ops right now! It's important!" Struggling out of bed, I glanced at my watch. Damn! It was only 4:00 o'clock in the morning! Are you kidding me? After slipping on my uniform and boots, I grabbed my flight gear and struggled over to the small building that was home to Flight Operations at Vinh Long Army Airfield.

Stepping inside, I was surprised to see Warrant Officer Jon Myhre, a senior Outlaw pilot with whom I had flown numerous missions, already standing there. Jon said, "Willie, we have a top-secret mission to fly, and I requested you as copilot. Are you OK with that?" The last remnants of cobwebs suddenly evaporated from my brain. I was wide awake! As in totally awake! Top-secret missions could get you killed! Plus, I had already learned: you never volunteer for anything in the military! But Jon was asking, and I respected him a lot. So, I said sure; and the briefing began.

A joint USMC and South Vietnamese amphibious landing was going to commence in the early morning hours the next day, January 6th, and the operation would continue through January 15th. Named Operation Deckhouse V, it had been planned for some time and was the only amphibious assault ever made outside the USMC assigned I Corp area up north. They were hoping to kick-ass, capture tons of clandestine supplies and just create all kinds of havoc in general! Our job was to fly to some small village located somewhere along the South China Sea coastline to pick up two special-ops guys, then carry them out to the fleet. As in offshore, in the ocean for God's sake! Those two guys had infiltrated into the local population and knew everything about the area; consequently, they were essential to the success, or failure, of the operation. That trickled down to Jon and me, because it meant the success or failure of that entire

massive operation had suddenly been dropped onto the shoulders of two lowly Warrant Officers that had to fly over lots of water to keep the mission on track! I was already regretting my decision!

After signing off for the Standard Operating Procedures (SOP) of the day, complete with codes, call-signs and countersigns, we picked up our flight gear and, using our flashlights to point the way, walked down to Jon's assigned helicopter, Outlaw 17, or Snoopy. The crew had already arrived and between the four of us, we soon proclaimed everything ready to go. It was pitch-black when we departed Vinh Long, and had a long flight ahead of us before hoping to find some non-descript little village located somewhere along the coast of the South China Sea. It took our best navigational skills to find it in the early morning darkness, but we did!

We circled once and sure enough, right on cue, a flashlight blinked on and off. Aha! Our boys were there, after all! At least we hoped so! Just in case, our gunners automatically slid the cargo doors open, then locked and loaded their M60 machine guns. After landing, someone whose face we couldn't really see walked up to Jon's door window and told us to shut down and standby. Damn! Some things never changed in the Army. It was always "hurry up and wait!"

The sun rose not long after shutting down and after a couple of hours of standing by, our bellies began to growl, as in "feed me!" We'd already missed breakfast, so we broke out our C-rations, then argued about who got the beanie-weenies, fruit cocktail and the small can of "pound cake". Canned biscuits were un-edible, so they were always used for fires to warm your meal, or brew hot chocolate. All it took was a jigger of JP-4 the crew chief drained from the fuel drain located just underneath the belly of the Huey. Pour the JP-4 in with the biscuit, light it up, and "voila", you had heat! Everyone was soon chowing down and the growling in our bellies ceased. Life was good!

Just about the time we finished our C-rations, our two special-ops guys walked up and announced they were ready to fly out to the fleet located about 40 miles at sea. "Are you kidding"? "Forty miles out to sea"? "Yes", they said; "To the U.S.S Iwo Jima". It was the U.S. Navy's first amphibious assault ship designed specifically for helicopters. Fortunately, they provided us a frequency to contact the Iwo Jima as soon as we went "feet wet", as they say in the Navy. Sure enough, the Iwo Jima responded to our request for directions by providing us vectors to find them. The entire crew tried hard not to think about the fact we had no life jackets, no floats on the helicopter, no nothing! Not even a rubber ducky to cling to! But luck was on our side and we soon made out several ships in the distant haze. The closer we got, the Iwo Jima became more distinct and we were soon given permission to land. Boy, they were taking a chance since neither Jon nor I had ever landed on a U.S. Navy ship! I guess we were issued a "combat" waiver.

We approached from astern and fortunately, the flight deck was clear except for a couple of old CH-34 helicopters, the Marines main helicopter in Vietnam. Jon was flying, and the helicopter's hovering wasn't very stable with all the wind he was having to contend with. But the grizzled old "master of the deck" really didn't care! He was holding a flag in each hand, directing Jon to land! Little did we know he was timing us to land in rhythm with the rise and fall of the Iwo Jima in the big swells of the ocean. When Jon finally touched down and lowered the collective, I could see beads of sweat on his brow. The old sea dog walked up to Jon's open window and we both thought he would be of assistance to us. I guess he was, sort of! His exact words were: "Son, if you don't do what I tell you next time, we're going to throw this son-of-a-bitch into the sea! Got me?!" He gave really good instructions, because we both said at once, "Yessir, we do!"

Before we could shut down or screw up anything else, a Marine officer ran over and told our passengers we needed to fly over to the U.S.S Canberra, a cruiser delegated to be the Flagship for the operation. When we were ready to go, Jon followed the old sea dog's instructions to the letter, and we were soon flying across the

China Sea toward the Flagship. The Canberra didn't have the big flight deck like the Iwo Jima, but it had a nice little helipad on the ship's stern. Landing on it was a "piece-of-cake" as Jon said, and we were soon secured to the ship. The two special-ops guys were whisked down below to see the Admiral, or Marines, and we were left under the guidance of an older Navy Warrant Officer, like us. He was a very pleasant fellow who asked if we had eaten. Jon and I looked at each other, then in unison, we said "no". Well, lucky for us! We were just in time for brunch! Brunch sounded really elegant!

We were soon down below in the Officer's wardroom for brunch, and it was very nice! We weren't dressed properly either, since everyone else was dressed in the Navy's immaculate white uniforms. Standing there in our dirty, mud-stained flight suits, Jon and I had all the presence of poop-in-a-punchbowl! A young Filipino waiter took our order and, even though we had just eaten, we ate again! We ate a lot - again! The Warrant Officer then asked if we'd like to watch their 5-inch naval guns in action. "Well, certainly!" Before long, we were standing somewhere up by the Admiral's bridge waiting for the action. I guess someone had requested naval gunfire, and they were going to provide it! How exciting!

Just about the time the Warrant Officer said "open your mouths wide and hold your ears", those big 5-inch cannons let fly with their first rounds. Oh my God! Fire belched a good 100-feet out from the barrel, smoke obliterated the sky and all I heard was ringing. I was totally deaf, and I was afraid to close my mouth because I just knew the recoil had taken out all my teeth! What the hell did he want us to open our mouths for anyway? Since I couldn't hear what he was saying, I tried to read his lips. The man wanted to know if we wanted to watch another round of cannon fire. Holy mother of God! No! I was speaking for both of us because Jon was still gasping like a guppy out of water. Plus, I thought his teeth were gone!

Before they could reload their 5-inch cannons, we scurried back to Outlaw 17 just as our special-ops guys returned to tell us "Thanks to you, our mission was a complete success!" Well, now, that was right nice to hear! We launched, flew back across the China Sea, dropped our spook-types off, then headed for Vinh Long. We arrived just before dark, totally exhausted after such a long day. As we were gathering our gear, several crewmen walked by and one of them asked where we'd been. Since we'd been given strict orders not to discuss that mission with anyone, Jon said "Nowhere in particular; same old stuff." The other crew chief spoke up and said "They took some guys out to the ships with all the Marines on board". So much for secret missions!

Operation Deckhouse V lasted from January 6 through the 15, 1967. It was a total disappointment by military standards. The results were 21 VC killed and a few weapons were captured. The price for such a victory was the loss of seven Marines and one South Vietnamese soldier. In fact, at the completion of the operation, it was reported at the daily military briefing in Saigon, commonly referred to as the "Five O'clock Follies", that the operation had "proven unproductive". At least they didn't lay the blame on two Army helicopter pilots who didn't like to fly over water!

Combat Assault at Vi Thanh, South Vietnam
February 15, 1967

Just hearing the name Vi Thanh struck fear in the hearts of veterans who had survived a previous combat assault that occurred just before I arrived in country. Located at the southern tip of the Mekong Delta, the area was a well-known Viet Cong (VC) stronghold that no one wanted to enter, not even the local military. I had already seen the horrors of Vi Thanh, but even though I had heard stories of that particular battle, I wasn't overly concerned. We were briefed on the evening of February 14th, and were informed the Outlaws would be supporting the 336th AHC Warriors, our sister company based further south in Soc Trang. The combat assaults would take place near Vi Thanh, and the mission would be dubbed The Second Vi Thanh. By the time it was over, I think we all preferred it to be called The Last Vi Thanh!

My assignment for that mission was to fly as copilot with Captain Ray Leuty as Outlaw Lead! I would be flying as pilot on the controls leading a formation of ten Outlaw slicks while Captain Leuty would keep track of a multitude of details he had to deal with. Naturally, we had another early morning takeoff to make the long flight to a staging area very close to the town of Vi Thanh. After arriving and attending the briefing, we learned there would be two flights composed of ten helicopters each. The Warriors formed the first flight, the Outlaws formed the second. The Thunderbirds, referred to as T-Birds, and Mavericks were assigned as gunships for the mission.

The T-Birds launched about 30 minutes prior to the slicks so they could recon the LZ, as well as the approach and departure paths in preparation for the arrival of the slicks. Not long afterward, two flights of ten helicopters each were winging their way towards the assigned Reporting Point (RP) as directed by the T-Birds. Each flight had formed up into what we called a "V Formation" consisting of two V's of five in tight formation. It was the typical formation that allowed all helicopters to land, then takeoff at the same time. The Warrior flight was approximately 5 minutes ahead of the Outlaws since they would be the first to go into the LZ.

Twenty Huey helicopters laden with 10 troops each, plus 5 gunships, made a lot of noise, so our assault force was not a surprise to the enemy! I'm sure everyone within 25 miles was instantly aware we were inbound, but it probably really didn't matter. The VC had a spy network that rivaled the FBI! Consequently, there were probably few real surprises when it came to flight operations. I heard Warrior Lead report in to T-Bird Lead at the RP, and he was provided all necessary information for him to take the Warriors into the designated LZ. I continued flying while Captain Leuty copied down the same information that informed us what direction to depart the RP, our altitude, our route, what to look for and avoid, what the LZ looked like and which direction we would land. T-Bird Lead also informed us they had received no enemy fire during the low-level recon of the LZ. It appeared the

LZ would be cold; no enemy fire expected. Great news, but things can change quickly in combat.

A couple of miles directly ahead of me, I could see Warrior Lead start turning his formation of Warrior slicks onto a descending right-hand turn that would allow him to reverse his heading a full 180 degrees so he could land into the wind. He had already passed the LZ that now lay a couple of miles outside my right window, at my 3:00 o'clock. It looked quiet and serene, much like the calm before a storm. It also looked small, which was a concern. We could get ten helicopters into the LZ, but it would force the lead helicopters very close to the tree line, and we didn't like tree lines on combat assaults! That's where the Viet Cong hid, and they were pretty damned accurate with those heavy automatic weapons accompanying them!

Before long, Warrior Lead called out that his flight was on final descent into the LZ. That meant he was only one or two miles out, descending and simultaneously slowing the flight in preparation for landing in the LZ. As Warrior Lead, his plan was to land in the LZ, drop the troops off, and get out of there! He didn't want to be in that LZ more than 60 seconds; if his flight began receiving fire and he stayed in there more than 60 seconds, it was no longer an LZ. It was basically a kill zone where your life expectancy was 30 seconds!

Even though I was leading our Outlaw flight in, at that moment I was literally sitting in Warrior Lead's cockpit, at least mentally. I knew what he was doing, even what he was thinking. I knew, because very soon, my own actions would be a mirror image of his. I was laser focused as I started a right turn in towards the LZ. I began to wonder, "will the LZ be hot, or cold?" A simple question, but oh-so critical! Warrior Lead was probably already slow enough to settle into the rice paddy. He was at his most vulnerable position. Not really flying, yet not completely landed. Somewhere in between. He was akin to a duck on a pond, a small pond! If luck was with us, the LZ would be cold. If not, there's an old saying about "just like shooting ducks in a barrel". Except on that day, it would be helicopters in a rice paddy.

Mirroring Warrior Lead's actions of just a short time ago, I had just started turning our Outlaw flight onto a long final into the LZ, when all hell broke loose! Warrior Lead called out "receiving fire", then ominously, it ended abruptly! Immediately, a chorus of voices from other Warrior pilots chimed in, all of them screaming "receiving fire!" In a heart-beat, our life expectancy had ratcheted down to 30 seconds! Another excited Warrior's voice advised us Warrior Lead had been shot down, along with two other helicopters! I was already on final descent into that same LZ where three helicopters now lay silent in the mud, their blades stopped. No longer capable of flying, they had become olive-drab colored coffins for their dead crewmen, a VC shooting gallery for the wounded. It was our turn to land in that little area of hell that lay just ahead of us! Just as I was thinking it couldn't get much worse, it did!

"Mortars! Receiving mortar fire!" cried multiple Warrior pilots. Damn! Our life expectancy just dropped another notch, to 8 seconds! T-Bird Lead was screaming: "Get the hell out the LZ!" Seven Warrior helicopters staggered out of the LZ, desperately trying to climb out of that kill zone! One of the Warriors was trailing smoke, unable to gain altitude. He'd been hit hard! I was on short final and saw multiple explosions in the LZ as mortar rounds dropped into it! I didn't want to continue, but I had no choice! Those downed airmen and troops were in a desperate battle for their lives, and we had 100 troops on board our helicopters! I saw the T-Birds roll in on what I hoped were the mortar positions. "Please, dear Lord, let them be successful in their efforts!" Ironically, I was praying for the death of others, so we may live. I don't know if God was playing favorites that day, but the mortar rounds stopped dropping in.

I keyed my "mike" to announce the Outlaws were on short final and planned to land behind the three Warrior helicopters sprawled at odd angles in the rice paddy. I also noticed how close Warrior Lead had gotten to the tree line! Damn! He was way too close when

he was hit! Like a duck in a barrel! The LZ was tight, but the Outlaws squeezed their formation a little tighter, then settled into the mud to disgorge their troops. Just a few more seconds and we'd get the hell out of the LZ! That's when mortar rounds began dropping in again! Voices began screaming in my headset about receiving fire and mortar rounds dropping in, but I ignored them. Hell, I already knew we were in trouble! My God! We were sitting in the middle of a kill zone with 8 seconds to live! The T-Birds were busy hammering everything in sight when I heard the Mavericks check in to join the fray! Just when we thought it couldn't get much worse, it did.

A mortar round hit extremely close beside my aircraft and our gunner screamed "I'm hit!" I knew we'd been hit hard because I could actually smell cordite from the blast! Captain Leuty and I were assessing the damage when Outlaw Trail, the last aircraft in our flight, announced all troops were out of the aircraft and the Outlaws were ready to depart the LZ. I immediately squeezed the mike and announced "Outlaws on the go!" as we departed! I wanted to say another prayer, but I was too busy! The Mavericks announced they were alongside us, and that was comforting! Not far from the LZ, I saw a Warrior helicopter sitting in a rice paddy with the blades stopped. Another shock; but then I saw one, two, three and yes; oh, hell yes! I saw the fourth crewmember! All four crewmen were OK! A small victory! One of our empty Outlaws peeled off and dropped in to pick up the crew. There were no rivalries today! We were all best friends, and would become even better friends because some of us still had a long day ahead! But for those Warrior crewmen and troops who lay silently in the LZ, their day had already ended.

At the staging area, there was no time to lick our wounds. Our gunner was tended to, a replacement found, and our helicopter was declared flyable! We were already back in action! The only thing, nobody wanted to go back into action; at least, not into that God-awful LZ that had morphed into a kill zone! But we did; we had to! Again, and again, and again! All day long! We also heard Warrior

Lead and his copilot were dead, as were several other crewmen in the downed aircraft.

During one of the flights back out to the LZ, we heard an American advisor in the midst of the ground battle making a desperate plea for help from the gunships. Every time he keyed his "mike", we could the ferocious battle going on around him. We could hear screams of men locked in mortal combat, plus the rapid staccato of machine guns steadily firing, and grenades exploding nearby! We could hear men screaming and dying over the airwaves.

The fighting was up close and hand-to-hand. Desperately fighting for their lives, the voice on the radio kept pleading for the gunships' assistance! The T-Birds and Mavericks were delivering everything they had! Swarming like angry bees, they were firing every weapon in their arsenal! Grim-faced men in every helicopter sat in their cockpits, silently praying for that unknown American desperately pleading for help! Aboard the Outlaw helicopters were another 100 troops that would soon join the horrific battle in and around the LZ. Maybe, just maybe, they would be in time to make a difference in that desperate battle! I fervently hoped they could help that incredibly brave soldier whose voice we had heard. Unfortunately, I never found out. I like to think our actions made a difference on that long ago day, and that brave warrior survived, as did his men. I can only hope my prayer was answered.

One of the heroes that day had been a young pilot from one of the helicopters that had been shot down. Leaping out of his mangled, twisted helicopter, he began directing his two gunners to set up a defensive position. Shell-shocked Vietnamese soldiers soon rallied around them! No longer able to fly, the downed crewmen had become the tip-of-the-spear in that LZ. The young pilot also used his hand-held radio to direct gunships and jets as to where to drop their heavy ordinance. His courageous efforts made all the difference, and the tide of battle slowly began to turn! The young Warrant Officer saved the day, and was awarded the Distinguished Service Cross (DSC), second only to the Congressional Medal of

Honor. It was a well-deserved award! He later said his best reward was simply making it back to his home in Texas. That, and getting back to his Mama's fried chicken!

As for the rest of us, it was a day when each of us resolved ourselves to the fact we were living on borrowed time. We couldn't possibly survive another flight into that LZ! Between flights, we simply rolled our helicopter's engines to flight idle so we could pee, eat cold C-rations, wash them down with tepid water, and refuel. Some of the pilots and crewmen were pacing back and forth, throwing up whatever they tried to eat; even talking to themselves! They were trying to talk themselves into making the next flight into that LZ! Some would say a prayer, then climb back inside their cockpits, ready to make one more flight. I prayed with them! "Please Lord! Let the next one be the last flight into that killing zone!" There were no atheists in the cockpits that day!

Mercifully, our long day ended just as the sun was setting. The Outlaw's day began at 04:00 AM, and lasted until 08:00 PM. That's when we finally made it back to Vinh Long. It was an incredibly long day, and my logbook entry shows I flew 12.8 hours with Outlaw Lead. It was a weary bunch of flight crews that finally got to bed around midnight. It was also a short night, because we had another combat assault somewhere the next day. Over the next two days, my logbook showed another 9.3 hours. After that introduction to Vi Thanh, I joined the ranks of the older veterans. Just the mere mention of Vi Thanh raised the hair on the back of my neck! It lasted until the end of my tour of duty.

Sgt. Freddie J. Williams, Long Range Recon Patrol
173rd Airborne Brigade, Vietnam
March, 1967

Sometime in January, 1967, I received a letter from my parents advising me that Freddie, my younger brother, had been ordered to Vietnam. They were concerned because they had no idea where he was stationed, nor had they heard from him. That bit of information

came as a shock because Freddie and I had discussed the real possibility of both of us being sent to Vietnam, and neither of us had wanted that. Knowing our parents would have a difficult time with both of us in Vietnam, I advised him to try to avoid an assignment to Vietnam. He, on the other hand, advised me to avoid Vietnam because I had a wife and infant son to look after. Now, that stalemate had become a moot point. Both of us were in Vietnam.

Since he had signed up for the paratroopers when he joined the army, I had a good idea he would be stationed well north of me because that's where the bulk of U.S. troops had been stationed. They definitely had the right guy because even before he joined the army, he was well-known around town as being utterly fearless, and tough as they come! It took me a while, but I finally found out he was assigned to the 173rd Airborne Brigade and had volunteered to join a new group of volunteers called the Long-Range Recon Patrol (LRRP), commonly referred to as LURPS. Those young men were hand-picked volunteers since they would be operating in such well-known areas as the Hobo Woods, Iron Triangle or some other forbidding name. Those areas were well-known, and feared, by U.S. troops who had fought there. They probably remembered them because they had lost so many of their comrades there.

Soon as I knew the general area where my brother was stationed, I set about getting a pass to visit him. As a member of the "brotherhood of Vietnam pilots", it was rather easy hitch-hiking helicopter-style to Saigon's Hotel 3 Helipad in early March, 1967. After landing there, I simply went from one helicopter to another until finding a Huey with the Cowboy logo on the nose. As luck would have it, that crew actually supported the company my brother's LURP team was assigned to! After explaining my intention of meeting up with my brother, they graciously invited me to accompany them to their base of operations so I could tag-along when they flew out to support the LURPs the very next day! Those Cowboy flight crews of the 335th Aviation Helicopter Company were top-notch guys who took good care of my brother and his fellow LURP teammates.

After spending a "boozy" night in the Cowboy Officer's Club bar, I was up early the next morning to hitch a ride on a Cowboy's Huey that was going directly to the LURP base before going on other missions. It wasn't long before I stepped off the Huey and began making my way across a rather primitive base camp that exhibited very few comforts of home. It was a tent-city surrounded with sandbagged walls to absorb shrapnel from exploding rockets or mortar rounds dropping in. Finally seeing a familiar looking silhouette, I casually walked over to a young three-striped Buck Sergeant who was giving out orders to a few soldiers. When he finally turned my way, he just about passed out when he saw his older brother standing there! It was a great moment!

After visiting a while, I determined my younger brother had already become a battle-hardened Vietnam veteran! Arriving in Vietnam at the age of 19, he had turned 20 in January and had already been promoted to Buck Sergeant; plus, he was the team leader of a LURP team made up of five of the toughest soldiers you've ever met, all of them younger than my brother! He had also been awarded the Silver Star, a medal for bravery surpassed only by the Distinguished Service Cross and Medal of Honor. He was an American hero who had not turned 21. Had he been back in the states, he would have been unable to buy a beer. There were many such heroes in Vietnam.

They lived rather primitively, and LURP team members seldom bathed because the Viet Cong had sensitive noses accustomed to jungle smells; an odor of soap or after-shave lotion could result in the loss of an entire LURP team! Each member carried up to 75 pounds of ammunition, radios, batteries and food, assuming you could call Vietnamese rations soaked with nuoc-mom sauce food! Nuoc-mom was made from fermented fish drippings with a pungent smell so strong it actually gagged most American soldiers! "Saddled up" with such trappings of war, LURP teams were typically inserted into specific jungle insertion points just before darkness set in. Silently as ghosts, they began infiltrating enemy-infested territory searching for enemy positions, communicating only by sign-language for days on end. My brother made more than 40 such insertions during his one-year tour, most of them deep into enemy territory.

I had a great day visiting with my brother, but just before dusk, I heard the familiar "wop-wop" sound that heralded the approach of three Huey helicopters inbound to pick up my brother and his team for their next mission. As the Hueys settled in and landed like a covey of quail, I silently watched the battle-ready LURP team troop out to the Huey and jump on board. Being the last man to board, my brother turned towards me for one last look, then gave a thumbs up as he clamored on board. Watching that Huey carry my brother and his team into harm's way was a very difficult time for me, and tears quickly filled my eyes. It was made worse when I heard one of his team members had been wounded on that particular mission. I'm still thankful my parents never knew how dangerous it was for their two sons in Vietnam. I also thank God for bringing both of us home.

My brother was a veteran in a company of heroes, and one of his closest friends, and fellow LURP team leader, was Staff Sergeant Laszlo Rabel. A tough Hungarian immigrant, he would later be decorated with the Congressional Medal of Honor for bravery after his LRRP Delta Team had come under attack while on a mission. Seeing an enemy grenade land inside his team's defensive perimeter, Sergeant Rabel immediately leaped on the grenade just before it exploded, killing him instantly. His heroic action saved the lives of all his men. Such was the caliber of men my brother served with.

Like most combat veterans, my brother doesn't talk much about his time spent in the jungles of Vietnam; perhaps it is too painful to remember. However, I assure you he most likely uttered a few prayers during his time in Vietnam. He might also have turned to his close friend, the Arbitrator. That was his specially-designed knife that could split hairs with its glistening, well-honed blade. Collectors of Vietnam war memorabilia have offered thousands of

dollars for a knife such as his. But a friend that accompanied you on every combat mission, and perhaps saved your life, has no price tag. It is priceless! The Arbitrator may well have been my brother's savior, but it displays no conscience or remorse. It now rests in a place of honor, in silent repose; but like a sleeping Cobra, it is still lethal.

Easter Sunday
March 26, 1967

Saturday, March 25, 1967, had been just another day in Vietnam. As a gunship pilot, I had spent a good part of it logging time in a UH-1B Maverick gunship in support of a mission somewhere up north, near Saigon. Later in the evening, we were informed there would be some operational requirements for a few of us the next day but, overall, we didn't expect to be that busy. That was a good thing, because the next day was Sunday; Easter Sunday to be exact. A day of peace.

The event that changed everything occurred sometime late Saturday night when a large VC company attacked an ARVN outpost, launching a battle that continued throughout the night into the early morning hours of Sunday, March 26th. The mere fact the battle had lasted so long was cause for concern because everyone instinctively knew an attack of that magnitude occurred only when it had been well planned, and expertly executed, usually by a large force of VC soldiers. We also knew an attack of that magnitude dictated a heavy-handed response from ARVN military units and their U.S. Army advisors, and they would need transportation. Transportation that would be provided courtesy of the Outlaws and Warriors.

Warrior slick helicopters and T-Bird gunships arrived from Soc Trang early Sunday morning for a briefing drawn up during the night to counterattack the enemy force. The Outlaws of the 175th AHC would be making the first lift into LZ Alpha, but the T-Birds would provide primary gunship cover for the first lifts. Maverick gunships would relieve them on station about one hour after their departure. It was a simple plan for another combat assault. We had done a lot of them. There was no reason to believe this latest one would be any different.

Interestingly, there was a different kind of helicopter in the 175th being utilized that day. Nick-named Smokey, it had a smoke generator that would pump out a heavy layer of smoke while flying just parallel to the tree lines normally surrounding the LZ, thus obscuring the vision of VC machine gun crews. Having flown with other smoke generating helicopters, I can testify that it was a special effect greatly appreciated by all slick pilots! Flown by veteran Outlaw pilots Mike Hershey and Ron Petty, the mission on Easter Sunday was its debut.

My own assignment for Easter Sunday was to fly as copilot with a new officer who was a relative newcomer to the Mavericks. Further complicating things was the fact he had also assumed the duties of Maverick Lead, as well as platoon leader of the Mavericks. In achieving that lofty status, he had deposed Captain Joseph Moffett, a well-respected combat leader of the Mavericks. I wasn't happy about having to fly as copilot with the new Maverick Lead, but Captain Moffett had made the aircraft assignments and had been adamant about me flying with him. His concern was the new Maverick Lead wasn't quite up to speed for what would be one of his first missions as Maverick Lead, even though he was on his second tour in Vietnam. Bearing all that in mind, I had concerns. The good news was the mission would probably be the type where there wouldn't be too much happening. The type of mission they called a "milk run", or, as Warrant Officer Jon Myhre always said, "A piece of cake, Willie; a piece of cake".

The morning of Easter Sunday wasn't unlike any other morning at Vinh Long. I recall it was a beautiful morning, and just another day I could X-out a date on my Date-Estimated-Return-Overseas (DEROS) calendar! After an early breakfast, I joined other flight crews for the early morning briefing, then we broke up into individual crews

and went to the flight line to preflight our assigned helicopters. All crews scheduled to make the first flights began strapping in while most Maverick crewmen made their way back to the company area for that all-important last cup of coffee. It was an odd feeling because usually, when the 175th was the primary company for the mission, the Mavericks would startup earlier than the slicks to be first on station. But not so on that day; that had been the T-Bird's job.

The T-Birds launched about 07:15 AM, the Outlaws and Warriors a short time later. Maverick crews had approximately an hour to kill before taking off to relieve the T-Birds on station. That gave us time to catch up on the latest scuttlebutt, or fill our bellies with whatever snacks we could find, then wash it down with black coffee. Some Maverick crewmen had already started ambling toward the flight line. Everything changed in dramatic fashion when flight operation personnel ran out and began screaming for the Mavericks to get airborne! Quickly transitioning from ambling to scrambling, we began racing toward our assigned gunships where our flight gear was already stowed in place. We were also trying to figure out what was going on!

When I arrived at the aircraft, I saw the crew chief and gunner were already there. Jumping into the copilot seat, I was just starting the engine when the new Maverick Lead arrived at the aircraft huffing and puffing, then jumped into the pilot's seat. Soon as we were ready for takeoff, I called our wingman, Jack Bryant, for an "up", acknowledging he was ready to go. Getting that, I switched over to Vinh Long tower frequency and requested an immediate takeoff. We were cleared and soon as we were airborne, we switched to company frequency, hoping to get some idea about what was going on! It didn't take long to figure things out, because various frequencies were alive with incessant chatter! We also heard Mike Hershey making desperate calls as he and Ron Petty attempted to nurse their badly shot-up Smokey helicopter back to Vinh Long. As it turned out, the only thing we didn't need was directions to the LZ. Even though we were still some distance away, it was well-defined by a large black cloud of thick, black smoke, courtesy of a dying helicopter's JP-4 fuel fire, spiraling upward into a clear, blue sky!

It all seemed so damned surreal! A few short minutes ago, we were all joking around, laughing and enjoying coffee, even talking about the cold beer and grilled steak we were going to enjoy that very evening! Sunday was the one day of the week when we could go to the Mekong Manor, grab a beer, select a steak, then proceed to grill it ourselves! Now, we were pulling the guts out of our heavily loaded Huey B model helicopter as we streaked toward that black column of smoke! Smoke that could very well be a funeral pyre for fallen Outlaws! We chose not to think about that.

Along the way, we flew low over a Warrior slick and T-Bird gunship, both sitting in rice paddies just a few miles apart! They had been hit hard in the LZ but managed to get out of the kill zone. Their crews waved at us to signal they were OK, a small victory! We could still hear Smokey's desperate radio transmissions. They had been hit hard on their first run across LZ Alpha! Outlaw pilots Mike Hershey and Ron Petty had their hands full flying the helicopter since the hydraulics were shot out, a testament to the accuracy of that deadly fire in LZ Alpha. It was a struggle, but Smokey made it back to Vinh Long. Both pilots then jumped into another helicopter and returned to the battle! The downed crewmen in LZ Alpha needed everyone's help, and they were prepared to risk it all!

As we got closer to the area, we checked in with Outlaw Six, Major Meehan's call sign as overall mission commander. Flying overhead the LZ, he had an excellent bird's-eye view, so he quickly gave us a Situation Report, or SITREP, and it wasn't good! The first lift of Outlaws had been hit hard by multiple heavy automatic weapon fires just as they had settled into the muddy LZ to disgorge their troops. Nine Outlaw helicopters managed to stagger out of the LZ, leaving one behind. It was Outlaw 17, assigned to Warrant Officer Jon Myhre. Jon, a guitar-playing, well-respected aircraft commander and Jim Martinson, the copilot, comprised the pilot crew that day; Joe Watson and Mike Redd were crew-chief and gunner. All

were missing. Making matters worse, a Dustoff medevac helicopter had darted in to rescue the crew of Outlaw 17, and they had also been shot down! The helicopter had been transformed into a pile of molten metal and ashes whose remains were marked by thick, black smoke! The same smoke leading us to the LZ.

Listening in to the incessant radio chatter, it seemed everyone was pleading for help! Outlaw Six wanted more gunships over LZ Alpha! He was screaming for that even as I was trying to spool my mind into fast-forward mode so I could digest the overwhelming series of events that had unfolded in the past 30 minutes or so. Trying to monitor multiple radio frequencies didn't help, either! I was jolted back into reality when we soared over LZ Alpha for our first birds-eye look into the deadly maelstrom below. It suddenly became very real! It wasn't a nightmare! Another booming voice over my radio shook the final vestige of uncertainty. It was Delta Six, Colonel Jack Dempsey, commanding officer of the 13th Aviation Battalion. He was also demanding our presence!

Our sudden popularity in a time of such urgency created an overwhelming task for me, made worse by flying with a new Maverick Lead! Even though he was on his second combat tour in Vietnam, it took an inordinate amount of time to adjust to the rapid fluidity of combat, especially as a gunship pilot. Whether flying as Maverick Lead, or functioning as a gunship fire team leader, he has a tremendous responsibility to absorb and respond to rapidly changing events on the battlefield! It required lightning fast, calculated responses to control the ebb and flow of life-or-death situations. He must also be able to fly his heavily-armed gunship while monitoring, and responding to, multiple radio frequencies, plus maintain constant contact with his wingman. Making gun runs to unleash 2.75" Folding Fin Aerial Rockets (FFARs) and fiery streams of machine gun fire into enemy emplacements was the easy part! Unless, of course, one strenuously objects to heavy automatic weapon fire being directed toward your gunship regularly, with horrific accuracy! Much like it was on Easter Sunday.

Unfortunately, the man filling the boots of Maverick Lead was not up to speed! He was functioning in a "one-potato-later" mode just to keep up with, and process, rapidly changing events while flying a heavily loaded gunship. Just as I was thinking "What am I doing here", I remembered Captain Moffett's advice of "Just do your job, Willy!" He had confidence in me, or I wouldn't have been there in that situation. I was more experienced in gunships, so I was better prepared to handle the heavy-duty life or death situation we had found ourselves in. Everything came into sharper focus, and I was mentally prepared for the next step, whatever that was! It was game on, for all of us!

The first order of business was to find out just what type of enemy forces we were dealing with. After seeing firsthand the battle-damaged and destroyed helicopters strewn about in the LZ, it was stunningly obvious we weren't dealing with some rag-tag group of Viet Cong! They appeared to be experienced combatants who were well prepared for battle! The question at hand was whether or not they were still manning their weapons, or had they melted into the thick jungle so prevalent along the rivers throughout the Mekong Delta, their typical method of operation. For them to stay and fight meant they were well prepared to do so! That's why we were really hoping they had melted into the jungle!

Maneuvering for a recon by fire attack, I advised my wingman of my intentions, which he acknowledged. With total trust in my wingman, I planned to roll into a steep dive to perform what we called recon-by-fire, then continue our attack by maintaining a tight cartwheel maneuver overhead, thus enabling us to keep firing into the enemy position. My intent was to keep the enemy's heads down if they were still there, thus allowing the downed airmen and ARVN troops to put some distance between themselves and the deadly fire coming from the tree line.

Just as I was planning the best way to roll in, Delta Six questioned again whether or not he was cleared to start inbound from his circling orbit to "get my boys out". Shaking my head vigorously

side-to-side, Maverick Lead responded by advising Delta Six "negative". We needed to recon that tree line first! Finally, in position for my first gun-run directly into the tree line, I nosed the helicopter over into a dive and started inbound! We held off firing at the beginning of the dive, intently listening for incoming rounds. There had always been the chance they had melted deeper into the tree line, and had become ghosts. It was ominously quiet at first, especially seeing the carnage laying beneath us. Just as I was thinking the enemy might have disappeared into the jungle, everything changed!

The entire tree line exploded with heavy automatic weapons fire! Accurate fire! Not only could we see tracers and hear rounds snapping by us, we could also hear a few well-placed rounds thumping into our gunship! We instantly responded with rockets and machine gun fire! My wingman screamed he was taking hits as well! Even though both gunships had unleashed heavy fire into their positions, there was no end to that accurate heavy automatic weapons fire! I called out "breaking left" into my "mike" as I initiated a hard turn, expecting my wingman to cover us, then execute the same maneuver himself. That was a critical time for gunships since we were exposing a big, flat belly toward the enemy and couldn't return fire except for the gunners. Having a wingman covering your break was critical, but soon as I made my break, the enemy shifted some of their heavy fire into him as well! Whoever they were, those guys were good! Damned good!

None of the gunners on either helicopter waivered in their efforts to cover us. All were unleashing torrents of machine gun fire from their hand-held M60's directly into the tree line to cover our less-than-glorious retreat! I thought it best to break off the attack and devise a better plan. I had no desire to join the graveyard below where Outlaw 17 now resembled a skeleton, and Dustoff a pile of ashes!

We instinctively knew we were fortunate to have survived that impressive show of fire power exhibited by such a determined enemy! We also answered our question about whether or not they were still there! They definitely hadn't melted into the jungle! They were still entrenched in bunkers, were well-trained and they sure as hell knew how to shoot! Very disciplined, too! They had just enticed two well-armed, normally much-feared gunships, to dive directly into the "mouth of the dragon" before unleashing that incredibly accurate machine gun fire!

It was an unusual scenario. There we were, in the early morning hours with lots of daylight left, yet had taken that kind of fire! It was very clear those guys had absolutely no intention of slipping away from LZ Alpha! They knew full well their comforting cloak of darkness was hours away, yet they seemed ready, willing and more than capable of mixing it up with the T-Birds, Mavericks or any other gunship fire-team that happened to pass through their area looking for a gunfight!

Out of range of the "shooting gallery", we began flying a circling orbit while assessing the damage to both gunships. After determining our new "air" holes didn't affect our ability to fly and fight, we began another turn in towards the tree line from another direction to see how wide their cone of fire was. As we were turning inbound, we noticed a strange looking helicopter making an approach into the tree line without a word being spoken to anybody! I finally determined it to be a Kaman "Huskie" utilized by the USAF as their search and rescue aircraft. I knew their call-sign was PEDRO, so I made numerous attempts to contact him and advise him of the situation. I certainly admired his intentions, but I knew he was a good candidate for induction into the VC's helicopter graveyard!

I made numerous calls, trying to deter him from his effort to land in LZ Alpha, but to no avail! He continued his slow approach into the LZ in much the same manner as strolling through a park! Then, when all that heavy fire was suddenly unleashed in his direction, he had a sudden change of plans! The pilot immediately banked away from the LZ, put that "Huskie" into overdrive and began putting as much distance between him and that tree line that

he could! Even though we hadn't talked to him the entire time, we all decided he was a pretty clever pilot! Soon as he realized the situation, he beat feet back to his base and never returned. Those were the actions of a pretty smart pilot! Unfortunately, the short, humorous break in action changed nothing; our flight crews were still trapped in a LZ the enemy firmly controlled! We still had a job to do!

Watching our would-be rescue helicopter disappear over the horizon, we received yet another call from Delta Six, inquiring as to whether or not he could attempt a rescue mission into LZ Alpha! My response was "negative", but I was quickly over-ridden by my copilot (Maverick Lead) who cleared Delta Six into the LZ! I was stunned! Knowing Delta Six had been impatiently waiting for some time, I couldn't reverse that ill-advised clearance into a hot LZ! Even worse, we had no visual contact with Delta Six even though he had been cleared into LZ Alpha! We finally spotted him just as he was turning onto a long final approach, and I warned him one last time about the heavy fire we had taken! He was firmly committed to rescuing his men. Though ill-advised, it was a very courageous act by a heroic crew!

Caught off guard by the Major's decision to clear Delta Six into LZ Alpha, I was not in the best position to escort him in, so I directed our wingman to assume the lead position since he was in a better position to do so. After getting a "Roger that!" from Jack, I maneuvered my gunship into a position alongside him so we could provide maximum covering fire into the tree line. Once Delta Six was in the LZ, we would establish an overhead cartwheel maneuver whereby we could keep pouring continuous fire into the enemy positions! I was still wishing Delta Six hadn't been cleared to make his valiant rescue attempt, but we were already on short final into LZ Alpha, and Delta Six had begun slowing down to land alongside Outlaw 17. Much like the pig that had provided our bacon for breakfast, we were committed.

When Delta Six initiated his landing flare, that single heroic act

committed all of us to a deadly game of high-stakes poker where the cards had been dealt, the bets were in, and the last player had just called to show the cards! It was a simple matter of who held the winning hand in a deadly game of life or death! The temporary lull was deafening as we became hyper-focused on what we hoped would be a short, intense gunfight, lasting just long enough for Delta Six to rescue our downed crewmen, then get the hell out of LZ Alpha! But luck wasn't with us that day! It had abandoned us on a day normally celebrated as a day of peace.

I heard the first of several desperate calls from Delta Six telling us he was "taking heavy fire", immediately followed by "we're hit! We're hit"! Those calls were over-ridden by my wingman screaming he was taking heavy fire! Another call said he was hit! The radio that had been so silent a scant few seconds ago was now exploding with bad news! Outlaw Six began yelling, ordering everyone to "get the hell out the LZ!" Then more desperate calls! "I'm hit, I'm hit!" "Delta Six going down!" "I'm breaking right!" "Cover me!" "Damn! I'm still taking heavy fire!" "Get the hell out of there Delta Six!" "I think Delta Six is dead!" It was an endless stream of radio chatter too horrifying to contemplate!

Both Maverick gunships poured continuous fire into the tree line, but the situation deteriorated so rapidly it became increasingly difficult for my mind to comprehend what my eyes were seeing! Despite the terrible series of events unfolding all around me, I seemed to be watching a slow-motion horror movie, complete with a voice track of strained, incomprehensible sounds of men engaged in a desperate life and death battle! Men who wanted nothing more than to get out of that god-forsaken, nondescript rice paddy! Men who wanted to scrub off the mud, the stink, and death! Men willing to sacrifice their own lives for their brothers who now lay dead, or dying, on the battlefield!

I saw Delta Six's helicopter slowly drop out of the sky, then slam down alongside the skeleton of Outlaw Six and the smoldering ashes of Dustoff, before slowly settling into a rice paddy that was some

Vietnamese rice farmer's prized possession. Unfortunately, an endless and deadly chain of events had turned it into a muddy gravesite of mortally wounded helicopters. The crewmen still trapped in LZ Alpha knew for certain there would be no more rescue attempts. Our last attempt had resulted in yet another crew of dead and wounded men. We desperately hoped they would be the last.

It was definitely a kick-in-the-butt moment for all of us, but a very determined enemy allowed no time for mourning our losses! They knew they had firm control of LZ Alpha and were determined to keep it! They were also quite capable of adding another helicopter to their impressive count of downed helicopters! Our Maverick gunships were still able to function as a fire team, even though both aircraft had taken multiple hits. We could still provide firepower even though we were hearing a new sound! A whistling sound normally attributed to a hole being punched through the main rotor blade by a bullet! A small hole, we could live with; a larger hole, maybe not. The strength and integrity of the blade might be compromised, and if we were forced to take evasive action, the results could be devastating! We chose not to think about that scenario. We still had work to do.

As we were deciding our next step, another Maverick gunship fire team arrived on station. The familiar voice of Captain Joe Moffett advised us he would take over from there. His gunship was what we referred to as a "hog", since his armament was 48 2.75 FFAR rockets! Twenty-four rockets on either side! They could do some damage, especially if their crews were pumped and eager to join the fight! And they were! Our fuel state was getting critical, so we turned our battle-damaged helicopters toward Vinh Long while I provided a SITREP (situation report) to the new fire team. We needed maintenance crews to take care of the damage, then refuel and rearm. The sooner the better! We would need all the gunships we could muster for another rescue attempt!

The whistling blade entertained us on our return trip before finally contacting Vinh Long tower, and they quickly cleared us inbound for a long approach to the Maverick ramp. After landing, we were somewhat shocked when we saw literally every man assigned to the airfield, regardless of rank, color, creed or job, had joined together on the flight line! They were all working side-by-side, driven by the single desire to rescue those men still trapped in LZ Alpha and bring them home!

The helicopter we were flying was grounded soon as we landed, due to the round passing through the main rotor blade; but it didn't stay grounded long! Everyone worked together to change the blade, rework the ride, then get us back into the air! Whatever it took! That's the kind of effort we had that day! Working together as one, nobody stopping until every one of those helicopter crewmen made it out of the LZ! Helicopter crews were known throughout Vietnam for never having left a man behind, regardless of whether it was a fellow crewman, or wounded soldier! It took all day before we finally got all our crews and wounded soldiers out, but we did. That day may have been my proudest moment as an American serviceman in Vietnam!

Crash of Outlaw 22
April 19, 1967

My good friend, Warrant Officer Gary Wesselman, had been aircraft commander and Warrant Officer Larry Reeves flying as copilot on Outlaw 22 on April 19, 1967. On that morning, they had been flying in a small flight of four Outlaw helicopters conducting what we call "ass & trash", carrying people and supplies, near Sadec, a small town not far from Vinh Long. They had completed the mission and were returning to Vinh Long when they encountered heavy rainstorms. Unfortunately, the pilot assigned as flight leader made the decision to descend through extremely thick clouds over Vinh Long. Three of the Outlaw helicopters broke out of the clouds at a very low altitude, but they lost contact with Outlaw 22. There had been no May Day call, and repeated calls resulted in no response. An Outlaw helicopter and crew were missing!

I had already transferred to the gunship platoon, so I was

assigned as a copilot on one of two Maverick gunships dispatched to help search for the missing Outlaw aircraft. Fortunately, the weather had lifted by the time we took off, so the first thing we did was retrace the flight path they had been flying when Outlaw 22 disappeared. Joining up with a couple of Outlaw slicks, we promptly spread out to begin our search, hoping to find Gary and Larry waiting impatiently for us to find them so we could all go enjoy a beer while they told us their story. It didn't take long before we spotted it. Actually, it wasn't so much seeing a helicopter as it was a crowd of Vietnamese villagers gathered around what looked like a charred pile of rubble in a rice paddy. It was apparent Outlaw 22 had plowed into the rice paddy at high speed, upside down. There had been no hope for survival. If there had been a "silver lining" in such a tragedy, it was the fact the two crewmen who normally flew with them had remained behind. They would live to fly another day. But we heard no stories from Gary or Larry that night. In fact, we would never share another beer, and talk of home. Their dreams of tomorrow ended tragically in a non-descript rice paddy that had already begun to fill in with mud.

The young Warrant Officer who had been flying as mission commander for the small flight, and made the fateful decision to descend through thick clouds, was a basket case. He tried to drink himself into oblivion to erase the guilt, but it didn't work. The base doctor had to inject a strong tranquilizer into him to provide his tortured brain some relief. He was grounded for a couple of weeks before returning to flight duty. I don't know if he ever got over the guilt of the loss of Outlaw 22, flown by our brother pilots. By my way of thinking, that certainly would have been a heavy burden to bear.

Another Mortar Attack at Vinh Long
April 21, 1967

Some of the Maverick flight crews, myself included, had flown for ten straight days with no time off, so we had been looking forward to a couple of cold beers, a good night's rest, followed by a "down" day with no scheduled flight operations. We had been in no hurry to close the Mekong Manor officer's club, but after the final beer call, several of us had retired to our hootch to play poker. Everyone was in good spirits because even though it was late, we had all day tomorrow to rest. We were really enjoying ourselves when we all heard a distant explosion. A very distinct explosion! As veterans of numerous mortar attacks, we knew the difference between outgoing or incoming. Everyone froze, trying to decide if that explosion was the latter. A split-second later, there was a tremendous explosion nearby that erased all doubts!

Every pilot sitting at the table instinctively knew what had to be done. As gunship pilots, we had to run the gauntlet! At the Vinh Long Army Airfield, everyone who was stationed there could run to the nearest bunker; that is, everyone except for gunship pilots! Even though the Cobra gunships of the 114th were also stationed at Vinh Long, the Maverick pilots' hootch was closest to the flight line. Just like the Maverick gunship crews had taken off during my first mortar attack, it was my turn now! A relentless, merciless enemy would keep dropping mortar rounds in until we did. So, my footrace toward the flight line in a mortar attack began!

A huge explosion in close proximity to pilots running a short distance away spewed thousands of red-hot, steel razor blades in all directions, each seeking out warm flesh to penetrate! Somehow, they were spared, so we kept running toward the flight line! Whump! Boom! Whump! Boom! Over and over! It's not unlike firework displays you enjoy on July 4th, except it's deadlier! Much deadlier! We had always been lucky at Vinh Long because no one had been hit. Not so at Soc Trang! They had lost a couple of pilots racing to the flight line, exactly as we were doing. We finally made it, and soon as we jumped into the cockpit and flipped the battery switch on, red anti-collision lights began flashing! We were telegraphing our intentions to every VC out there the Mavericks were coming after their ass! No doubt they had assigned an invisible, black-garbed observer to watch for our beacons. The mortar rounds already in the air when our beacons first flashed continued on their trajectory, but they would be the last! By the time the Mavericks launched, the mortar crews were already dismantling their weapons and slithering off into the darkness. But they would be back! They always came back! We averaged three mortar attacks a month during my time at Vinh Long.

It's horrifying to hunker down in a bunker while mortar rounds drop in all around you! It is horrifying beyond belief when you're racing through mortar rounds dropping in all around you! If that red-hot shrapnel traveling at warp speed ever hit you, it would most likely end all your earthly concerns, thus earning you a trip home before your short-timer's calendar projected date. No, thank you! I would much prefer the mortars hit other targets; like helicopters! On that particular night, three helicopters were destroyed, and numerous others were badly damaged. A cheap price to pay as far as we were concerned!

We also found a deep hole gouged out of a small marsh-like, muddy area directly alongside the well-traveled road leading to the flight line. Had the mortar round dropped on the hard surface of the road, it would have wiped out whoever happened to be unlucky enough to be racing by. But luck had been with us. The marsh-like, muddy area had absorbed the shrapnel before it had a chance to unleash its devastation and death. It had simply rearranged the mud. It became a sacred area. It was a place where good luck had taken refuge that night.

Landing in a Minefield
April 24, 1967

We had been staging out of an old airstrip near Chi Lang, an area that pilots always avoided if they were flying single-ship missions. The airstrip was in VC country, and a dangerous place to be in with no security. Since we had so many helicopters and troops involved in a large military operation, it was deemed safe for our planned combat missions. The Mavericks had already completed a couple of missions, so we had flown in and landed on the dirt strip behind a couple of refueling Outlaws to wait for our turn to refuel. After refueling, we would then reposition to the opposite side of the runway prior to shutting down, and several Outlaw slicks had already accomplished that. Since the airfield was so small and crowded, other helicopters behind us had to pick out a spot to land wherever they could, then await their turn to get in line for refueling.

My gunship was sitting on the ground about 2 or 3 helicopters behind the refueling pad when I noticed a helicopter hovering at my 3:00 o'clock position, searching for a clear spot to land. I recognized it as a Red Knight helicopter from the 114th AHC, our sister company at Vinh Long. Watching it, I saw no problem since the pilot had selected a large vacant field some distance from us where he could land to await his turn at the fuel pump. He made a smooth landing and the helicopter was just settling into the soft ground when there was an extremely loud WHUMP, followed by a huge fireball that engulfed the entire helicopter! Thick, acrid smoke fed by JP-4 fuel began billowing over the area.

Every pilot sitting in line, myself included, thought we were being mortared by some unseen enemy and desperately began rolling our throttles back to a flight setting! Everyone was extremely anxious to get out of there before the next round dropped in! However, by the time we had everything ready to go, we realized no other rounds had rained in on us. Apparently, it was a "one-of-a-kind" event, so we rolled our throttles back to idle, then sat and watched, riveted by what was unfolding at the Red Knight helicopter!

How that entire crew survived that massive explosion and subsequent fireball, I haven't a clue! All four crewmembers scrambled out of that burning maelstrom and began sprinting like gazelles straight toward my helicopter! When they got close enough, I recognized the aircraft commander as one of my closest friends, the pilot with whom I had travelled to Vietnam! There was a 4-strand barbed wire fence that separated us but with Olympic-like grace, all four crewmen managed to get over it! Since my helicopter was the closest, it represented safety, exactly what they needed to calm shattered nerves! I quickly jumped out of the cockpit to greet them, but they were too out of breath to talk! They had made the mad dash wearing their heavy, burdensome chest protectors.

After everything finally settled down, we walked up to the barbed wire on which old, rusty signs were hanging. Bearing skull and cross-bone markings, they warned of death and danger. The entire area had been mined! The French forces that fought there so long ago had done the honors. The pile of ashes that was once a helicopter, now sat at the epicenter of a mine field. The Red Knight flight crew had raced through that mine field like gazelles, then cleared a barbed-wire fence to safety with nary a scratch! They did have frayed nerves, as we all did! That night, my good friend Bill Whitlow and I, along with several others, shared a few cold beers to salute their close call! I think we also wanted some of his luck to rub off on us.

Combat Assault at Moc Hoa
June 28, 1967

Moc Hoa was home base for not only a Special Forces team, but also for a tough crowd of combatants referred to as Mike Force, short for Mobile Strike Force Command. Mike Force was a component of Special Forces made up entirely of highly trained and well-motivated indigenous soldiers, all of them hand-picked from the ranks of Chinese Nungs, South Vietnamese Special Forces and Montagnards, to name a few. They were well-known as ferocious fighters and would fight to the death on behalf of their American advisors. The calm, peaceful look of the small tree-lined hamlet of Moc Hoa belied itself, because it was a place of intrigue, spies and maybe even counter-spies who crossed back and forth across the Cambodian border. There always seemed to be a lot of action in that area and I can recall supporting them both as an Outlaw slick pilot, and flying Maverick gunships, such as I did on June 28, 1967.

I had already flown a mission in the early hours of the morning,

but was scrambled out later in the day to support a combat mission gone bad. It had been a well-planned operation contrived by a Special Forces "Mike" team based in a distant, dangerous part of the Mekong Delta near the Cambodian border. It had certainly sounded simple enough! A blocking force would sneak out under the cover of darkness to set up an ambush site, then the main strike force would push through an area well-known as a hotbed of VC activity, thus driving the crafty VC into the trap! Unfortunately, the enemy had read the same book of military techniques, albeit in Russian; or Chinese!

While the main strike force was maneuvering into their launch position, a strong enemy force attacked the Special Forces blocking force. Cutoff and surrounded, the beleaguered group was soon in very serious trouble! They desperately needed help, or they faced total annihilation! A Maverick light fire team composed of two gunships were scrambled and were soon racing toward Moc Hoa, leaving a second fire team of two Maverick gunships behind to relieve us about an hour later.

As soon as we got in range for radio communication, we immediately realized the seriousness of the situation. As a gunship pilot with so much fire power at our command to diffuse most situations, it was always difficult listening to desperate cries of men in serious trouble, pleading for someone, anyone, to help them! Even worse was when they somehow knew they were about to die! Such as it was in that instance. The enemy had sensed the Special Forces troops were "on the ropes" and were preparing to launch a final assault to annihilate them! That was their plan, but the Maverick fire team arrived on station at just the right time! After listening to the horrors of what was happening down there, we were ready. No quarter asked for, none given!

Screaming in at treetop level just over the heads of the surviving Special Forces team, I saw the stunned faces of a confident enemy looking up at us! The snorting Maverick logo painted on the nose of our gunships was the last thing many shocked enemy

soldiers saw when my copilot began firing our mini-guns, spewing out 6,000 rounds per minute into their midst! It was a weapon that could literally destroy everything within the perimeter of a football field, and that was just for my copilot! I had 14 rockets to launch wherever they would do the most damage! Meanwhile, our gunners were wielding their M60 hand-held machine guns with expert efficiency. Like the merciless enemy below, we could also be merciless, especially when American and ARVN soldiers' lives were at stake!

Everything was happening at break-neck speed, and everyone on board both gunships were experienced veterans of many hours of combat. There was very little chatter or wasted motions. Everyone knew exactly what to do, and did it well! We were expending ammunition much faster than normal, but we had to breach the enemy positions to clear an escape route for the friendlies, whatever was left of them! The other fire team flying out to relieve us was still too far out! We had to make it happen quickly! Facing what had seemed like certain death just a scant few minutes ago, the survivors were screaming on the radio now, urging us to create an escape path for them! With just enough ammo on board for one more run, we had to do our best!

It was a determined Maverick fire team that rolled in for one last attack to break through the enemy force. Even though low on ammo and rockets, we were still formidable, at least for one more attack! Halfway into our gun run, one of our gunners who had been laying down such deadly volleys of fire abruptly quit firing. He announced he was out of ammunition! That told me he had used up 2,000 rounds of 7.62 ammunition! He'd also changed a gun barrel sometime during the exchange. The other gunner kept hammering away, but I knew he was almost out as well! When I glanced back over my shoulder, I saw the gunner who had expended his ammo raise his machine gun over his head, preparing to hurl it at the enemy below! He appeared to be in a trance brought on by the constant din of battle! To him, it seemed simple enough; he was out of

ammo, but not out of the fight! He would hurl his useless weapon at the enemy! My scream in the intercom stopped him just in time!

I was trying to think what my next step would be when the beleaguered blocking force let us know they had punched through a gap! They had escaped! I exhaled heavily, then shifted back into my seat. The adrenaline-high that had been present for so long left my body, and it was like the air had suddenly escaped a balloon! But we had done it! Thank God! Even better, the second Maverick fire team arrived on station just in time to relieve us.

There's an old saying about how you can never let it rest until your good is better, and your better is your best! We didn't let it rest, and we did our best. Now we just hoped to fly back to Vinh Long without any problems. If we had been forced to land anywhere in between there and Ving Long, two Maverick gun ship crews would've been reduced to throwing mudballs at the enemy.

Low Level Across the "Y" July 13, 1967

The "Y" was the confluence of several rivers, surrounded by thick, jungle-like vegetation. It was an area that lay almost directly between Vinh Long and Can Tho, the city where the 13th Aviation Battalion Headquarters was located. It had also been deemed a "free-fire" zone. That meant any gunships, or slicks for that matter, assuming they liked living dangerously, were authorized to open fire on any suspicious activity without first obtaining clearance to do so. It was a dangerous area full of thick tree lines, canals, tributaries and rice paddies, all excellent areas for an elusive enemy! It was the perfect hiding place for determined VC who wanted nothing more than to kill U.S. invaders, especially if they were flying helicopters! Needless to say, we normally tried to avoid it, even though we were in gunships. It was that dangerous!

On that particular day, I was flying a mission as copilot with one of the Mavericks' most respected fire team leaders, who also happened to be flying his last mission in Vietnam. We had been scrambled to Can Tho in the early morning hours and were returning to

Vinh Long after being released from our mission. We were the lead gunship and covering us in the wingman position were two wild and crazy guys who were not only good pilots, but best friends as well. We had been released as soon as we had put in our last strike, and even though we were low on fuel, we figured we would have just enough if we flew home as the crow flies; like in direct, as in a straight line. A straight line that would take us directly over the "Y"! It was a decision we would soon regret.

Rule number one for a VC gun crew was never, ever fire at the lead gunship! The lead gunship always had a wingman flying behind him, ready to pounce on whoever fired at the lead aircraft. As it turned out, we just happened to fly over an experienced machine gun crew because they let us sail right over them in the lead aircraft, then cut loose with a machine burst directly into our wingman! It was an accurate burst of machine gun fire that penetrated the chin bubble, hit the copilot's leg, literally exploding it, then splattered blood and chunks of body parts all over the cockpit! The stunned pilot screamed out "taking fire, taking fire! My copilot's hit bad!"

We responded by rolling into an immediate high-rate turn and I began raking machine gun fire all around the area we had just flown over. Seeing our wingman climbing out of harm's way, we darted in beneath him to draw any subsequent fire while our gunners laid down covering fire! After doing all we could, we broke off our attack and rejoined our wingman racing toward Vinh Long. The wounded pilot was pumping blood out of a lacerated artery and needed medical assistance immediately!

It was a relatively short flight to Vinh Long and we called ahead to request medical personnel to meet us at the flight line. After landing, we shut down our aircraft, then sprinted over to our wing man. Jerking the door open, we were stunned at what we saw! It looked like a butcher shop, or a horror movie where a madman takes a bucket of blood and splashes it all over the cockpit! We knew the copilot was alive when he started yelling: "Did you get those SOBs who shot me?!" The camp doctor arrived, and it wasn't long before

we managed to get our pilot out of the cockpit, then tie a tourniquet on his upper thigh to stop the flow of blood. A 114th White Knight helicopter pulled up alongside us, so we whisked the wounded pilot, the Doc and his able assistant into the cabin and they launched immediately, headed for the medical facilities in Saigon.

After seeing our comrade depart, the adrenaline rush began to recede and we had trouble just standing up; leaning against the open cargo area of the blood-covered cockpit, we just stared in disbelief! The cockpit was covered in blood, gristle, bone, muscle and God only knows what else! It was even hanging from the roof! We finally left the flight line and began working our way back to our quarters to drop off our flight gear, then head toward the Mekong Manor for a beer. Food was out of the question! It just didn't sound too appealing.

Later in the evening, I had retired to my hootch and was surprised when the lead pilot of our morning mission walked in and sat down. We started out having a few drinks and friendly chit-chat, but his conversation inevitably turned to what had happened. He began to beat himself up over coulda, woulda, shoulda – but didn't! No, flying over the "Y" hadn't been the best idea, but we all agreed to do it. Other pilots had joined us in the hootch by then, and we all did our best to dissuade him from shouldering all the blame. Unfortunately, along about midnight, he lost it! His remorse was simply too great. We had to wrestle him down, while one of our pilots ran over to get the Doc for the second time that day. He showed up shortly and pulled out that big syringe full of really potent stuff. Just one injection and boom, the pilot was out! After Doc left, a couple of us picked up the stricken pilot, put him in bed, then remained with him throughout the night. Had he awakened; he might have needed us. Amazingly, he slept soundly throughout the night, not waking up until late the next day. Whatever Doc shot him up with, we needed some! Not all the time, just those times when we had experienced a really bad day. Maybe a day like the one we just had!

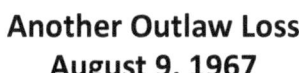

Another Outlaw Loss
August 9, 1967

We had just returned from a tough mission when we heard about losing another Outlaw helicopter. Major Charles Latta, Executive Officer of the 175th AHC was killed along with three other crewmembers, plus two passengers, when his helicopter was involved in a mid-air collision near Bien Hoa, located in the Saigon area. He had just departed the 24th Field Hospital heliport and was climbing to 1,500 feet MSL when an Air Force RF-101 reconnaissance jet smashed into the helicopter, shearing off the entire main rotor system! The helicopter immediately began tumbling out of the sky in the same manner of a brick! There were reports some crewmembers were either thrown out, or maybe jumped out, in a desperate attempt to survive! The Air Force pilot managed to eject but was severely injured. The loss of so many crew members from the 175th AHC in one accident was a terrible shock for everyone at Vinh Long. It was yet another long list of names for our chaplain to read and pray about during the memorial held in our small chapel.

Anyone for Ice Cream

Warrant Officer Ron Petty and I had flown an Outlaw slick to Saigon on a "scrounging" mission to procure hard-to-come-by parts for our mini-guns and rocket pods. Maintaining parts to keep armament systems on our Maverick gunships was a never-ending process! Since Ron's extra assignment was to fulfill the duties of an Armament Officer, I was always glad to help out whenever I could. When you go out on such missions, it always helps if you have a few "souvenirs" to offer in hopes of getting the best deal possible. A bona fide VC (Viet Cong) or NVA (North Vietnamese Army) flag, or some other such item would almost always result in obtaining armament parts.

On that particular day, we had arrived at Hotel 3, a large soccer field turned into a heliport, just before lunch. Standard Operating Procedure (SOP) for getting into Hotel 3 was to make your approach to the center of the field, then taxi directly into the nearest available space around the perimeter of the field. It was akin to parallel parking all around the old field. After shutting down the helicopter, we released our two crewmembers to take a long lunch break so they could take advantage of some good food at the nearby restaurants, go shopping or both. After giving them explicit instructions regarding our departure time, we sent them on their way. Ron and I commenced doing our job of scrounging parts. We also intended to take advantage of our close proximity to the nearby USAF Officer's Club. Soon enough, we had accomplished everything we had planned, then headed back to our helicopter.

As always, it was very hot and humid, so the first thing we did was to open up the cargo doors and stretch out. Since we had arrived a little earlier than our planned departure time, we had about forty-five minutes to kill, so why not get comfy? Ron and I were doing our best to stay cool, a difficult task in Vietnam, especially from Saigon southward toward the Mekong Delta. Not wanting to waste much energy, we were sitting and chatting when we noticed a tow-motor carrying a very large crate, enter the large gate across the field, directly opposite where our helicopter was parked. After driving onto the field, the driver began passing by each helicopter, probably hoping to locate someone to whom the large container belonged. As it turned out, Ron and I were the only crewmembers sitting in a helicopter at Hotel 3! Finally, the young GI on the tow-motor pulled up alongside our helicopter and asked, "Are you waiting for this, sir?"

"What is it?", asked Ron.

"50 gallons of Ice cream".

Ron quickly responded, "Hell yes! Is that all you have?"

The young GI was so happy to find the owner he quickly replied, "No sir! I have another box; soon as I unload this one, I'll go get it!" Ron told him to hurry because he had already caused us to delay our takeoff back to our base. In his haste to satisfy us, he turned that tow-motor and shot directly across the middle of Hotel 3! As

soon as he began moving, Ron and I jumped into action! We didn't have the first clue who that ice cream belonged to, but we saw an opportunity to make a lot of folks happy at Vinh Long! We got busy untying the main rotor blades and got everything ready to go. We also hoped our crewmembers would make it back in time!

It wasn't long before the young GI reappeared and, in his haste to make up for being late, he once again shot directly across the airfield! It wasn't long before we had the second crate loaded and tied down. We thanked the young GI and told him we weren't upset anymore! Relieved, the young man waved goodbye and took off back across the airfield. Just as we began thinking about starting up, we saw our two crewmen running towards us. Seeing our blades untied, they were afraid we were going to leave them! Before long, we headed toward Vinh Long with 100 gallons of ice cream! When we checked in with Vinh Long tower, we told them to have someone meet us with a truck because we had a surprise for everyone. Ice cream was a rare commodity in Vinh Long; but not that night!

UFO in the Delta

Warrant Officer Jon Myhre and I had been conducting what we called "ass & trash" missions in Outlaw 17 when we received a call directed to "any helicopter in the area, please check in." After listening a few minutes and not hearing a response from anyone, Jon told me to respond to the request and provide our location. Following his instructions, I responded with our Outlaw 17 call sign and gave our location. The voice came back on the radio and explained what the call was about.

An American advisor located at some little nondescript village along the South China Sea coast was desperately seeking assistance to transport a Vietnamese child who had been critically wounded when the bus he was riding in hit a land mine. Following protocol, the officer had already tried to get a South Vietnamese helicopter to pick the child up but had been turned down because the area was known to be dangerous. Plus, they had said, it would soon be

dark! Again, too dangerous! The critically wounded child would almost surely die without medical assistance.

We were under no obligation whatsoever to assist that young officer located in some lonely, dangerous, remote outpost that now harbored a critically wounded child! Sensing the desperation in his voice, Jon and I exchanged glances, then I turned to our two crewmen sitting behind us in the cargo area to get their opinion. Their response was two thumbs-up! I keyed the "mike" and told him Outlaw 17 was heading their way soon as we could drop off our cargo which would complete our assigned mission. We could sense the relief in his voice when he responded to my call with "Roger that, Outlaws." I began to plot a course to our new destination.

While the U.S. Army advisor had been relieved and happy, Jon and I were not so positive! Knowing a young child's life was in the balance made our decision an easy one, but we also knew it was the type of mission that could "go south" in a heartbeat! Flying into a small village where we had never been before, in total darkness, was the kind of situation that was fraught with danger. It could also be a trap, but we knew somewhere out there in the darkness was a small child who desperately needed our help. It was a life-or-death situation, and we had to go.

It took a while, but we finally saw a small flashing light just outside the small village now bathed in total darkness. It was crunch time, because neither of us had been there before, and had no clue about obstacles such as power lines or antennas! Enemy gunners could also be hidden in the darkness, patiently waiting for the perfect time to shoot as we settled into the crosshairs of their machine guns! Holding our shared breath, we began our approach to a single beam of light emitting from a flashlight being held by the young American advisor who had seemed so grateful for our offer to help.

Directed to land adjacent to the road, Jon gingerly lowered the collective until the helicopter settled firmly onto the ground. The beam of our landing light provided a soft glow around the nose of our aircraft, and an officer quickly walked up to my side of the

helicopter. Opening the door, he stepped up on the skids and we finally saw the face behind the pleading voice. The sincerity in his voice and handshake convinced us we had made the right choice! An old, beat-up white ambulance with a faded red cross pulled up just outside the main rotor diameter, and two Vietnamese nurses carefully lifted up the critically wounded child, then carried him to our helicopter.

Accompanying the small group was a very concerned young Vietnamese mother whose child, through no fault of his own, was fighting for his life. Surprisingly, the child was conscious, frightened and reaching out for his mother! Consequently, when the advisor asked if his mother could make the flight with us, we responded by inviting the entire family to go, but he declined that offer. He also thanked us profusely for responding to his pleas for that "mission of mercy" for such a badly wounded child no more than three or four years old. His genuine concern for the small child elevated his stature as both an officer, and human being! I can only hope he survived the war and made it home. We can always use his kind.

After everyone was safely loaded and secured, Jon nursed Outlaw 17 off the makeshift landing pad and into forward flight, soon leaving the small village behind in the stifling darkness. Since we were flying alone in a remote, dangerous area of the Mekong Delta, Jon wasted no time in quickly climbing to a flight level of 2,500 feet, then turned onto a heading that would take us directly to My Tho, a little city that lay along the Mekong River about 50 miles east of Vinh Long. It was a pleasant little city that not only had a hospital, it was home to several U.S. Army advisors assigned to a local ARVN unit. The down side was the fact it was located just across the river from a massive jungle that was one of the most dangerous free-fire zones in the Mekong Delta! The entire area belonged to the VC.

After leveling off at 2,500 feet, I made a radio call to Delta Center to file our time-off, direction of flight and estimated time of landing at My Tho. That bit of information provided critical information for search and rescue in the event we might have gone down on our flight to My Tho. We'd already had a long, tiring day, so after all the menial post-takeoff tasks had been accomplished, Jon and I settled back to relax for what we hoped would be an uneventful flight to our destination, then head home to Vinh Long for a cold beer and dinner, not necessarily in that order!

It was a moonless night and flying over the inky darkness of the Vietnam jungle is almost overpowering in its enormity! In the words of one pilot who hailed from the farm country of Georgia, it was like "flying around inside a cow's belly!" Even though Jon and I had relaxed a little, both of us still had that "sixth sense" all combat veterans know well! Like something wasn't quite right! Much like that well-known Murphy's Law, we instinctively knew "if there was the slightest chance something could go wrong, it would!" And it did! Just not in the way we thought it would in a South Vietnam combat zone!

I think Jon and I first saw the light about same time, but sat quietly observing it, trying to figure out what it was. At first, we both thought it had been just another aircraft like ours. Maybe some other aircraft flying in the vast emptiness of the Mekong Delta. Perhaps it was "Puff, the Magic Dragon", so-named because it could spit red, fire-like streams of death in much the same manner the dragons of yore could spit flames to roast their enemies! That old, much-respected, armed DC-3 could take off at dusk, then fly around an assigned area until either getting a desperate call from someone pleading for its death-dealing flames of death, or sunrise – whichever came first! Truth-be-told, I think the sun's appearance was always a welcome sight to every soldier stationed in the Delta. That's when an elusive enemy would once again disappear into the jungle, signaling the end of yet another long night. Problem was, there were too many long nights!

Jon finally broke the long silence by saying, "Willy, have you been watching that light at about our 2:00 o'clock position?" My response was, "I have."

It clearly hadn't been a red anti-collision light because that would've been very distinguishable; nor had it been an aircraft position light. Those two lights are red and green, located on either side of an aircraft. If seen from a distance, you could determine which direction the aircraft was traveling. What we were seeing was a white light, much like an aircraft landing light, or taillight. It seemed to be far off in the distance, but then it began to get closer. I made a call to Delta Center to see if they had other aircraft traffic in our immediate area and their response had been "Negative; I have no reported traffic in your area". They suggested we call Paddy Control to see if they might have a blip on their radar screen. It was a good suggestion since Paddy Control was located in Can Tho and had radar coverage over most of the Mekong Delta.

Acknowledging their recommendation, I switched to Paddy Control's frequency and made a call to them. After assigning us a transponder "squawk code" that would identify us, they acknowledged they saw "something else" near us, but had no contact with whatever "something else" was! They also made repeated calls such as: "Unknown aircraft in such-and-such coordinates, please identify yourself!" The silence was deafening. Whatever, or whoever, was out there had chosen to remain silent, ominously so! They also decided to get a closer look at us because they seemed to be getting closer!

What had simply been an unknown light in the distance began to take on a more definite shape in the form of an orb, or globe. A large one! It also seemed to have the luminescence of a pearl. Now more nervous, Jon made a gentle left-hand turn away from the orb and it followed us, remaining what appeared to be no more than a couple of miles away. Then, when Jon began a right-hand turn toward the object, it seemed to maintain its position at first, then actually seemed to close the gap between us! It was getting to be a little unnerving, especially on the part of the two crewmen sitting behind us. They were requesting permission to open the cargo doors and fire some warning shots with their M-60s! Jon and

I both agreed that wouldn't have been the best approach in trying to identify what we now referred to as a UFO. We certainly didn't want to appear hostile since their intentions seemed to be based more on curiosity, rather than hostility.

Repeated calls to Delta Center and Paddy Control resolved nothing! They both knew we were there, and they knew something else was with us! They just couldn't identify who had joined up with us in that dark sky over the vastness of the Delta! Whenever Jon had made an attempt to turn away from it, it followed. Then, when we turned toward it, it refused to back off! Finally, in sheer frustration, Jon made a rapid, almost aggressive right-hand turn towards the bright ball and what happened next got our attention!

It was if it had taken Jon's sudden movement as one of aggression and actually seemed to accelerate towards us! Jon let out an expletive, then made a sudden left-hand turn away from the orb and began an immediate descent! Seeing lights off in the distance, I scanned my map and determined them to be the lights of My Tho, our destination. Knowing that, Jon continued our rapid descent at the fastest speed we could attain in our now-protesting Huey! Huey's didn't particularly like high speeds, and they let you know it by increasing vibrations, but we didn't care about vibrations that night! We just wanted to put some distance between us and that damned light!

We were coming up from the south towards the Mekong River, and My Tho lay on the north side of it. Since VC-infested jungles lay along the bank of the south side, no helicopter pilot ever made their final approach over that area! A low, slow-flying helicopter would provide too tempting a target! We always flew low-level up the Mekong River from either east or west, then made a sharp ninety-degree turn up the canal to the helipad located alongside that same canal inside the city perimeter. On that night, we both decided it would be best to continue our high-speed descent out toward the west of the city, make a right-hand turn, then low-level

up the Mekong River from the west before making a hard left-hand turn toward the heliport.

Jon wasted no time in his haste to overfly the VC-controlled south bank of the Mekong River! It was almost as if our sudden acceleration and descent had taken our UFO buddy by surprise and we managed to increase our distance from it. In fact, the crewmen sitting behind us declared they had lost sight of it, a good feeling! Before long, we had safely crossed over VC territory, then dropped in just above the Mekong River a few miles west of My Tho before making a right-hand turn, then racing towards the lights of My Tho! We were so low our helicopter's skids were literally skimming the surface of the water! The crewmen opened the cargo doors, locked them open and readied their M-60 machine guns in case they were needed. We were only a mile or so away from My Tho! Everything was going great at first! Then we saw the light again! It was directly in front of us!

Whoever, or whatever, was flying that giant, glistening ball of light had literally executed a mirror-image of what we had just accomplished! We had descended to the west of My Tho, while "they" had descended to the east! We were now racing up the Mekong River toward each other from opposite directions! Jon swore, then pulled in more power to increase our airspeed, hoping to make it to My Tho before the glowing ball did! None of us could take our eyes off that object! It was like we were flying toward a mirror that was reflecting the beam of our landing light into our eyes! We had become locked into a race of death with the finish line being the heliport! The giant, glistening orb seemed to fill our windshield when Jon made an aggressive hard left-turn towards the heliport! That's when it got worse!

Jon's hard left-turn was literally a 90-degree banked turn so tight, Jon and I both lost sight of it! The gunner who sat on the right-hand side of the helicopter was a genuine badass! The kind of gunner you wanted when you were in combat being shot at! Brave and tough as they come, he began screaming like a child! Since the cargo door was locked open, he had an unobstructed view of that huge ball of light and he screamed "My God it's going to ram us! He's going to hit us! He's going to hit us!" Then he did what any of us would have probably done, he dropped his M-60 and threw his hands up to cover his eyes, still screaming! Then, suddenly as it had begun, Jon rolled the helicopter back to level flight and directly in front of us was the helipad! Like the pro he was, Jon gently settled the helicopter onto the helipad, then rolled the throttle back to ground idle. We could scarcely believe what had just happened to us!

After shutting down, it was a shaken crew that crawled out of the helicopter and just stood there taking in the sounds of the night, trying to figure out what we had just encountered. Making it even more surreal was how beautiful and serene everything was there on that heliport, especially when just a few short minutes ago, we had physically prepared ourselves for a collision we hoped would never happen! We even questioned the advisor when he pulled up in his jeep to greet us, but he had seen nothing, nor had he heard anything. It was an eerie moment. Shortly afterward, the child and mother were carefully placed inside the ambulance and carted off to the hospital. A thankful mother's smile was our reward.

It had been a long and harrowing day, even without that unexpected encounter with the unknown! Our adrenaline had peaked in those few short seconds just prior to landing, then bled out in the same manner air departs a punctured balloon! Suddenly dead-dogged tired, we were reluctant to leave the safety of the heliport, but we knew we were expected back in Vinh Long for another early morning get-up for another combat assault somewhere in the Delta. Soon enough, we were safely back at Vinh Long, preparing for tomorrow's operation. We hoped it was just the standard things to be concerned with in combat. Simple things like bullets, bombs, or pissed-off VC! We could accept that easier than another confrontation with something completely unknown, unexplained and unwelcomed!

Courage of an Enemy Soldier

We had accomplished our mission of escorting the Outlaw slicks into the LZ one morning, and had been flying support missions for the troops on the ground. It was an area where several battles had been fought throughout the years because there had been numerous bomb craters, plus evidence of Agent Orange damage in several areas. There had been some action earlier when the friendlies had engaged a small group of VC troops that had managed to slip further into the jungle, so we had begun the arduous task of trying to find a clever enemy that had gone to ground deep inside the jungle! We had just about given up when one of our gunners spotted a couple of enemy soldiers in typical black VC clothing. Quickly rolling my gunship back around, sure enough, we saw two well-armed VC soldiers scrambling at the edge of numerous bomb craters!

I ordered my gunners to take them under fire while I maneuvered my gunship into a better position to launch an attack. Just as I rolled in to make a firing run, I saw one of the enemy troops go down, hit by our door gunner. The remaining soldier managed to escape into the cover of the thick jungle surrounding the craters. Seeing the unmoving black-clad enemy soldier in the bomb crater, I made a low pass to check him out and it was obvious he had been badly wounded but was still alive. As the fire team leader, I was trying to decide whether we should kill him outright, or direct ARVN troops to him and try to take him prisoner. Before I could come to a conclusion, I noticed the enemy soldier who had escaped leave the safety of the jungle and, un-armed, run back out to his fellow combatant! I was literally stunned by his action, because with two heavily armed gunships on the prowl, it had taken courage to do that. A great deal of courage!

He made it to the crater where his buddy lay, then picked him up and began to assist him on his journey back toward the safety of the jungle. He had to know with great certainty his chances of survival were slim, just as I did! With the firepower of my two gunships armed with miniguns, the entire route from where he was, all the way to the jungle, could be turned into a virtual killing field! Yet, he had taken that risk to save a badly wounded comrade. My wingman and I circled the pair as they painfully made their way to the relative safety of the jungle. We remained silent while watching that act of bravery unfold. Even the gunners were quiet! That was amazing, too! Normally, our gunners would have been begging for the chance to "waste" those guys! They had almost made it to the jungle when I made my decision.

Circling wide, I dropped down low over the jungle, then slowed down just as I passed close beside them. Looking down, I nodded to the man who would live another day, and perhaps his brother-in-arms would as well. His actions had shown him to be someone willing to risk it all for his comrade. Had we been locked in combat that day, we would have killed him without a second thought! There was also the chance we might meet up with that same soldier again under far different circumstances, perhaps on a day when he may not be so lucky. But on that day, that single act of courage restored some semblance of faith in my fellow man in a very deadly war.

Snake in the Cockpit

A flight of ten Outlaw slicks had launched out of Vinh Long in the early morning hours and were flying toward some distant abandoned airfield from which yet another combat assault would be initiated. It had been just the beginning of what would be another long day as a helicopter crewman, and having to get up at such an early hour didn't boost one's morale! Grumpy and morose, everyone had gone about their duties, trying to sleep with one eye open! There was never much conversation during those early morning repositioning flights. The gunners were trying to sleep off last night's beer, as was the non-flying pilot. That all changed as the sun began peeking over the eastern horizon!

A blood-curdling scream of "there's a snake in the cockpit!" reverberated through every cockpit in the Delta! It was as if an electric shock had jolted the entire flight of Hueys, and everyone who had been sleeping or dreaming of home, had been jolted by it! It took a second or two before Outlaw Lead could muster a "say that again" over the radio. "There's a big f-ing snake staring at me from between my tail rotor pedals!" That sounded serious! When asked if the snake was poisonous, the frightened pilot's response was: "How the hell should I know? Oh, God! Now he's sticking his tongue out at me!" Without a word being said, every crewmember in every helicopter began looking for snakes in their own aircraft! No one wanted an uninvited guest of that nature! Outlaw Lead asked the pilot if he felt OK to continue to our destination, since it was only another ten minutes away, rather than attempt a landing in what well might be some VC battalion's back yard. He elected to continue to our destination.

As soon as we landed and shut down, literally every crewman raced over to the Huey with the uninvited guest! The snake had disappeared back into the bowels of the helicopter, and the pilots and crewmen were contemplating their next step toward dis-inviting him! Finally, someone came up with the idea of "smoking" it out with smoke grenades, which everyone agreed might work. Without further ado, the mechanic pulled the pins from two red smoke grenades, tossed them inside the helicopter and closed the door. Within a matter of seconds, smoke was so thick you could barely see the helicopter! Even though everyone was carefully watching for a coughing, hacking snake to come crawling out of the helicopter, spewing red smoke, he never appeared! More smoke grenades, but again, no snake! Since we had to start engines in another 30 minutes, the doors were all opened in hopes of clearing the smoke before start time. Everything went as planned for the rest of the day, and the snake never appeared again.

We don't know how it escaped, but we never saw the snake again; consequently, we had no clue of what type of snake it was. As far as we were all concerned, it had been a Cobra! Maybe even one of those "spitting" Cobras, and nobody wanted one of those staring at us from between our feet resting on the tail rotor pedals!

I assure you, after that experience, every crew chief, gunner and pilot accomplished a thorough preflight on their helicopter!

Improper Dress for Saigon's USAF Officer's Club

A Maverick light fire team, consisting of two gunships, had been scrambled out in the early morning hours in response to an urgent call from a remote Special Forces outpost under attack and in danger of being overrun. The weather hadn't been the best, but we elected to go and fortunately, the weather improved not long after we had departed Vinh Long. It wasn't long before we were able to contact a DC-3 "Spooky Bird" gunship circling overhead trying to help the besieged outpost. After observing all the tracers crisscrossing in-and-around the outpost, plus ongoing explosions, it was evident the soldiers in the outpost were in big-time trouble! Amazingly, the outpost commander came online to inform us the entire force was pulling back from their outer perimeter and "digging in" around the bunkers in the center of the compound to prepare for what might well be the final assault if we couldn't stop it! We hadn't fired a shot, and just like that, things had just gone from bad to worst!

After coordinating a plan with "Spooky", he continued dropping flares to light up the area while our Maverick gunships would make continuous gun runs utilizing mini-guns and rockets to break up the attack. As we prepared to make our first gun run, the first of many flares dropped by "Spooky" lit up the entire area and we were stunned by what we saw! The outpost itself was surrounded by a jungle, and there were so many black-clad VC attacking it, it looked like an ant-bed! There appeared to be swarms and swarms of them already clamoring over the outer defenses while an unending stream of VC funneled out of the jungle! I quickly advised the survivors to withdraw into the bunkers because we were going to roll in on firing runs and we couldn't pick and choose targets! There had been too many of them! We would simply fire into the mass of enemy soldiers!

Rolling in for our first firing run, we knew the focus of those automatic weapons would probably shift upward toward us and we didn't want to be shot down into the middle of that "ant bed"! We dropped down to treetop level roaring in from over the jungle and firing literally at point-blank range! When I made my break away from target, my wingman immediately began firing into the enemy to cover my break. I pulled up into a hard left turn away from the target, then spun around to be in position to cover my wingman during his break. We were in synch! We were at our deadliest! Dodging, darting, firing, breaking off, spinning, then rolling in, all the while being cognizant of everything going on around us. Spooky continued doing his job to perfection with those flares! He even fired some bursts from those flame-spitting mini-guns. Even though only every fourth round was a tracer, it looked like a firehose of red water pouring into the enemy soldiers!

The battle seemed to last forever, but all that firepower took its toll! There were no more targets to shoot at, at least none that were moving. Everything had been so intense we had not even noticed the sun just beginning to creep over the horizon. Circling over the battlefield, it was suddenly very quiet, ominously so! We could see VC bodies, but had no idea if any of the Special Forces troops and their ARVN soldiers had survived their ordeal. The silence was suddenly broken by the commander of the outpost. In his southern drawl, he announced "he certainly would like to buy us a beer one day for saving everyone's ass!" Oh my God! What a wonderful moment that had been! They had survived! Most of them, anyway.

Our Spooky Bird guardian angel came online to tell us he had run out of flares, fuel and bullets so he was departing for his home base near Can Tho. We responded to his call, and offered to buy him a beer some day! But we knew what everybody else knew. None of us involved in that night's combat would ever meet up again. We were like ships passing in the night. We had been summoned to that place in time, and had done our best.

Almost out of fuel and ammo ourselves, we broke off and

headed for Saigon's Hotel 3, the nearest refueling pad. After we landed and refueled, we released the crewmen so they could head over to the non-com mess hall while the four pilots went over to the USAF operated Officer's Club for a hard-earned breakfast! Standing inside the "lobby" of the Officer's Club, we had been busy taking off our caps and shoulder arms when an Air Force Colonel approached us to ask what our intentions were. "To have breakfast!" we responded. Glaring at us, he informed us we weren't eating in his club "dressed like that!" I had to admit, we hadn't been much to look at. Scrambled out at an early morning hour, we had on various flight suits, pants and shoes, one pair being tennis shoes! Plus, our faces and clothing were black from all the gun smoke after fighting throughout the pre-dawn hours to save lives.

We tried to explain our situation, but it was a losing battle. The pompous-ass Colonel really didn't give a damn that we had just been in a fire fight! Living the good life in Saigon, his sole job had been to serve as the club officer. He probably even lived in an air-conditioned condo! Finally giving up, we went over to the Army's mess hall and ate breakfast. They didn't seem to care what we were wearing. We had been in two battles that morning and lost one. Fortunately, we won the battle that counted most!

A New Pilot on Board

Early one morning, we were re-positioning from Vinh Long to Soc Trang where our sister companies, the 121st and 336th Assault Helicopter Companies, were stationed. It was still dark when we had taken off and climbed to 2,500 feet, then settled back for the one-hour flight to Soc Trang. Our course was to fly directly over the city of Can Tho enroute to Soc Trang. About ten-minutes from Can Tho, an Outlaw aircraft broke radio silence by announcing he had to land As-Soon-As-Possible (ASAP)! Sounding somewhat concerned, Outlaw Lead asked if he was having some sort of mechanical problem. "Negative', said the Outlaw. "I have a new pilot on board". "Roger that", said Outlaw Lead. "See if he can wait until we get to

Can Tho and you can break off and land there". "Roger that", came the response.

Soon as the Outlaw pilot had said "I have a new pilot on board", everyone knew exactly what he was saying. The new pilot had diarrhea! Everyone had it sooner or later, some even multiple times! The Outlaw flight kept humming along until we passed over Can Tho at which time Outlaw Lead told the Outlaw pilot to break off and land at Can Tho. The pilot responded with "No problem sir. We don't have to land anymore". "Roger that" said Outlaw Lead. Montezuma had his revenge at about 2,500-feet over the Delta.

Fortunately, almost all the little airstrips we used as our base of operations were always surrounded by rice paddies. The new copilot was able to strip naked, wade into the middle of the paddy and wash his "dirty ditties"! The crew chief even wanted him to clean out his own seat! Such was life for a "newbie"! For everybody else, it had been just another day!

Saying Goodbye
November 16, 1967

As slow as each day seemed to be, it was amazing at how fast my tour of duty in South Vietnam went, especially looking back at it. It just seemed to be that one fine day I put the final X in my 90-day DEROS (Date Estimated Return Over Seas) that I had dutifully kept up-to-date. It was an exciting moment, but I still recall how difficult it was to say goodbye to all those young men to whom I had become closer than brothers.

But for young Warrant Officers Philip Reichard, Lloyd William Whitlow, Jr. and myself, it was a journey home we would never forget! We reported in to the 90th Replacement Battalion in Long Binh on November 8, 1967, our flight home departed the next evening at 7:30 on November 9, 1967. There was a loud cheer from every veteran on board when the Continental airliner broke ground, followed by another round of cheers when the captain announced we had cleared South Vietnam's airspace!

Little did I know, that even though I made it home, the memories of the time I spent in Vietnam would never leave me. Like so many other Vietnam vets, I am still haunted by the futility of it all, and the deaths of so many young men, friend and foe alike, who shall remain forever young. I salute them all.

Warrant Officer Jon Myhre and me during our secret mission for Operation Deckhouse IV on U.S.S. Canberra in South China Sea just off coast of Vietnam (01/05/1967).

My brother's LRRP (or LURP) team members are (left to right): Ray Freeman, Rick Brooks, Art Silsby, Freddie Williams and Jeffrey McLaughlin. They're all prepped to go out on recon mission (03/1967).

Remains of Outlaw helicopter totally destroyed in early morning mortar attack on Vinh Long Army Air Field (04/21/1967).

U.S. Army recovery team preparing to remove Outlaw 17 from LZ Alpha after Easter Sunday (03/28/1967).

Warrant Officers Phil Reichard (on left) and Bill Whitlow at Long Binh, Vietnam, the day before we all departed for the U.S. (11/1967).

CHAPTER 6

From Vietnam to Fort Wolters, Texas

After completing my one-year tour in Vietnam, I was reassigned to Fort Wolters, Texas, the place where it all began. I was thrilled to get that assignment because my parents lived only an hour away, plus my in-laws were about four hours away. We had a year of catching up to do, so our time at Fort Wolters was a good time for our small family of three.

After spending a one-month leave reconnecting with both my family and life in the United States, I reported in at Fort Wolters, Texas in mid-December 1967. Since Fort Wolter's training was shut down during the Christmas holidays, there was very little to do regarding military duty, plus I was given time to relocate my family. When everything finally returned to normal, I began the U.S. Army's Method of Instruction (MOI) course, a mix of flight instruction and ground school, in early January, 1968 to become a flight instructor in a Hughes TH-55 helicopter. Due to a shortage of helicopters, my first flight in the TH-55 wasn't until January 11, and the course continued into mid-February. Fortunately, every student in my MOI class was a Vietnam combat veteran and two of them, Warrant Officers Lance Fogde and William (Bill) Jenkins, had been my classmates as a Warrant Officer Candidate (WOC). We were thankful to have survived Vietnam and had pleasant times in MOI sharing stories of our experiences.

On February 1, MOI training had been put on hold so everyone in my MOI class could go on search and rescue missions trying to locate two Iranian students who had disappeared the previous evening while on a night cross-country training mission. It seems the young captain who was the flight commander had allowed the

students to fly together even though foreign student-pilots flying together had been strictly forbidden! A missing helicopter with two students on board was a serious matter! Consequently, each of us had flown multiple search and rescue missions searching for it. Finally, after several days, all search missions ceased, and MOI training resumed on February 5. A few days later, Fort Wolters' authorities received a call from a farmer who advised them he had located a bright orange helicopter with the remains of the two missing students still in the cockpit.

We had searched a very large area based upon the students assigned flight course, but the location of the crash site hadn't even been close to where we had searched. It was obvious the students had taken off, flown over the first check point, then turned the wrong direction and had flown until they ran out of fuel before crashing into the farmer's field. We knew that because even though the helicopter had been badly damaged, it had not burned. It was certainly sad to think of what it must have been like for those two lost, confused students as they desperately searched for some place to land before draining their fuel tank, then falling out of control into a black abyss!

I finally completed my MOI training after a successful check ride on February 13, 1968, and was immediately assigned to C-Branch, in flight C-12 at Downing Heliport on February 16. That was where new students began Presolo/Primary I training, their initial phase of flight training which entailed 50 flight hours. C-12, or "Charley 12", would be my military home the entire time I was at Fort Wolters. I remember that time well and have many good memories of being an Instructor Pilot (IP) at Fort Wolters.

The flight commander of Charlie-12 was Major Art Conlon, a career military man who had served time in the Korean conflict as an enlisted man with the rank of Sergeant, then received a direct officer's commission. He was a large man, standing 6-foot, 4-inches tall with a wide girth that could probably be contributed to the consumption of many beers throughout his military career. He might

not have exactly looked the part, but he was a good commander, and a fine man. He loved his Warrant Officers and always stood up for us! It has always been my great pleasure to have served with him.

My First Class as an Instructor Pilot (IP)

Even though I had just returned from Vietnam, nothing could prepare me for my first class of students as an Instructor Pilot (IP)! Typically, each IP was assigned three students at the beginning of each class, and it was his responsibility to teach each of them the basic fundamentals of flying a helicopter. Considering the fact none of the students had ever even seen a helicopter before being assigned to flight school, it had been a difficult task! So much so, that an IP rarely kept all three of his original students throughout the Presolo/Primary I phase of training. He would be lucky if he had kept even one, and some IP's would eventually swap out all three of their students with other flight instructors! It was a "musical chairs" scenario that had come about because the Army method was to swap students among flight instructors after five hours if they had problems flying the helicopter. That solved the problem in many cases, but not always. Meanwhile, I was becoming nervous since I was the only IP in Charley-12 who had kept his original three students after soloing them, then preparing them for their flight checks with standardization instructor pilots.

Not being an experienced IP, I had no idea whether my students were that good, or if I simply couldn't recognize a below average student. Several of the more experienced instructor pilots actually gave me a hard time about my instructing abilities because "nobody" could take their original three students through the entire phase of training, especially a newbie IP, as I had been. In fact, it even elevated to a confrontation with two of the senior IP's who carried it too far one day. They were very vocal about it until I finally ended the confrontation rather quickly when I told them to stay the "f–" out of my business, or I would sure as hell get in the middle of their business, either there inside the classroom or outside! I was vindicated at the end of Presolo/Primary flight training when my three young WOC students ended up being among the top candidates in our class. Those young men made me mighty proud and that was the last time anyone ever spoke to me about how my students were doing!

The Dreaded Annual Written Exam
September – 1968

If you were a U.S. Army aviator, you were expected to take, and pass, a written exam every year. That requirement had been waived in Vietnam, so I was quite shocked when the Flight Commander announced we would all be taking our annual written exam the next month and everyone should begin preparing for it. Taking the Major at his word, I paid a visit to Fort Wolters' Publications Department to pick up the applicable books to study and was rather stunned when I could barely carry the manuals to my car! Oh, my Lord! All I could think about was the written exam was a scant one month away and I had approximately two dozen thick books to study! There was no way for me to adequately prepare for my first ever written exam for U.S. Army aviators!

The next day, I was explaining my dilemma to the flight commander and he simply told me not to worry about it because it was always an open-book test. That was all well and good, but an open-book test wasn't going to be any easier if I didn't know where to find the answers in the book! I also thought it rather strange that I seemed to be the only IP in the entire flight department that seemed to be worried about it. Perhaps they had all known something I didn't. As it turned out, they did!

The annual written exam was scheduled to be taken on a Saturday morning, and every pilot at Fort Wolters was provided a time to report to one of the large WOC mess halls. Early on a Monday morning just prior to the test on Saturday, there appeared a single, mysterious sheet of paper laying atop each IP's desk. As I

sat down, I glanced at it and it appeared to be answers to four separate tests, as in Test I, II, III and IV, and 50 answers on each test. I quickly glanced around and saw other IP's quietly fold the paper up, then put it in their pockets. Not a word was spoken as every sheet silently disappeared into flight suit pockets, mine included! Just like that, I was prepared for my first ever, annual written exam!

On the following Saturday morning, there were probably 100 pilots who showed up at the mess hall to take the written exam at the assigned hour. The folks overseeing the exam had cleverly moved all tables in such a manner that not only were they were spread apart from each other, but each seat was approximately 5-feet apart. Plus, pilots sitting opposite each other were seated in staggered seats, so no one sat directly in front of each other. The message to everyone was "there will be no cheating!" But, hey – everyone had a secret piece of paper stashed somewhere in their pockets! We were also allotted four hours in which to take the test. What could possibly go wrong?

First of all, after being seated, I noticed the two pilots sitting across from me were both Majors, definitely not a good situation! I had an answer sheet, but how was I going to use it with two high-ranking officers sitting directly across from me? After making a few half-hearted attempts to dig through a stack of books looking for answers, I knew I was in trouble! There was simply no way I could find the answers to fifty questions in four hours, nor could I use my answer sheet. Somewhat distressed, I leaned slightly across the table, cradling my forehead in my hands and began to ponder my dilemma.

The two Majors sitting opposite me were expertly whizzing through books, plotters and navigation computers, then writing down answers! I was both amazed and impressed with their prowess until I saw something rather intriguing. I noticed one of the Majors periodically flip through one of his books, hesitate just long enough to memorize a few answers from his sheet, write the answers on his IBM card, then continue flipping through his other

books. How clever, I thought! After further checking, sure enough, the other Major was doing the same thing. Aha! Maybe it wouldn't be so bad after all!

After my on-the-job training, I began doing the same thing and after an hour or so, I was finished. Looking around, I could tell most of the other pilots had completed their tests as well. At that point, it simply became a waiting game, because nobody wanted to complete a four-hour test in just 45-minutes or so. As we killed time, everyone continued doing a masterful job of looking busy. Finally, after a couple of hours, the first brave soul, naturally a Warrant Officer, stood up and began confidently striding towards the desk where the officers-in-charge of the test were seated. They would grade all tests on the spot so we would all know immediately whether we had passed, or failed. The ice had been broken and other pilots began heading towards the desk until it literally became a tsunami of pilots rushing to join them, myself included!

By the time I got in line, there were already quite a few pilots in line, and standing directly in front of me was one of the IP's from C-12, my own flight. Since he was holding his IBM answer card behind him, I idly began looking at it and happened to notice his test was labeled TEST I, the same test I had taken. Since we had both taken the same test, I began comparing his answers to mine and was startled when they didn't seem to match up with each other! I whispered over the IP's shoulder to get his attention, then asked if he had taken Test I. He looked at it, then responded with a "yeah" as he turned to face me. I quickly informed him of what I had seen, so we began a thorough comparison of our answers and found none of the answers matched up! At all! He looked at me and said: "You really screwed up!" My response was: "Hey, maybe you screwed up", but I was also thinking we could be involved in a "sting" to ferret out cheaters by handing out a bogus answer sheet! A not so good situation for my military career, or his!

Since we had no clue which answer sheet was correct, we began passing the word up-and-down the line for all pilots who took

Test I to gather around us so we could resolve our dilemma. Soon, there were about five or six pilots who had taken TEST 1 in line with us, so we began comparing answers. Every answer card was spot on regarding our answers except for one pilot; my fellow Charley-12 IP! It was apparent he had used the wrong answer sheet for his test. Plus, he couldn't correct it either, because we had all been given black ballpoint pens to mark our answers on our IBM card. With eyes the size of saucers, my fellow IP asked in a quivering voice what should he do? Unfortunately, none of us had the slightest idea what to tell him.

Meanwhile, the line was steadily moving toward the desk where all tests would be graded. In the same manner as Michael Jackson's epic "moonwalk" moves, my fellow IP began working his way backward even as the rest of us were steadily moving forward. By then, all IP's had completed their tests and were now joining us. As each of them lined up, my buddy steadily worked his way backward, toward the end of the line that snaked all the way to the rear door exit. It was quite impressive how well he finally made it all the way to the door, and with one last scared look followed by a quick wave of the hand, my fellow IP disappeared out the door! Shortly afterward, my own test was graded and after receiving a high score, I could finally breathe a sigh of relief. The world was right again!

Needless to say, word got around quickly about my buddy's royal screw-up, so he was the butt of many jokes for several days afterward! However, with so many students to fly, we were all too busy to give it much thought and after a week or two, it had become ancient history. However, the military had a long memory, especially when it came to annual written exams for U.S. Army aviators, and they didn't forget! We had that hammered home to us about a month later when two Majors, a Captain and a six-striped Sergeant paid Charley-12 a visit.

All the above-named gentlemen sauntered into our classroom and the ranking Major called out the name of my buddy who had skipped out of the mess hall! Our Flight Commander responded by

asking them: "Who the hell are you?" Their response was that they were the team in charge of all aviators' annual written exams and my fellow Warrant Officer IP had NOT taken the written exam! He was to be immediately grounded, and maybe even shot at a later date for such a horrendous crime! The IP at the heart of it all turned ashen-faced and began struggling to stand upright. Things got so loud the commander of our entire Flight Department entered our classroom to see what all the ruckus was about. The rest of us were trying to make ourselves small and insignificant! Sitting at the rear of the room, I was quite happy to have distance between me and the combatants at the front of the room! Unfortunately, my relief was short-lived because my Flight Commander bellowed for me to come up there! I quickly stood up, marched up front and said: "Yes sir, how can I help?"

My red-faced, teed-off Major responded by telling me the officers in charge of written exams had made some outlandish accusations that one of his Charley-12 IP's had not taken the written exam! Since I had been scheduled to take my exam the same time as my fellow IP, he knew I had to have seen him and if I could verify that, he could then tell our unwanted guests to, as he put it, "get the hell out of my classroom!" Consequently, he asked me one question:

"Did Warrant Officer Smith take the annual written exam as scheduled?"

"Yes, sir," I responded, "Warrant Officer Smith took the annual written exam as directed. I saw him there."

The officers-in-charge of annual flight exams were told to leave and that there would be no further questions about a missing exam because "as incompetent as they were, they had probably lost it!" As soon as the unwanted guests stormed out of the room, my ashen-faced Warrant Officer buddy walked up and said: "Thank you for saving my ass! I thought my career was over!" I said, "No need to thank me. They simply asked the wrong question. They asked if you had taken the test, and you did. They didn't ask me if you'd turned in your IBM answer card for grading." In retrospect, I didn't lie to a direct question from a superior officer, my fellow IP hadn't

been grounded and everyone knew our Major was a stand-up officer who had our backs. To my knowledge, none of Charley-12's IP's ever failed another annual flight exam!

1968 – A Year to Forget!

On March 7, 1968, our flight was conducting flight training at My Tho, a remote stage field located just south of Mineral Wells, Texas. I had just completed my first training period, had landed, and was in the process of shutting down when my student and I observed a TH-55 helicopter being flown by a solo student attempting to land his helicopter in a strong tail-wind condition. Suddenly, the student seemed to lose control and allowed the helicopter to spin while the nose pitched up, thus allowing the tail boom to hit the ground. The helicopter itself then slammed down hard onto its right side with a tremendous boom, immediately followed by its engine screaming since it was no longer turning the main rotor.

Knowing the helicopter could explode into flames at any minute, I jumped out of my helicopter and raced over to help pull the student pilot out of the aircraft through the cracked plexiglass, just about the time a fire truck arrived to spray the helicopter with foam to prevent a post-crash fire. A couple of other pilots also rushed over and we desperately tried to resuscitate the unconscious pilot while waiting for the rescue helicopter to arrive. Unfortunately, we were unsuccessful in our attempts, and the young student never regained consciousness. The next day we heard he had died later that same evening. Everyone in our flight took his loss pretty hard. He had been a good student who would have made a fine pilot.

On September 23, 1968, my student was hover-taxiing our TH-55 helicopter into takeoff position from one of the departure panels at Downing Heliport when I heard Lance Fogde, one of my flight school classmates, call for takeoff just prior to my own call. Lance's TH-55 was cleared for takeoff just prior to us receiving our own takeoff clearance. I saw his helicopter accelerate through translational lift, then begin climbing out just as we received our takeoff clearance. I turned my attention inside to check the instruments just about the time my student began initiating our own takeoff.

We had just taken off when I recognized Lance's voice screaming out a "mayday" call on the radio and was "going down"! I quickly looked out just in time to see his helicopter plummet from the sky, hit the ground and literally explode! I lost sight of him as we continued climbing out to avoid colliding with other helicopters taking off behind me. Even though it had been a horrible sight, we had to continue to our assigned stage field to conduct flight training. It had been a bad day that became much worse when we returned from our training period and were informed both Lance and his student had been killed in the crash.

On November 20, 1968, I lost another flight school classmate when Bill Jenkins, crashed at an adjoining stage field while in the process of giving a student pilot his standardization check flight. Again, there were no survivors. It was another tough loss of a good friend and classmate. I was glad to put 1968 behind me!

Bravest Pilot I Ever Knew

Most helicopters at Fort Wolters were outfitted with dual controls so either the IP or student pilot could control the helicopter, as necessary. However, there were also a few helicopters that had a single set of controls, and they were strictly for solo flights only. After a student had soloed after 12-15 flight hours, they could start accumulating solo flight-time as called for per the flight training matrix, and that was what solo-only helicopters were for. They definitely weren't dual flights, because no IP in his right mind would crawl into an aircraft without dual controls, especially with student pilots! However, there were those occasional exceptions.

Every stage field where we trained had a large, metal control tower. It was rather rudimentary in comparison to what they have at large airports, but it had a radio, a light gun, binoculars and a telephone for emergencies, and that was enough. Typically, the Flight Commander would be the one who occupied the tower so

he could control flight operations, but many times, whenever an IP had spare time, they would climb up into the tower to assist him. On that particular day, that's exactly what I did. The Flight Commander could always use the help, plus I enjoyed doing it. Everything was going well until he noticed one helicopter being flown in an erratic manner, so I assumed duties as a controller while he grabbed the binoculars to zero in on the young student. After a minute or so, he advised me to tell the student to hover over near the control tower so he could meet with him to see what his problem was.

Following orders, I called the helicopter and ordered the student pilot to hover over near the tower where he could meet with the Flight Commander. As soon as he landed, the Flight Commander walked up to the helicopter and began talking to the student. Shortly afterward, he pulled himself up into the helicopter. He was a large man, so the helicopter seemed to groan as it squatted a little closer to the ground as he settled into the seat. That done, he called the tower requesting takeoff clearance, which I granted. The little helicopter struggled to get into the air with the added weight, but it was soon in a hover, albeit in a somewhat herky-jerky manner. Looking at it through the binoculars, I also noticed something else a little different about the helicopter but thought I would say nothing until they returned for landing. Everything seemed to be going just fine, so why mess it up?

After flying a traffic pattern, they hovered back over near the control tower, landed and the Flight Commander climbed out and clambered back up into the tower. As soon as he was seated, he began telling me how that ignorant student pilot had forgotten to reduce the control frictions, thus making the helicopter harder to fly. Aha, that explained the "herky-jerky" manner! I acknowledged as to how some students can sure do dumb things! Then I asked him if he had noticed anything different about the helicopter he had just flown in, such as maybe it had only one set of flight controls?

Suddenly, he had such an incredulous look on his face that I had to chuckle out loud! He had that "deer-in-the-headlights" look

before finally squeaking: "Are you kidding me?" "Nope", I said, "it was one of those solo helicopters that only had one set of controls for the student!" Then I added, "Now I know you're not near as dumb as that student pilot, so I'm just going to tell everybody you're the bravest pilot I've ever met!" I don't think he appreciated my joke, but he was a good sport about it, and all of Charley-12 IP's had a good laugh about it. As far as I know, he never again crawled into a solo-only helicopter, at least not on my watch!

The Price of Soloing

As previously mentioned, the vast majority of students who showed up at Fort Wolters, Texas had never even seen a helicopter, much less flown one, and the first benchmark of success was to demonstrate the skill to solo. The typical time it took for a student pilot to solo was approximately 12 - 15 flight hours, which was usually in the third or fourth week of training. Naturally, it was always a big event for the student, but the IP's enjoyed it as well. The IP's enjoyed it, first of all because it documented their teaching skills, and every IP took great pride in their student's accomplishments. Secondly, it was also the time their bar was re-stocked!

I don't know when, where or how it started, but the ritual was for each student to present a bottle of liquor to the IP who soloed him. Not just any brand either! It had to be a bottle of the IP's favorite brand! As you might imagine, a lot of liquor exchanged hands during the third or fourth week of training since that was solo time, and that was just for the young American student pilots! Later on, when we started training Vietnamese students, they would present the lucky IP with an entire case of his favorite drink! Those were very lucrative times, at least in regards to liquor! In fact, I cannot recall having to buy a single bottle of liquor the entire time I was an IP at Fort Wolters because I soloed a lot of students! Great days!

Launch and Recovery at Fort Wolters, Texas

During the Vietnam war, Fort Wolters might have been one of

the most important bases in the entire United States military establishment, not to mention one of the busiest! I can make that statement because that's where 100% of new U.S. Army helicopter pilots were trained. The scope of such an undertaking was astronomical, especially when the demand for helicopter pilots ramped up dramatically in the 1967-68 timeframe, whereby 600 students were graduated each month!

To accomplish such an undertaking, there was the insatiable need for training helicopters. The Army utilized several different models, but the helicopter most of us recall was the bright orange TH-55's, or in accordance with the Army's history of naming helicopters after Indian tribes, the Osage. They had 760 of them, far more than any other model. I think it's safe to say there were probably 1,000 helicopters of all make and models at Fort Wolters, and that's a lot of helicopters to manage, especially when approximately 800 helicopters were launched, twice a day, to accommodate all the students in flight training.

There was also the need for training areas and offsite facilities to support the three main bases where helicopters were stationed and from where they would launch. Then, after launching, there were 22 stage fields of various shapes and sizes in an area encompassing more than one million acres, all generously donated by patriotic farmers and ranchers. To indoctrinate the student to Vietnam names, most of the stage fields were named after some of the larger cities in Vietnam such as, My Tho, Vinh Long, Soc Trang, etc. Getting to and from those locations during peak launching and recovery times was definitely not for the weak of heart!

The morning flight training classes stretched from 6:00 AM to noon, the evening classes from noon to 6:00 PM. Helicopters were literally in the air throughout the daylight hours, and the sound was not unlike thousands of bees filling the skies! Each helicopter, whether dual (with IP) or solo (without IP), would depart a helipad, then follow a series of painted markers on the ground to exit the three main heliports. As each helicopter passed over the

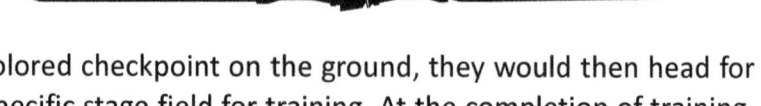

final-colored checkpoint on the ground, they would then head for their specific stage field for training. At the completion of training, everything was reversed, and I think that was the most dangerous time!

All 800 or so helicopters launched, had to be returned back to their respective heliports and each one had to approach in a specific route and air speed. Doing anything else could be dangerous and there were several mid-air collisions to prove that point! As dangerous as it was on a daily basis, Fridays were always the worst, especially in the afternoon, because everyone was anxious to get to happy hour at the officer's club! Somehow, we all survived, and I remember those days fondly. We worked hard and played hard, but I don't think I could ever do that again!

Memories from Fort Wolters

I have a lot of great memories from my days at Fort Wolters, plus I also have two special gifts. I was in flight school when my son, Keith Wayne Williams, was born on February 25, 1966. I was an Instructor Pilot when my daughter, Nicole D'Ann Williams, was born at Beech Army Hospital on June 11, 1969. They have both made me very proud.

Another special memory was Neil Armstrong's moon landing on Sunday, July 16, 1969. An emotional Walter Kronkite, famed CBS news anchor, and the world watched as Neil Armstrong stepped out onto the moon. No one wanted to fly on Monday! Everyone wanted to talk about that exciting moment!

U.S. Army instructor pilots assigned to Flight C-12 (I'm 3rd from left, front row): Major Arthur Conlon is receiving flight safety award at Fort Wolters, TX (1969)

I'm standing with two of my students beside a TH-55 training helicopter at Fort Wolters, Texas (1969).

CHAPTER 7

Petroleum Helicopters, Inc. (PHI)

In the late 1940's, Bell Aircraft Corporation certified a new kind of machine that eventually became an integral part of offshore oil operations. That new machine was called a Model 47 helicopter, but when it was first certified, there was no market for it; consequently, one had to be developed! Seeing a future in the swamps of Louisiana, some innovative men saw a need for helicopters and formed a company known as Pet/Bell. It started out small, but eventually grew into what became known as Petroleum Helicopters, Inc. (PHI). It grew so quickly, when I left the military in 1969, PHI was the world's largest civilian helicopter operator! They were certainly the big boys on the block in the Gulf of Mexico, plus they also had ongoing world-wide flight operations. They were, and still are, a very reputable company. That was the company that offered me, and countless other Vietnam helicopter pilots, a job to fly helicopters all over the Gulf of Mexico!

After separating from the U.S. Army on October 10, 1969, I loaded our small family of four into the family auto, drove through the gates of Fort Wolters and headed southeast for Lafayette, Louisiana. It was the beginning of a new adventure for two reasons: first of all, it was the first time my wife and I had been out of the military as a couple since we were married on August 12, 1965; secondly, it would be my first job as a commercial pilot. In the military, you literally have everything you need to exist; medical, dental, commissary for groceries, service stations, etc. Leaving all that behind to become a civilian was a difficult decision to make! It was rather scary for both of us because we had no idea what life on the "outside" would be like.

Getting hired by PHI had been a difficult process. Not only did I have to possess the right qualifications, I also had to sit before a five-member board to explain why I wanted to work for PHI. The war in Vietnam had created an insatiable appetite for helicopter pilots, and most of them were choosing to separate from the military soon as their required time was up. Due to the large pool of pilots, PHI could afford to be picky in their quest to hire only the best pilots available. A job offer a few weeks later signified I had met all requirements and my starting salary would be a whopping $725.00 per month! It would be the first non-military pay check of our marriage! I just hoped it would be enough!

My wife, two small children and I arrived in Lafayette full of hope, yet somewhat apprehensive of what the future would hold. We checked into a small motel that would be our new home for a couple of days until we could find a rental house to move into. Rising early the next morning to begin training at PHI, my first order of business was to order a cup of coffee to start the day and I darned near choked when I gulped down the first sip! God was it strong! It was definitely high octane and that was my introduction to "Cajun coffee" laced with chicory! It took a lot of cream to tame that beast and I never learned to like it! Fortunately, most offshore platforms had three coffee urns, one each with mild, medium and regular. The regular contained chicory. I never graduated above the mild coffee, but I have to admit, that Cajun coffee did a good job of getting me hyped-up and ready for training on my first day at PHI, October 20, 1969!

There were four new pilots in my introductory class taught by Merlin Bute, a fine gentleman who was an experienced offshore pilot. The course lasted a week and the first portion consisted of what it was like flying in the Gulf of Mexico. The second portion would consist of five hours training in a Bell 47G model helicopter flying in an offshore environment. Merlin passed on a wide variety of tips that would hopefully keep us alive flying offshore. He also taught us a very basic method of navigation since the only instruments on our helicopter were a clock and a magnetic compass, commonly referred to as a "whiskey compass". Time, distance and heading was all you needed to navigate offshore! Merlin explained that unlike the roads, rivers, water towers or other handy check points on land, there were very few oil platforms and drilling rigs in the broad expanse of the Gulf. Plus, the drilling rigs moved around – a lot! Consequently, you couldn't rely on those for navigation. And most important, there was lots and lots of water! It was almost like he was telling us it was far easier to get lost than find the right platform! And oh yeah, always keep track of your fuel! FAA regulations required all pilots to have enough fuel on board to fly back to the shoreline, regardless of your destination offshore, or wherever you were located.

When I actually started flying operationally for PHI, I became a "floater", in that I had no assigned job. I would simply fly, or "float", wherever, or whenever, they needed me. It could be at Morgan City, Cameron or any of the numerous bases located from Texas all across Louisiana. PHI ruled the Gulf at that time! They even had "service stations" for refueling on all the Gulf Oil platforms that stretched across the Gulf. They definitely had their act together! There was always a lot of flying to be accomplished during my seven-day work shift. At the same time, there was also a lot of boredom. There really wasn't much glamour flying repeatedly over vast expanses of water in the Gulf of Mexico. However, in between the blurry lines of day-to-day humdrum, there were many isolated instances that kept you either scared out of your wits or rolling on the floor with laughter! As a pilot, you had to have a good sense of humor and accept things as they came, whether good or bad. The one thing you could not be was thin-skinned! You had to be willing to laugh at your own screw-ups, because if you couldn't, you would be called out mercilessly!

Beach In, Beach Out

"Feet wet, or feet dry" is still the typical phrase utilized by Navy and Marine pilots when they were either crossing the beach heading out over the ocean or had crossed the beach were returning

to dry land. At PHI, we simply used "beach out or, beach in". It's a simple phrase but means oh-so-much when you're not used to flying offshore, especially for someone like me! I had only one experience going offshore and that had occurred while I was still in Vietnam. We had to fly quite a distance offshore to find the U.S.S. Iwo Jima, then over to the U.S.S. Canberra. The thing is, I didn't like flying over water then, and I'm not sure I was ever comfortable flying over water, even though I did it for almost 4 and a half years when I was with PHI!

I still recall with great clarity the first time I called "beach out" over the radio as I flew offshore in a Bell 47-G4. The beach disappeared behind me and there was nothing but water to see in all directions! Flying solo, it was very disconcerting at first, but I soon adapted to it. You really didn't have much choice, because flying offshore to support the oil industry was the bread and butter of PHI's vast, worldwide operation. When you hired on with PHI, you were reconciled to the fact you were going to fly offshore. PHI had other operations literally all over the world but, in the end, you always came back home to the Gulf of Mexico. PHI had been in business since the late 1940's and by the time I arrived in October 1969, they had gotten pretty good at flying offshore to support oil operations, or in support of any other offshore activity for that matter! They had a vast array of offshore refueling and "flight-following" radio stations located on Gulf Oil Company's platforms that spanned the entire Gulf of Mexico. They were indeed the best in the business in those days, and had practically no competition!

Weather was another factor to contend with. The only time I can recall when the prevailing weather was crystal clear was maybe the next day or two on the heels of a cold front. Ironically, that was also the day when everybody seemed to get lost! That was because you could simply see too much! There were too many drilling rigs and production platforms you didn't recall seeing before, because the norm was always much less visibility due to haze, fog, mist, low clouds, etc. There was also one area to avoid at all costs.

It was actually named Vermillion Bay, but everyone referred to it as Vertigo Bay, and it was well named! I don't care what the weather was, you simply didn't want to navigate through that area, especially with the miserly few instruments we had back then. For some reason, there was never a visible horizon to use as reference to keep the helicopter flying on an even keel! I don't know why, nor did I ever learn how Vertigo Bay received its name. I simply avoided it and that was good enough for me!

Another Bay, referred to as Mosquito Bay, was also an area to avoid! You never, and I mean never, wanted to land at Mosquito Bay! Inhabited by millions of mosquitoes, it was literally a place of death, where neither people nor animals could survive! I still recall the only time I ever had to land there to drop off a part for another helicopter that had made a precautionary landing. Even though I sat in the cockpit with blades turning, I could see swarms of mosquitoes! It was a lesson well learned for me, just as it was for the pilot who had made the landing. We both agreed we would never land there again! In fact, we wouldn't even fly over it after that!

Newspapers in the Gulf

Every morning when I arrived at whichever base I was assigned to, the first thing I did was to pick up my "allowance" of the TIMES-PICAYUNE newspaper. Every pilot did the same! You gathered up newspapers for every stop you were scheduled to make amid the multitude of drilling rigs, barges or platforms in the Gulf! It might've been Knute Rockne, Notre Dames' famous football coach, who was once quoted as saying, "There are some folks who think football is a matter of life or death, but they're wrong. It's far greater than that"! I think that quote was also applicable to how much those papers were worth! Getting a current paper was far greater than "life or death" to some of the folks who worked off shore in those days! Every paper was also read multiple times; it never went to waste!

In those days, communication was not instantaneous as it is now, plus television reception offshore was horrible! It was like watching a snowstorm! Additionally, trying to call in via the old ship-to-shore routine was not the best way to call home to your wife. I still recall one young stud who would call his girlfriend the same time every night and commence having phone-sex, totally unaware he was extolling his sexuality to everyone in the Gulf! Since it broke the monotony to the great glee of everyone listening in, no one ever bothered to tell him! To some, it was better than watching a snowy TV screen, but probably not as good as a newspaper!

    I learned the worth of a paper on one of my early flights with a grizzled old PHI pilot who was "showing me the ropes". We had landed on a drilling rig that served the best steaks, shut down and walked down the flight of stairs before entering the office area. Passing by the "pusher's" office, the pilot tossed him a newspaper, then we proceeded down to the galley where fine steak lunches were being served. After eating one of those steaks with great relish, it was time to be on our way.

    Let me explain PHI's generous per diem reimbursement plan for all employees who worked away from home. If we worked away from home at a base, we were reimbursed $2.00 for breakfast, $2.50 for lunch and $4.50 for evening meals, for a grand total of $9.00 per day! If you should eat a meal at an offshore platform, the protocol was to sign a "chit" for that meal, and PHI would reimburse that company for your meal. You could not turn in an expense account for the meal you signed a chit for! That would be double-dipping! The pilot just lost whatever amount that meal would've been, and in those days, a couple of dollars was significant! In reality, none of those offshore rigs ever required a pilot to sign those chits. At least, as long as you took them a paper!

    The grizzled old pilot and I started to walk out the galley door when the cook's galley hand stopped us to sign a chit. Stunned, the old pilot jerked the pen out of the helper's hand, scrawled his name across the chit, then bellowed to everyone in sight, "that's the last newspaper you'll ever get from me!" He then stormed out, went directly to the pusher's office, stepped inside, jerked the newspaper out of his hand, then reversed course and began climbing the steps to the heliport! The pusher asked what in hell was going on, to which I responded by saying "we had to sign a chit". The pusher then screamed for the cook to "bring me those g--damned chits!" Which he did! The pusher grabbed them, caught up with the pissed-off pilot, then ripped them up in front of him! "That's better", said the grizzled pilot, "here's your paper back!" No one ever had to tell me again how much those papers were worth, and I was never without one! If I were going to land at 7 different locations, I had 7 papers - always! Papers were more important than free meals in the Gulf of Mexico oil fields back then.

Have You Eaten Yet?

I had made my last stop of the day and was heading for my home base where I could take a well-deserved rest. It was a beautiful Sunday afternoon and I was looking forward to a snack since I had missed lunch. Since it was summer, I was cruising at an altitude of about 4,000 feet in the cool air on a course that would take me directly over Union-26, a small platform located just a short distance offshore. Just before I flew overhead, I received a radio call by someone with a very Cajun accent:

"This is Uniform-26 calling helicopter passing overhead, come in!"

"This is Sierra-41. Go ahead", I responded.

"Had anything to eat?"

"Negative", I replied. Uniform-26 then asked, "You like smoked spare-ribs?"

"Affirmative! Love them!"

"Got a newspaper?" he asked.

"Got one with your name on it!"

"Come on down! We got the best spare-ribs you ever done ate!"

I landed, handed over the paper, and was handed the best plate of spare-ribs "I ever done ate"! In fact, I think I ate two plates that

day! That became a ritual that lasted until I moved on to another job assignment about a year later.

Carlos Marcello, the Don of New Orleans

I hadn't been at PHI very long when I checked the assignment board one day and noticed 3 or 4 small helicopters were scheduled to go to New Orleans Moisant Field (now Louis Armstrong Airport) to pick up passengers. I didn't recognize the name of the company leasing them, but one of the more senior pilots nonchalantly said: "That's Carlos Marcello's group." Since I was the new kid on the block, I had absolutely no idea who Mr. Marcello was! Several pilots jumped at the chance to tell me all about him, because it seemed he was a very famous gentleman in those parts.

Carlos Marcello had been known as either Godfather, or the Big Man, of the New Orleans crime family. He had been involved since 1947, and that lasted well into the 1980's before he was finally arrested and spent a few years in prison. Interestingly, there are still many who are convinced that he, along with several others, masterminded the assassination of John F. Kennedy in retaliation for federal prosecution of their organization. But to me, old Carlos was just a blank sheet of paper when those three other pilots and I flew over to Moisant that day.

We were flying old Bell J-Model Rangers (not the popular turbine powered Jet Rangers) and after landing, we patiently awaited the arrivals of our customers. They arrived at various times, so when one of us had a load of two or three customers (depending on size), we departed and headed out to the duck hunting camp marked on our maps. Once everyone was there, we all shut down at the camp to await further orders. Expecting to set out there in the heat and misery of southern Louisiana, we were quite surprised when a "guard" approached and asked if we'd like anything to eat or drink. We were shocked, but very pleased at his generous offer! When we said yes, he went back to the cabin and true to his word, returned with quite a spread! Some of the other guys came

out to visit us as well. They were very interested to hear we were all Vietnam veterans. In fact, they even thanked us for our service! Wow! They were really nice guys! When we got back to PHI's old Morgan City facility, we all agreed they may be bad guys, but they were some of the nicest customers we could ever hope to fly for!

The Old Southern Gentleman

I can't recall exactly what type of mission I had been on when I found myself deep in Mississippi one beautiful day. Most likely, it was searching for oil leases or something like that, but I can't swear to it. What I do remember is being directed to land at an absolutely gorgeous southern mansion right out of "Gone with The Wind". It was the spitting image of Tara! You almost wanted to look around for Scarlett, and maybe even Rhett! It really had that look!

Not wanting to blow anything up, out or away from such a place, I parked way out on a paved area, hoping not to interrupt anyone. As I was sitting in the cockpit at flight idle, waiting for the engine to cool down, I noticed what appeared to be someone sitting on the porch, watching us. As soon as I shut down, my customer walked over to meet some gentlemen standing near some automobiles while I waited beside the helicopter. It wasn't long before the person I noticed up on the gigantic porch stood up and began walking toward my helicopter.

As he approached, I could scarcely believe what I was seeing! That old gentleman was right out of Hollywood casting! An old southern gentleman reincarnated! Dressed in a white linen suit with his head topped by a straw hat, he had been quite a sight! He was ramrod straight and had a graceful stride I admired. In his right hand was a drink; it just had to be a mint julep! In his left hand, a cigar! He approached to within 4 or 5 feet from me, then stopped. Looking squarely at me, he half-bowed and said: "How do you do, suh." I responded in kind. I also shook his hand, then extended a welcome for him to look around. He looked at the helicopter for a minute or so before turning to me, then said in a deep southern drawl: "Suh, that's

a mighty fine-lookin' vessel!" I had never heard anyone use that description for a helicopter, but I readily agreed with him. "Yessir, it is indeed a mighty fine-looking vessel!" He had been a real, "genteel" southern gentleman! I'm sure he's long since passed on, and I doubt there's any elderly southern gentlemen of his kind still around in this day and age. That's a shame. He had been one-of-a-kind, and we could certainly use more of his kind, especially today.

My Welcome to SMI 58

In January 1970, I got my first regular assignment as a regular pilot. It was on an offshore platform sitting approximately 60 miles out in the Gulf of Mexico. In that particular block, there were three Shell Oil platforms; SMI 58 A, B and C. SMI 58 B was home, or at least it was for 50% of the time, since we worked a 7-day on and 7-day off schedule. The pilot who preceded me was a fellow you might have heard of. He had been a U.S. Army Aviator like the rest of us, and even a Rhodes Scholar, so, he was pretty smart! He also knew how to play a guitar and write songs! In fact, when I first showed up as their new pilot, the first thing some of them asked was: "Can you play the guitar?" I think they were pretty disappointed when I said no. I wasn't overly concerned, because there's not too many people who could play the guitar, write songs and sing like Kris Kristofferson!

Book on How to Fly Helicopters

When I was first assigned to SMI 58, like everyone else, I read paperback books or magazines to pass time. That's because there always seems to be a lot of standing by when you're a helicopter pilot! After a couple of months, I began to think I should do something more worthwhile, like getting an FAA Airline Transport Pilot (ATP) rating for helicopters. There was very little use for me to obtain such a rating that was typically required only for airline captains, but I thought it would be more industrious than just reading books and magazines. It took a lot of preparation on my part, so I

began gathering all the pertinent material I could find to assist me. Since I knew FAA exams always had some of the most basic questions, one of the books I took to the platform to study was the FAA's book entitled "How to Fly a Helicopter".

Between flights to the various platforms, I had been spending a lot of time in the break area on the main platform studying my material. On one particular morning I was sitting in one of the comfortable chairs and reading "How to Fly a Helicopter", when one of the Shell employees walked through, saw me, then walked over and said: "What are you doing?" My response was: "Studying for a FAA test". I was holding the book where he could easily see the cover, but thought nothing of it. However, during the course of the day, I had never seen so many of the Shell employees casually stroll through break area, take a look over my way, then continue through the door. I really didn't think too much about it other than it must have been a slow day for those guys!

That evening, right after dinner, the platform manager stopped by and asked if I could please step into his office. He was a very pleasant fellow, and one of the few Texans on the platform, so I said "sure!" When I walked in and took a seat, he started asking me questions about my past flight experience, which I thought was rather odd. After a few minutes of chit-chat, he finally said, "Dwayne, didn't you fly helicopters in Vietnam, and have a lot of flight time?"

"Yes", I answered. "I was in Vietnam and I have well over 2,000 flight hours in a helicopter".

"Then why are you reading a book about how to fly a helicopter"?

Aha! The light bulb just got brighter! "So that's why all your hands kept walking through the break area!"

He then explained how concerned everyone was about me reading a book about how to fly a helicopter, and they were getting nervous about it. I promised to never be seen reading that book again! The solution was to insert my book inside a Playboy and continue studying. Case closed! I never heard another complaint!

Friday, February 13, 1970

One of the most important things they beat into you when you first join PHI is to "tie your helicopter securely to the platform!" Then, when in doubt, "tie your helicopter securely to the platform!" They were adamant about that, and for good reason; winds can blow hard in the Gulf of Mexico! In fact, I guess that's true about all big bodies of water, since there's nothing to slow the wind down. But on that February day in 1970, it was a wind such as I had never experienced before!

As I climbed the stairs up to the heliport located on top of the living quarters, I was surprised at how strong the wind was, especially at such an early hour. I was flying an old Bell 47-G4 model, on large inflated floats, and it wasn't very stable flying in high winds. Consequently, I always paid attention to the wind velocity. I started the helicopter and flew a few of the hands over to the other platforms so they could do whatever it was they did. Somewhere around mid-morning, I advised the supervisor I needed to pick up all the guys from the outlying platforms, because I was going secure flight ops for the day. The winds were already bouncing off 35 knots, and that was my limit since it's difficult to startup or shutdown in high winds. In fact, we actually had brooms secured on the platforms, in case we had to push up against the blades when they began slowing down to prevent tail boom strikes.

But that day was different! Since the winds seemed to be increasing rapidly, the supervisor concurred with my decision to gather everyone up and bring them back to the main platform. I made the rounds, picked everyone up, then secured the helicopter to the helipad. I used every tie-down PHI provided to secure it, and even tied the main rotor blades with extra rope. Satisfied everything was OK, I went down the stairs into the living quarters to wait out the weather.

It was going to be a long day, so I settled back, dug out a paperback book and began reading, and listening to the wind. Shortly after lunch, I checked the wind velocity and it actually peaked at

60 to 70 knots. My God! You could actually feel the platform shake when those high waves hit it! That's about the time the standby boat called and said he had broken free from the platform and, in spite of using full power, was drifting further away from the platform, unable to make any headway back toward the rig! So, there we sat! My helicopter was grounded and our stand-by boat gone! Even though I was on a platform, I felt all alone in a raging sea!

I had always heard the term, "fury of the seas". On that day, I learned what it meant. It was horrifying for me, because I had never experienced that before! Even some of the old-timers on the platform admitted they were a little concerned. Well, if they were a little concerned, I was a whole lot concerned! I just kept going to look out the windows at the sea. It was raging! Then the rain hit! It was like a fireman spraying the windows. I couldn't even see out. Reading my book was out of the question, so I tried my best to look calm. Nobody wants to see a nervous pilot. But I fooled no one that day. It didn't matter, because no one was calm. Everyone was nervous! Then, just when you thought it really couldn't get much worse, it did.

There was a huge noise! Not a boom, just an extremely loud noise that came from right outside the door. Even though SMI 58 was a production platform, it had what they called a "work-over rig" that looked much like a drilling rig, since it stood about the same height. It occupied the entire work deck, and everyone was absolutely certain it had toppled over. So much so, everyone gathered together to figure out our next move. It's interesting how, when the shit-hits-the-fan, everyone wants to be with someone. They don't want to be alone. So, we're all together, talking at once, trying to figure out what the huge crash and horrific noise was right outside the door. The front door! The one we were all watching. Thing is, we had forgotten about the back door.

While we're watching the front door, the back door opened. As one, we all turned to face the back door just in time to see an apparition rush through it! An apparition that looked very similar to the

275-pound foreman for the work-over rig, except he was dripping wet and his eyes looked like saucers! Looking straight at me, he said in his best Cajun drawl:

"Your helicopter done crashed! I done seen it levitate, den it crashed! Den when I seen it lift up, I ran!"

He paused for a minute, then added:

"Well, I thought I be runnin'! Den sumpin hit me in da ass! Den I be runnin'!"

It was a very good description I thought; in fact, we all did! Even in the the chaos around us, we all had a good laugh. It certainly helped break the tension.

After we had settled down, I tried to open the front door again, but the wind was just too strong. I could crack it open an inch or so, but all I could see was my smashed-up helicopter just a couple of feet away from the door. There was nothing else I could do except call the lead pilot located in Morgan City and inform him of what just happened. His order was to take photos of the accident, then standby for a team to come by and investigate. His last question was: "Did you tie it down properly?" Yes, I did! My next call was to my wife on the ship-to-shore frequency to tell her to hold off on unpacking. We had just moved to Lafayette in October and now, four months later, I had just lost a helicopter! I was sure I would be terminated.

The next morning was an absolutely beautiful morning. The wind had blown through, done its damage, then departed. My beautiful Model 47-G4A helicopter was totally destroyed, since it had pitched nose-first between the work-over rig and the crew's quarters building. It would never fly again! As I was contemplating my future employment elsewhere, the foreman yelled and said someone from PHI was on the phone. Quickly grabbing it, I started telling the lead pilot the situation when he cut me off.

"Don't worry about it", he said. "We had four more helicopters totaled just like yours! We're sending a barge around to pick all of them up. I'll try and get you a replacement helicopter out there later today."

"Does that mean I'm not fired?"

"Of course not; you're doing a great job!"

Wow; what a relief that was! My wife and kids would be happy, plus I was feeling pretty good about it myself. At least until the replacement helicopter arrived! The N-number emblazoned in bold numbers on its side was N1313X. The date was Friday, February 13, 1970!

Help! I'm Locked Up

I had just flown the foreman from Morgan City back out to SMI 58 in the aftermath of a hurricane that had been a near-miss. As you might imagine, there were those usual things strewn about, but production platforms are sturdily built. After a bit of tidying up, everything was soon back in order, at least on the main platform. But not so much on an unmanned platform that was about a thirty-minute flight to the west of us in my BH 47-G4A model helicopter.

Shell had been in the process of upgrading the platform prior to the storm and had a single Shell foreman to manage approximately 20-or-so roustabouts to make it happen. All personnel had been evacuated, but since it was only a "small" storm, everyone had been advised to standby in the local area in hopes of getting everything back up and running post-haste. Sure enough, that's what happened. Even though everyone was told to stay put, they weren't told how much they could drink, not that it would have done any good. Naturally, the rowdy roustabouts tied one on! So much so, when they loaded up on the crew boats to take them back to the unmanned platform to get back to work, they decided to take a bottle or two with them. As they put it, just to clear their brains of the last vestiges of their massive hangovers the following morning.

Since it was an all-day affair to sail out to the remote platform, the "rowdies" all fell asleep on the boat ride. Just before dark, they had arrived at the remote platform. The Shell foreman I had flown over earlier in the afternoon was on hand to welcome them aboard. However, refreshed from their long snooze on the boat, the young

rowdies decided there had been no good reason to stop until all their booze was gone. Work could come later!

Sometime around midnight, the night-shift foreman on the main platform received a desperate phone call from the beleaguered Shell foreman overseeing the construction on the lone platform. The old "one riot, one Ranger" trick didn't work, at least for that Shell fellow. When he tried to intervene with their drinking party and put them all to bed, they had responded by throwing him into the small office building, then grabbing boards and a handful of nails, they sealed him in his own office. Even the windows were sealed! That's about the time he had made his desperate call for help.

Since the platform was a good distance away, the first order of business for the field foreman was to roust me out of bed so we could fly over to rescue the captive foreman. "Excuse me", I quickly informed him, "PHI pilots don't fly at night", especially over the darkest stretch of ocean known to man! I also explained all I had for instruments was an airspeed indicator to tell us how fast we're going, plus an altimeter to tell me how high I was above the ocean, neither of which would do much good at night. He finally accepted the fact we weren't going anywhere until daylight. We took the time to prepare our battle plans for the best way to invade that remote platform just at daylight. That was when we'd have the element of surprise.

At sunup, we were making last-minute preparations for our invasion to rescue our petrified Shell employee. One of my passengers was the field foreman; the other was the biggest, meanest Cajun boy he had in his crew! That was good! I liked big dudes like that on my side! I just wished our foreman had been comparable in size rather than being a pipsqueak, like me. Not long after sunrise, we were off. My plan was to make the first run "out-of-the-sun", then when they were blinded while trying to find me, I would land, let my warriors off to join them in battle, take off and circle in the event my intrepid warriors got tossed overboard.

I was very proud of how well we performed our plan-of-attack. It proceeded just as we had planned. The only thing I didn't have to do, was take off to circle and watch the battle from afar. That's because when we touched down, we were all rather shocked at what we saw! Bodies were everywhere! Some sitting up, some laying down, some even laying on top of each other! Sometime in the wee hours of the morning, our rowdy boys had run out of booze. With nothing to keep them fueled up, they had simply collapsed wherever they happened to be standing. The battle was over before it even began.

Fortunately, Shell had dispatched a crew boat to sail out soon as the mayday call had been received, complete with a replacement crew and maybe even a company of Marines. They arrived not long after we did. The beleaguered foreman was rescued, and the roustabouts were literally tossed on the personnel net and lowered to the waiting boat below. I think they slept all the way to shore, where they were all sent packing. I asked the field foreman if they would all be sent to jail for their "uprising". "No", he said, "We'll probably see them all back out here next week. Hard-working roustabouts are hard to find".

How to Wear a Life Vest

One day, I carried an engineer out to a production platform so he could assist in changing out a water pump that pumped in sea water for the toilets. Since he was the only passenger in my BH 206, he obviously was deemed to be very important in regards to resolving the problem. After all, when you couldn't use the potty, you had a big problem! Upon landing, I was told to stand by in the event I had to "hot-shot" a part back to the shore base and pick up a replacement. The engineer disappeared below while I went to do my usual thing; get a cup of coffee and relax. After a couple of hours went by, he came in and told me I needed to make a flight back to the beach to take the old pump he had just removed, plus pick up a replacement pump. Okey-doke for me. No problem!

When I got back up to the heliport, I was rather shocked at the size of the failed pump! It was about 4-feet long, and fat! Like a fat cigar! The engineer advised me a roustabout would fly back with me to assist in unloading, loading, etc. Not a problem. The first thing was to load the pump. Since it was rather heavy, I decided we needed to lay a piece of wood on the floor to distribute the weight, rather than allow the concentrated weight to punch a hole in the aircraft floor, not a good thing! Looking around, we located a 2 X 6-inch board about 3-feet long. Perfect! Just what we needed, except we had to cut it in half. I asked the young roustabout if he could take the board and cut it in half. Sure, no problem! He grabbed it, then dashed off to make it happen! Soon he was back with two pieces of board. He had done exactly as I had asked. He had cut it in half! The only thing, he had cut in half lengthwise! Duh! Luckily, we found another board to cut in half. I told him "this time, cut it crosswise!" He did.

Soon, with everything loaded, we were preparing to depart, and I began briefing the young roustabout about emergency procedures. Since it was common procedure to wear life vests at all times when flying, I pulled one out and handed it to him. Just as I did, the engineer yelled at me about something and it distracted me for a second. When I turned back around, I saw the young man trying to pull the life vest down over his head, just as you're supposed to do. However, it was much easier to do if you removed your hard hat first! I yelled out: "Take your hard hat off first!" Looking somewhat confused, he pulled his hard hat off, looked at it quizzically, then thrust the hard hat through the life vest, then proceeded to try and pull it up his arm! OK! Let's start over!

Somewhere, I recalled hearing a story about how the famed Green Bay Packers had gotten severely beaten one Sunday, a most unusual event for Coach Lombardi. The following Monday, he was standing in front of his dejected players, and in his hand was a football. Holding it up, he pronounced to everyone: "This is a football". Someone in the back spoke up and said: "Can you slow down,

coach? You're talking too fast!" I think I found him. He had escaped to the Gulf of Mexico!

Hell, Anybody Can Fly This Thing

PHI's uniforms never seemed to fit, or look right; at least, for pilots. They were a deep forest green that looked very much like those uniforms the old Texaco full-service gas station employees wore. They just weren't very classy for us pilot-types! Sometimes you never knew if the guy wearing it was a real pilot or not, unless he was wearing that distinctive yellow cap. One of PHI's more illustrious pilots proved that point one day.

PHI's large facility located on Lake Palourde was much like an airport terminal, in that passengers would arrive early in the morning of their scheduled crew change days, find the rig they were going to, then check in. A sign-in sheet was created each day for every drilling rig or production platform doing crew changes. In many instances, the passengers might not even know each other, especially if they were a "roustabout" crowd. Sort of like it was that morning. The aircraft they would be flying on was a Sikorsky-62 model, a large single-engine helicopter built to float and operate on water. It had a keel and everything it needed to do that. Since the cockpit sat so high above the ground, the pilot had to enter the aircraft with the passengers, then make his way forward to climb up into the cockpit.

That particular pilot checked the roster, saw all passengers had checked in, and noticed there was one passenger seat left open. Never one to miss an opportunity to mess with people's minds, he had the dispatcher announce over the loudspeaker that "flight so-and-so was ready to depart for Zapata 6 (or whatever drilling rig it was) on the aircraft parked on spot so-and-so." Eager to get the long flight over, everyone grabbed their bags and headed out to load up! The pilot, dressed in his normal looking uniform without his yellow cap, did the same thing. What he didn't do was continue into the cockpit. He sat down in the remaining seat and, like everyone else,

began waiting for the "pilot" to show up. After several minutes of impatient waiting, the passengers started getting restless and began to gripe about such things as: "Where's the f-ing pilot! We're wasting time! What the hell is going on!" As soon as the griping hit its crescendo, the pilot announced to one and all: "Hell! Anybody can fly this damned thing!" He then unbuckled his seat belt, angrily stomped down the aisle into the cockpit and proceeded to "fire her up!"

Since the sun had already peeked above the horizon, creating heat and humidity, most helicopters had already departed so there wasn't a lot of activity on the flight line, which had been a good thing! I was busy pre-flighting my own BH 206 Jet Ranger just a few pads away when an explosion of passengers burst out of the S-62 I heard starting up! Since that helicopter held quite a few passengers, it was quite a scene! A couple of them stopped just long enough to scream and yell for someone to "get that crazy f-ing roustabout out of that helicopter before he kills everyone on the flight ramp!" They then turned and joined the rest of the group whose legs were moving like pistons, since they were busy trying to put as much distance between themselves and that crazy "f-ing" roustabout who thought he could fly a helicopter!

It took quite a while to get things sorted out since no amount of pleading, or even a confession on the part of the pilot, could convince the passengers to get back on that helicopter! They wanted a new pilot! One of those REAL pilots who wore a yellow cap! They finally changed pilots, but it took a while since there had been very few S-62 pilots on the staff. The culprit who started it all apologized profusely and said he would never do that again! I don't remember him getting into much trouble, either. It was pretty funny, and I think everybody got a laugh out of it! Everybody except those passengers!

Frozen Fish That Floated

Oil-field hands were notorious for understating how much their "carry-on" luggage weighed back in those days. In most cases, it

shouldn't have been much, because most of the rigs provided them personal lockers, and usually, they even had a mini-washeteria. The thing was, many of them liked to fish. It had been a good way to pass the time of day, plus, it had also been profitable! For instance, one well-known restaurant would buy all the red snapper they could deliver for fifty cents a pound, a real fortune in those days. What made it even better was the fact they only had to gut'em, wrap'em in all the newspapers they could find, freeze'em solid, then carry them to shore on the helicopter! The restaurant folks grabbed the box of frozen goodies, weighed it, then doled out the dollars in cash, no questions asked! It was simply too good to pass up, even if they did have to fib a wee bit to their pilot about how much their luggage weighed!

I saw passengers staggering up the stairway to the heliport, veins threatening to rupture as they strained to carry all that weight! Then, when asked how much their luggage weighed, they had boldly announced: "twenty pounds!" On that particular day, the crusty, old pilot was flying a S-62 while making crew changes to a barge located just offshore on the Louisiana coastline. For quite a while, he had been making those same crew changes and he had been getting a little disgusted with them, since they were outright lying about how much their baggage weighed. So, that day was the day he had chosen to make believers out of them.

Sure enough, every passenger chirped the same "twenty pounds" when asked what their baggage weighed, even though their extended veins said otherwise! The crafty pilot solemnly accepted their word about the weights, loaded them up, then fired up his S-62 helicopter with a bottom that looked like a boat. Sure enough, when he picked the helicopter up to a hover, the power needle was bouncing off the red line, when it should have been sitting comfortably in the green! Even though the takeoff had been no problem, the pilot jiggled the controls enough to make it look a bit dicey, just to get everyone's attention! Then he leveled out and began screaming at the top of his lungs: "It won't climb! We're

overloaded! Prepare for a hard landing! We're going in!" And with that, he proceeded to pull full aft on the cyclic stick, pitching the nose of that helicopter to where it had pointed almost vertically, directly towards the heavens, then suddenly pitched the nose forward again into a level position, splashing gently into the shallow waters of a marshy area! It had all been safely, and expertly done, and those Cajun boys didn't have a clue they'd been had!

The crafty pilot later said he had never seen so much jettisoned flotsam in his life! And he had been a retired Navy pilot! Boxes and boxes of frozen fish went floating by! He later said he wished he had snatched one or two boxes of fish, because he really liked red snapper! He should have, too, because he never saw any more frozen fish floating alongside his helicopter that could float like a boat!

Just a Wire Brush

One morning I was on top of the helipad, getting ready for the day's activities, when surprisingly, the field foreman and two of his biggest, toughest-looking employees, climbed the stairs then stepped onto the helipad to await the arrival of a Bell 205 crew-change helicopter. Their timing was perfect, because I could already hear the familiar "wop-wop-wop" that was so familiar to every soldier who survived Vietnam! That was because the 205 was simply a commercial version of the famous Huey. That "wop-wop-wop" sound had been the cue for every roustabout who had spent the last two weeks on the platform to race up the stairway. That day was a great day for them; they were going home! Just like me, they were quite surprised to see the three amigos standing there. Three amigos with very stern-looking faces!

The foreman motioned for everyone to step over to where I was standing and wait for the 205 to land and shut down. When the incoming crew stepped off the helicopter, they were ordered to standby on the platform until they were released. Everybody had that puzzled look on their faces! No one had the slightest clue what was going on except for the three amigos!

Stepping up to the first roustabout in line, the foreman asked him if he had any Shell Oil property in his luggage. I have to say, it took guts to do that, because that roustabout was a big, rangy-looking dude with long muscular arms and big hands! Definitely not the guy you wanted to get in a bar room brawl with! Sitting beside him was a military-style duffel bag. A full one! The roustabout looked him squarely in the eye and said: "A wire brush". "Let me have it", demanded the foreman. The man reached down, unsnapped his duffel bag, picked it up, then turned it upside down. Bam! The wire brush slammed to the deck! All I could do was stare, dumbfounded. Hell, we were all dumbfounded!

He had not lied. It was definitely a wire brush! A circular wire brush! The kind of wire brush that had a large, one or two horsepower electrical drive motor attached to it! After the foreman finally comprehended what he was seeing, he just raised his head and looked at the roustabout with his mouth gaped open. The roustabout simply returned his gaze, shrugged his shoulders and said: "Like I said, a wire brush." Without another word, every hand standing beside him grabbed their duffel bags and began dumping them! I had never seen such a wide assortment of tools, nuts, bolts, food, fish or just about anything else imaginable that began tumbling out of their bags!

The incoming crew was finally dismissed and allowed to go downstairs, no doubt extremely happy they were incoming and not outgoing! They would've been caught red-handed just like the guys still being interrogated upstairs. The outgoing crew was finally allowed to leave with dire warnings about never coming back to a Shell Oil platform for as long they lived! Truth be known, they were all probably back out there soon as their two-week break was over.

What prompted the surprise search had been the disappearance of just one of two brand new walkie-talkies the foreman had received the previous day. In fact, I remembered how I had seen the foreman and one of his cohorts talking back-and-forth the previous evening. They had been doing the usual "Can you hear me?

Can you hear me?" routines that kids do. I can't imagine what idiot would have stolen just one! I mean, what good was that? The lone walkie-talkie was never found. In Mafia-speak, it probably "still swims with the fishes"!

My good friend and PHI pilot Wayne Brown is sitting on the float of a Bell Model 47 helicopter on Vermillion 39, a very old Union Oil Company platform in the Gulf of Mexico (1971).

CHAPTER 8

Mining in the Clouds

Hidden in the vast jungle wilderness and misty forests of what is now known as Irian Jaya, lay a mountain of pure ore. Like a predator lying in wait for its next victim, it had laid silent for millions of years because it was simply too inhospitable and inaccessible! It was hidden in a land that still is one of the most unexplored regions of the world. The land itself is so high and damp that vegetation grows to unheard of sizes! In the area surrounding our camp, I saw massive elephant-ear looking plants, along with an assortment of plants, all of them huge! No doubt, early explorers most likely considered the possibility of meeting prehistoric creatures when they first dared to venture into such a place. Further complicating things was a large number of tribes known to inhabit the area that had very little contact with each other, plus they each had their own language. None of them liked each other, and some of them had even developed an appetite for human flesh.

The large island was formerly known as Dutch New Guinea, and in 1936 a young Dutch geologist set out to climb a little-known mountain known as Carstensz Top. Even though he had been an employee of Shell Oil Company, the geological studies he carried out along the way were secondary to a mountain climbing expedition for his small team. They were deep inside the mountain range when they happened to stumble upon a massive black outcrop of copper which he named the Ertsberg, Dutch for "Copper Mountain". The young geologist didn't think it was too important, but he did take time to write a report.

Everything changed when a gentleman by the name of Forbes Wilson, manager of exploration for an American firm by the name of Freeport Sulphur, came across that same report while searching for references to potential mineral deposits. After reading the

geologist's report, he decided to undertake the difficult journey to see what the young geologist had written about for himself. Seventeen days into his trek, on June 16, 1960, Wilson saw the Ertsberg for the first time. Stunned by what he saw, he told me he "could feel the hair rising on the back of his neck!" The next day, he cabled a pre-arranged code back to base camp that described what was basically a mountain of copper ore; he also added the words "access egress formidable". In other words, it would be a difficult task to mine due to its inhospitable location deep in the mountains.

So formidable, they would have to wait until technology advanced sufficiently to make it a viable proposition. At the time he said those words, he probably had no idea what technology they were waiting for; but in September 1967, the first phase of development was initiated. An integral part of that first phase was two Bell 204 B helicopters leased from Petroleum Helicopters. Those helicopters were the new technology necessary to tackle the "copper mountain" of ore.

The initial shore base was located at Timika, a former World War II Japanese airbase. The plan was to push inward from there. The journey inland was difficult, but slowly and surely, they marched inward by building roads, bridges, camps, and tunnels just to reach the Ertsberg. The helicopter fleet soon increased to three, and in January 1971, six helicopters were required when the project moved into phase IV, the final phase. The project would require 11 pilots and 10 mechanics to keep up with the huge demand of transporting people, fuel, food and parts to keep all the machinery humming! The call went out at PHI they needed more pilots and mechanics, and I was one of those selected to try my hand at mountain flying.

Traveling to Australia

In early May 1971, I left a teary-eyed wife, son and daughter as I boarded a small airliner at Lafayette's airport and headed for Darwin, Australia. It was very hard having to say goodbye to my family! It was a long trip across the Pacific, and the only positive thing about it was being able to travel first class. That took some of the "sting" out of it. Traveling with another pilot, I still remember we were the only two first class passengers on a PAN AM 747 jetliner traveling from Honolulu to Sydney, Australia. The stewardesses, which is what they were called back then, took good care of their two lone passengers. In fact, so much so that we both simply passed out from all the food and drinks our tortured stomachs could handle! It was probably a good thing because it was a series of very long flights from Lafayette, Louisiana to Darwin, Australia!

After many hours of flying and layovers, we finally arrived at Sydney's International Airport, totally exhausted after such a long trip across the Pacific Ocean! All we wanted was to go to bed and catch up on our sleep! So much so that I remember very little of our time spent in Sydney. However, I do recall our flight from Sydney to Darwin, or at least a snippet of it. I remember the flight being four and a half hours of having to listen to a young Australian engineer complain mightily about having to travel to Darwin. He was distraught about being forced, once again, to travel to Darwin on business, and he was not happy! He proceeded to tell me about all the bad things one had to cope with; such as extreme heat, no decent bars, terrible food, etc., etc. etc. He was convincing enough that I was beginning to re-think why I had volunteered to come to Australia in the first place!

After landing in Darwin, I quickly concluded the young engineer had been spot on regards the extreme heat; however, I couldn't complain about the young secretary who had been assigned to pick up the other PHI pilot and me, nor could I complain about the Foster's beer we were treated to enroute to the office. It was a new brand of beer for me, but it was pretty tasty and very cold! The perfect antidote to the hot weather! Everything considered, it wasn't a bad way to be introduced to a place I'd never been before! The one thing that did catch my eye was the fact everyone drove on the wrong side of the road in Australia, the exact opposite of what I'd

always been taught in the U.S. It felt a bit odd sitting on the left seat of the car as a passenger, since the steering wheel was situated on the right side. But we soon adapted, and it wasn't too long before we were putting around like it was the most natural thing in the world. The next part of our journey would begin in a couple of days after we had time to overcome the doldrums of jet lag. It had been a very long flight to the "land down under"!

A few days after our arrival in Darwin, we found ourselves standing on the Darwin International Airport tarmac anxiously waiting to board the F-27 turboprop airplane that would take us to Timika, in West Irian Jaya. It was a place that would become our home away from home for 20 days out of every 30, since we would amass one day off for every two days worked. The F-27 was our sole means of transportation between West Irian and Darwin, and we would become very familiar with it during the upcoming months. The flight over was pleasant enough; it wasn't too crowded, and it was a smooth ride. A good start for what would be a six-month adventure.

The flight to Timika was uneventful, but I must admit I was somewhat surprised when I first saw the international airport! It had been an old Japanese base during World War II and old anti-aircraft weapons still sat starkly on the beach. They were once hidden among the palm trees, but years of receding beaches and erosion had laid them bare for all to see. There were still a few old Japanese Zero fighter aircraft that lay hidden in the surrounding jungle. Long forgotten, they now lay in various stages of rot, rust and deterioration, and will remain so until they will become part of jungle itself. Any human remains from that long-ago war have already done so.

The airport had no paved runway and the customs office was what appeared to be a shack constructed of whatever happened to be available to construct it. Whoever built it had utilized logs, nails, wire, plus maybe even glue and tape to erect a loose resemblance to a shack, skillfully topped off with huge palm leaves! The customs officers had the appearance of having come straight from

the surrounding jungle, but they were pleasant enough and we had little problem in getting our passports appropriately stamped. Just about the time we cleared customs, we heard the familiar "wop-wop-wop" of a Bell Huey helicopter inbound to pick us up. A familiar black and gold Bell 204B model helicopter landed close to the F-27; grabbing our suitcases, we climbed on board to be whisked away to our new home at Camp 22, so named because it was a mere 22 kilometers inland from the port site.

Our new home was a nice, well-built barrack type building called a Port-A-Camp. It consisted of a small room just large enough for two pilots located on either end, with a bathroom in the middle. With four pilots sharing one bathroom, it was not exactly the comforts of home, but certainly adequate. I later learned we were extremely fortunate to have such quarters because no one else in the camp had such "luxurious" homes except for the camp manager. I certainly couldn't complain about the food either, since they had a very nice screened in mess hall that was under the control of South Koreans. It was clean, neat and the food was really good!

PHI's Chief Pilot for the operation was a pilot by the name of Don LaFreniere, a very pleasant, debonair gentleman who was the real deal! He was a superb pilot who had once flown for President John F. Kennedy and was, at least in my opinion, the perfect choice to lead such an operation. He had arrived with the first helicopters in 1967 and as far as I know, was the last to leave when the PHI's contract was completed. I have always considered it a privilege to have known and worked with him.

My first flight in the mountains of West Irian Jaya, New Guinea was on May 15, 1971. It was an orientation flight up into the mountains with Dave "Matey" Simmons, an experienced PHI pilot. It was my first experience flying in mountains and I have to admit, I was dumbstruck about what I observed on that hour-and-a-half flight! Not only was I in awe, but I actually began having serious doubts about volunteering to fly in such an awesome, forbidding environment. Plus, making things worse, I had a serious ear block due to a cold I had picked up on the long flight over. We had climbed to an altitude of 12,000 feet so we could get a close-up glimpse of one of the only tropical glaciers in Asia. It rested on a massive mountain called Puncak Jaya, and although it was a spectacular sight, I paid the price when we initiated our descent. I honestly thought my ear and sinus cavities would rupture! It was so painful, I actually embarrassed myself by letting out a muffled yelp I tried to stifle, but couldn't! Fortunately, we resolved the problem by leveling off at different lower altitudes to let the pressure dissipate enough to continue our descent to the next level. It took so long we began to worry about having enough fuel on board to make it back to our base camp! Fortunately, luck prevailed and we made it home just in time for a late lunch.

I was signed off to fly missions on that single flight, but I did have to take a day off to clear my sinuses before flying again on May 17. I remember thinking had I been in the military, I would've undergone at least two weeks ground & flight training just to qualify as copilot for such difficult flying, then flown perhaps another week or so with an instructor! Then, probably a check ride! Conversely, commercial operators utilize the old "time is money" methodology. They simply don't have an unlimited budget for such niceties! Their attitude was I had experience in the helicopter, met all prerequisites for the job and had volunteered for it! Oh yeah, one last thing: Don't break the helicopter! Any other questions?

A Project Made for Helicopters

Of all the projects utilizing helicopters that I've been involved with, the Ertsberg project was made to order for helicopters! In fact, I seriously doubt it could have been accomplished without them. I only played a short 6-month part in the multi-year project, but it was a critical one. It began in September 1967, when Don LaFreniere, PHI's Lead Pilot for the project, made the first flight in a Bell 204B helicopter. In January 1971, helicopter operations moved into Phase IV, the final act of what had begun four years previously.

Now requiring six helicopters, the call went out for more qualified pilots, and Vietnam had qualified me with more than 1,000 hours in Huey model helicopters, the type they were flying. That's when I showed up.

The mountains and conditions we flew in daily were intimidating! In fact, a couple of pilots came over, took a few flights, then departed for home, and no one blamed them. Weather and mountains present two major problems to cope with, and any mental lapse could end in tragedy! We had one pilot who made a mistake by trying to land slightly downwind on a very narrow saddle stretching between two mountains. The aircraft hit going too fast, bounced off the pad, then tumbled down a very steep slope, ass-over-teakettle! When the dust settled, the helicopter had disintegrated, but the pilot was still strapped in his seat, perched about 100 feet down the 60-degree slope. He was lucky. Had he not been stopped by a projection of rock, he would have been part of the wreckage several hundred feet below. A very brave crew worked their way down the slope, freed him, then pulled him back to the crest of the slope where we he was medevacked, first to Tamika, then on in to Darwin. We never saw him again.

Each day began by loading personnel, food, parts and anything imaginable that we could get inside the aircraft, then heading up into the mountains! Once we finished that, we then began sling loading everything that was either too big, or too heavy to fit inside the helicopter!

Caterpillar specialists were flown in to carefully "dissect" D-6 Cats and front-end loaders into manageable sizes to sling load. The heaviest loads were the D-6 mainframes and we carried two of them to a mountain camp that was 8,700 feet high! We carried just enough fuel to sling load them to the camp, then literally autorotated back down the mountain to conserve what little fuel we had on board. The same specialists who cut them up, were then carried up into the mountains to re-build them.

One of the most interesting sling load operations was conducted

by Don, whereby he latched on to a spool of cable, then began a vertical climb, dragging the cable from Camp 72, located in a high mountain valley just below the Ertsberg, up to the Ertsberg itself. Once that was accomplished, they simply kept pulling bigger and bigger cable until they had the correct size needed to set up a tramway to ferry personnel and copper ore from the Ertsberg to the newly built work camp that was named Tembagapura. I remember picking up a British gentleman one day to carry him up to the Ertsberg and along the way, I casually asked what his area of expertise was. He explained he would be wrapping, twisting and tying the cables together in the "proper" manner. Turns out he was one of only two people in the world who was capable of such a thing,

They did have trouble with the tram in the first stages of operation, though. It would begin swaying back and forth when they carried the more heavily loaded ore cars that weighed several tons. A host of remedies were tried, including speed reduction, reducing loads and even changing centers-of-gravity, but nothing worked! When all else fails, read the manual! Or, in that case, canvass the world and see if there's anybody out there who could help! As it turned out, there was. An experienced Swiss engineer took on the challenge. He made some calculations and a few other modifications which greatly reduced the swaying. He said it had to be tuned, just like a violin. Who knew?

The Helicopters

The helicopters utilized for that particular job were six identical Bell 204B model helicopters, the same type I had flown in the Maverick gunships in Vietnam. They also just happened to be one of my favorite helicopters to fly! They were painted the familiar black and gold, even though some were somewhat faded due to so many hours in the sun. I was also somewhat surprised when I began my preflight because a few things were missing. Whatever the FAA said PHI could remove from the helicopters and still fly, was gone! There were sporadic holes in the instrument panel from random gauges that had been removed to save weight; for instance, the main generator was gone, and only the starter/generator remained. I can't recall what else was removed, but weight was absolutely critical flying high in the mountains, so most aircraft were barebones empty!

The helicopters also had emergency kits, just as every aircraft in the world has. However, that emergency kit had a most unusual item in it. It had a twelve-gauge shotgun! It was carefully secured and strict rules and regulations were always adhered to. The gun was there because that part of the world was probably the wildest and most dangerous place in the world to fly, at least at that time. We flew over triple-canopy jungles, vast mountain ranges and in some of the most treacherous weather I have ever flown. Plus, waiting down below in the thickest jungles on the planet, were wild tribes of cannibals who were always looking for a tasty snack! My main desire was never having to use that shotgun, and I never did!

When flying sling loads up to the 10,000-foot level, as little as 200 pounds of fuel was pumped in for the climb up. We didn't need much, since the return trip was a descent and pilots literally kept the collective on the bottom as they conducted needles-joined autorotation back down the mountainside to the operational area (comparable to a car racing down a steep hill with foot off the accelerator). There were literally hundreds of loads to be flown to the bases up in the mountains, so we would land, roll throttles back to idle, tighten down frictions, then climb out to stretch or relieve ourselves while another 200 pounds of fuel was pumped into the fuel tank. Then we repeated the entire process again, and again and again. We never ran out of loads to sling!

Cassowary Airlines.

Our little fleet of six Bell 204B helicopters was jokingly referred to as Cassowary Airlines, so named after a huge bird similar in size to an Ostrich, but almost looked prehistoric. It had a huge "comb" on its head like a rooster, yet it appeared to be made of the same material as the horn on a rhinoceros. They're also considered to

be one of the most dangerous birds on the planet! After having an opportunity to meet one up close and personal, I no longer considered the name of our little "airline" to be a joke!

It occurred when I was doing my daily routine of walking several miles up the new road. In reality, the road appeared to have been barely wide enough for a truck, plus there was no shoulder or ditch. In fact, on either side of the road was jungle so thick you couldn't see through it. I had walked several miles along the road when I heard an unusual sound somewhere just inside the thick jungle. Even though I thought I heard something just off to the side of the road, I really couldn't see anything more than a foot or two into the thick jungle-growth. I was on the alert for anything unusual and kept walking until I made a turn at a sharp bend in the road. What I saw caused me to suddenly put on the brakes!

Standing in the middle of the road, no more than fifty feet away, was an extremely large bird! I felt like we were looking eyeball to eyeball! It was a large beautiful bird and he showed no fear at all! Perhaps it was because I was on his turf and he had no reason to fear me. He seemed more curious than alarmed, mirroring my own thoughts! He did take a step or two toward me but, after deciding I wasn't a threat or a tasty meal, he slowly turned and ambled back into the thick jungle and disappeared. It was an amazing moment with an extraordinary creature! Not wanting to antagonize my new friend any further, I turned and retraced my steps back to the base camp. The name of our little "airline" was no longer a joke, at least for me! That magnificent bird was a throwback to prehistoric days, and it was my good fortune to have shared a few minutes of his time.

Our Chief Refueler

At Camp 22, the majority of employees were specialists from literally every country in the world! Each was an expert in their respective fields, and might have been anything from cooks to miners! However, local natives were employed to handle more menial tasks, such as refueling helicopters. I remember the native who considered himself as our "chief" refueler, and he took his work very seriously. His name was Pieter (pronounced Peter), and I don't recall ever hearing his last name. It was always simply Pieter, but I'm sure that was not his original name. He was probably christened with that name when he became a Christian. I always thought it was a good thing he finally accepted Christianity, because I'm sure he needed all the help he could get whenever he found himself standing before the "Pearly Gates". That's because the pleasant and hard-working Pieter had been a cannibal in his youth!

I found that fact out one afternoon visiting with Pieter while we were involved in flight operations. Even though he wasn't a PHI employee, we all considered him as one of our crewmembers because he was always there, eager to work, regardless of what we asked him to do. He also seemed to be very proud we all considered him to be our "chief refueler". One day, I was standing alongside Pieter while he was refueling my helicopter and since all PHI employees were American, I nonchalantly asked him how he liked Americans. I assumed he liked us and thought it had been a simple question, one to which he could quickly respond. I was quite surprised when he seemed to be thinking very seriously how to answer what I had thought was an easy question.

"Well", he said, "not too much. They have a bad taste." I was totally stunned by his remark! I think I even looked around to make sure I wasn't alone as I quickly climbed back into the helicopter. As I was flying back up into the mountains, I remember thinking I needed to find out what made us Americans taste so bad! Whatever it was, I intended to make it a regular part of my diet! Maybe garlic! Or better yet, maybe broccoli! Even President George Bush didn't like broccoli!

Later that same evening, I made a point of meeting up with the camp manager during dinner in hopes of getting a plausible reason for Pieter's strange answer. After explaining what had occurred, the manager looked me square in the eyes and said Pieter had most assuredly practiced cannibalism as a young man. He also told me Pieter was much older than he looked. He was 54 years old! My God! We all thought he was probably only in his mid-thirties! He could've sold his secret of eternal youth to women! He then told me Pieter's tribe had most likely practiced cannibalism probably until he became a Christian in the early-to-mid 1950's.

In fact, he also felt reasonably certain that some of the more militaristic tribes still practiced cannibalism, especially the Asmats! Young Michael Rockefeller, aged 23 and son of Nelson Rockefeller, had disappeared just a short distance from where we were standing. The young man had swum to shore, looking for help after his boat capsized. The manager felt certain he had most likely ended up being slain and eaten! The date of his disappearance was November 19, 1961, a scant 10 years before I was there.

The Witch Doctor

As I stated earlier, our camp was made up of specialists from all over the world, but most of the manual labor was accomplished by local tribesmen who appeared as if they were not far removed from the ice age! Some of them had come from local tribesmen in the surrounding area, while others we had to go and find. I remember one day when we flew deep back into the triple-canopy jungle and landed in a small clearing in our quest for more workers. I was advised not to shut down while one of the local managers, who was well-versed in local customs, stepped out of the aircraft to meet some of the local tribes. Just about the time he was moving back into the jungle, a very strange looking fellow all dressed up in feathers, mud and paint stepped out of the jungle! He wasn't armed but he was wielding a club-like "wand" made of what appeared to be a leg bone. He was also pretty fierce looking and seemed a bit put out with such rude outsiders who had invaded his domain. He went into wild gyrations that could be called dancing, while simultaneously throwing dust, feathers and God knows what else towards the helicopter! I wasn't feeling too comfortable with all that, and the way the young camp manager raced across the clearing and leaped back into the helicopter, I got the impression he wasn't feeling all that lucky, either! As he was buckling-up, he motioned upward with his finger and said: "let's get the hell out of here!"

I have to say, by the time he was finishing his sentence, we were already airborne and, in his words, "getting the hell out of there!" He then went on to say that "bloody bloke" was a local witch doctor who wasn't pleased at all with our visit! Consequently, he was putting a "hex" on us and our helicopter! He also went on to say there were some fierce looking warriors who seemed to be stalking us just outside the edge of the clearing, hiding in the jungle in case the "hex" didn't work! "I made a mistake", he said. "Those guys are from the Asmat tribe and they're warriors. Working is for women!" Well, hello! We sure as hell won't be doing that again! In fact, had I seen those warriors lurking in the jungle, my Ace might have found himself in a foot race just like Indiana Jones sprinting to outrun a band of bloodthirsty natives! Then he'd have to catch up with my helicopter!

Malaria "Ice Bowl"

When you work in the tropical jungle, everyone is susceptible to a wide variety of disease. There are also lots of bugs, beetles, snakes and other slithery critters that thrive in such jungles. But one of the deadliest is mosquitos! Especially the females! Always on the lookout for fresh blood, they pounce on unsuspecting victims, then literally spear them with very small knives. Then, after sucking out your blood, they sometimes leave their calling card that often leads to death. It's called malaria. And late one night we got an urgent call from one of the other pilots urging us to get help immediately!

One of our senior pilots had developed uncontrollable shakes, was vomiting and had extremely high fever! The kind of fever that can boil your brain! Fortunately, the camp had a very capable Australian doctor and he knew exactly what to do. Get him on ice!

While two pilots got him in a bathtub, a couple of us raced over to the mess hall and filled buckets with ice, then raced back to the cabin where the deathly-ill pilot was. We dumped the ice in on him in much the same manner you would ice down beer! And it continued until dawn when we could medevac the ailing pilot to Timika for transfer to Darwin. Prayers, ice and dedicated work saved his life. He was back at work a few weeks later, and he had the foresight to hide a case of beer in his luggage. We had the beer; we just needed ice! Fortunately, we knew exactly where to find some! I also think we put the beer in the same bathtub in which we had iced down the ailing pilot! We were veterans when it came to icing things down.

Changing Beacons in the Jungle

Even though we were in "ice-age" country, there were several antennas standing between some of the camps and the port site. They were necessary for radio communications, and even though there was very little flying activity around them, they all had beacons that flashed incessantly to warn any unsuspecting pilot of their existence. Since our small group of pilots represented the majority of area-based pilots, we remained acutely aware of their location in the event we were called upon to make an emergency flight at night. Sometimes, their bulbs burned out and replacing them wasn't all that difficult. However, getting there to do it was an entirely different story! Most were standing like sentinels in a sea of green, lush jungle, some of which was triple canopy. One day, one of the more prominent towers had gone dark, thus signaling a light change was due. Since all flight activities occurred from sunup until the rain formed in the mountains (which was literally every day), all maintenance support flights took place in the afternoon after flight operations in the mountains had ceased. Since it only required one helicopter, we had a rotating schedule to ensure every pilot had his turn at flying such missions.

On that particular day, several pilots, myself included, were hanging out at the flight ramp when a maintenance support request came in from the port site facilities for a helicopter. Since all flight operations had ended for the day, due to heavy rains in the mountains, it was already a slow day. So much so, that four of us clambered aboard and headed towards the port site base. Two pilots were up front in the cockpit, while another pilot and I sat in back acting as crew chief and "gunner" on the short flight. Upon arrival, the aircraft was shut down so we could be briefed on what our mission would be. We soon found out we'd be transporting one of the young local natives out to one of the antennas where we would lower him on a cable to the ground so he could climb the tower and change the bulb. It sure sounded simple enough!

We were all a little surprised when the young man appeared, because he wasn't all that big, and he just didn't look too enthusiastic about the job at hand. I sort of understood his feelings when one of the managers produced a harness he was required to wear. Made up of leather straps, fasteners, hinges and maybe even duct tape, it literally engulfed him! It took a magician to even get it on him! Good thing it was long after Houdini's time, because he would NEVER have gotten out of that particular harness. It was difficult enough just getting it on! We also noticed when each additional piece was snapped on him, the more apprehensive he became. But, then again, maybe it was just our imagination.

Soon enough, our young warrior was in full battle dress, ready for action! I must admit, he did look dashing, at least what you could see of him. It was so bulky, he had trouble climbing aboard the helicopter, so the two delegated crewmen (another pilot and I) assisted him up into the cabin. It didn't take too long to get to our "drop" site and as soon as the pilot at the controls established a stabilized hover about 100 feet Above Ground Level (AGL), he advised us via the intercom he was ready to lower our young warrior to the ground. Well, he might have been ready, but our young warrior wasn't! It was beginning to become quite obvious he didn't have a clue what he had gotten himself into! Someone had simply said,

"Hey! I need you to go change a lightbulb", then let it go at that! But the way he had braced himself against the rear firewall, it was evident he wasn't going anywhere! In the meantime, the pilot was getting impatient because it's pretty hard holding a steady hover at such an altitude. Soon he's screaming "Get him f-ing out! I can't hold this hover forever!" Desperate to come up with a solution, I had a brainstorm!

    I grabbed the hook on the cable and yelled for the pilot to hit the over-ride and spool off the cable for me! Our young warrior didn't know it, but he had a large ring on his harness just for this purpose. Both myself and the other pilot reached over to him and I snapped that hook in the ring and yelled into my intercom "suck him out of the aircraft"! As soon as our ace saw what was happening, he tried to unhook, but it was too late! Out the door he went and was just dangling there, screaming at us! Expletives probably! But we couldn't hear much with all that rotor and wind noise! "Lower him!", I bellowed to the pilot. Down he went. Not too fast, just enough. As soon as he was on the ground, he disconnected from the harness. He was a quick learner! We began to circle so the pilot could turn the controls over to the copilot and give his aching muscles a break! Like a real trooper, our young ace went right to work and began climbing the tower. Not wanting to leave him alone out in this vast wilderness, we hung around the area and enjoyed the scenery. It was a rare break for all of us. Four pilots just enjoying life and yakking it up! It didn't get much better than that, and we thoroughly enjoyed it!

    It wasn't long before he signaled he was ready for us to winch him back up into the safe, comfortable confines of the helicopter. Stabilizing at 100 feet AGL, we lowered the cable and soon as it reached him, SNAP! He snapped it in place and waved to come back up. Damn! He was already a pro! Not knowing what he would do when we got him back in the helicopter, we were watching him closely. But soon as we got him level with the helicopter, all we saw was a big smile and eagerness to get back on board! And we were glad to oblige! We were soon on our way back to the port site and

we had ourselves a real warrior on board! Even better, soon as we landed, he bailed out and was greeted by a huge crowd of adoring fans! Probably every native in his tribe knew what he had done! He waved happily back to us, we waved at him, and they all waved back at us! It was actually pretty cool! I'll bet his journey through the sky had become a lasting part of the legend and lore of his tribe! He had certainly earned his 15 minutes of fame. Sometimes you just have to drag them to it.

Flying Nekkid

When we flew in the mountains, we normally wore light jackets because it stayed pretty cool in the high altitudes, while the weather at sea level was rather mild. In fact, it could get quite warm at sea level, and in some instances, downright hot! Normal dress for mechanics was khaki pants or shorts, a T-shirt, or sometime no shirt when they were carrying out maintenance tasks on the six busy helicopters. Pilots wore PHI's standard issue uniform which was typically khaki pants and shirts and they were quite adequate while flying at sea level. Half of our flights were flown at lower altitudes since the mountains were typically socked in completely by noon. At that time, we shifted gears from carrying sling loads up into the mountains, to sling-loading cargo from the port site to the staging areas located at the base of the mountains.

One day, we were all busy sling-loading everything from the port site and I'll admit, it did become tedious sometimes, since it was steady back and forth flying. On that particular day, it was also hot! When it gets hot and humid, it soon ceases to be fun. That's when pilots start grumbling and griping, not that they ever needed much urging to do so! So, there we were, back and forth, back and forth, when finally, one of the pilots called out over the radio that he "wasn't hot no more!"

"What'd you do" one pilot asked, "finally take a cold shower?"

"Nope", he said. After a short pause for effect, he said "I'm nekkid!"

"You're what?"

"I'm flying nekkid! I feel free as a bird! I think I might be a hippie"!

As soon as we landed for refueling, a couple of us rolled our throttles to flight idle, then went over to check him out. Sure enough, there he sat, buck-naked! He said he'd always wanted to do that! At least once, just to feel unhindered! Like a bird! We all admitted he'd probably make a good bird alright. Maybe a loony bird, or a Dodo! That night over beers, the lead pilot advised us nonchalantly there would be no more "free-style" flying. That was the end of that; at least, no one ever admitted doing it again.

Crash Site in the Mountains

It's hard to describe what those mountains were like. Forbidding comes to mind! Some soared to well over 16,000 feet and were home to one of the world's only tropical glaciers in Asia! They were also shrouded in clouds most of the time, at least in the afternoon. Further complicating the problem was not many of them had been correctly charted. Even the aviation charts we used were inaccurate in some places. That could be deadly, especially when flying in clouds! We saw proof of that one day when we saw the wreckage of an aircraft still protruding from the face of a sheer granite wall. It literally resembled a dart protruding from a dartboard. A team was flown in as close to the crash site as possible, but it still took hard work and climbing to get to it. There were no bodies, just a couple of ID tags and the serial number of the aircraft. The fuselage had very little rust, and the tires looked like new. One still had air in it. That was all proof of the constant cold temperatures at that altitude. If memory serves me correctly, we found out it had disappeared sometime in 1942 while carrying a planeload of troops wounded in battle to the safety of a medical facility located in the rear area. It was also determined to be a Dakota, the British designation for a DC-3. It was a sad, stark reminder of what map errors can do. As a long-time pilot, I can attest to how luck, be it good or bad, can also be the determining factor.

Arming the Enemy

While walking around the compound one day, I noticed some of the local natives near the river, honing their skills with bows and arrows. Very primitive bows and arrows. They were made with local material and were seriously lacking in the finer arts of bow making! After careful observation and practical application, I realized their bows were not very flexible. It was sort of like trying to utilize a two by four pine board as a bow because it was that stiff! The arrows looked like arrows alright, but they had no feathers that are so critical to accuracy. They were having trouble hitting anything more than 20 or 30 feet away. I decided they needed help, so I spent the next few days trying to gather feathers so I could upgrade their weaponry. It was my good fortune I never found any.

I say that because I happened to make a comment one day to the camp manager about my plan to "upgrade" their weapons into something more efficient. Alarmed, he quickly and patiently explained the error of my ways. He told me that under no circumstances was I to enhance their armory with "upgraded" arrows! It would be akin to arming Tahiti with a nuke! There were many tribes in a small geographical area and the status quo allowed no one tribe to gain an advantage over the other, especially in terms of weapons! None of the tribes liked any of the others, nor did any of them even speak the same language! Any one tribe would love to take over another tribe's turf, and something as simple as making an arrow shoot further and straighter could greatly affect such a fragile "eco-system" made up of so many different tribes. I took his word for it and abandoned my project. Besides all that, I never could find any feathers!

Enemies on the Helipad

The local natives hired as workers had come from a wide variety of tribes, and there were a lot of tribes to choose from. None of them spoke the same language, plus they also disliked each other! And I can't overstate that, either! They just flat-out hated each

other! I found that out one day while transporting personnel to the various camps scattered throughout the mountains.

We always started our day by landing at the operational staging areas, where we were handed a schedule sheet directing us to pick up internal cargo and personnel, then deliver everything to the various mountain camps as per the schedule. As the pilot, I had absolutely no idea who I was flying, much less which tribe they were from. On that particular day, I dutifully dropped off three natives at the designated mountain heliport, then took off, heading to my next scheduled stop. I hadn't been airborne more than five minutes when I got a call from our base advising me to make an immediate return to the helipad I had just left. Hearing how desperate his voice sounded, I made an immediate 180-degree turn. The voice on the radio then increased several octaves in pitch when he screamed "they're killing each other!"

Even though I was flying fast as I could at that altitude, I seemed to be navigating in syrup! Finally, the helipad was in sight. A large crowd had gathered to watch the combatants going after it! My arrival seemed to diffuse the situation and the camp manager and his assistants helped the three wounded warriors onto my helicopter, then urgently waved me off! After lifting off, I assessed the situation and determined my young warriors had received more than they gave in terms of bumps, bruises, cuts, abrasions and Lord only knows what else! But they also seemed ready to have another go at it if provided the chance! But that wasn't going to happen, not on my watch! I advised the base controller of my assessment, and he directed me to go ahead and take them to another camp where their fellow tribesmen were. A crowd had already assembled to greet them, and they were given a hero's welcome!

By the time I lifted off, the young warriors were probably already describing their feats of fisticuffs and heroism, and might have embellished it a little. I probably would have! Since my charges were safe, it had become just another event in the course of another long day. I also found out later I had been lucky! The same thing had

happened to another pilot and his young warriors had been killed before he could get back and rescue them.

No More Two-Steppers

After we had finished flying for the day, another pilot and I gathered up our flight gear and began trudging back toward our living quarters when we noticed a group of workers with a giant crane picking up large steel pipe casings and loading them onto an equally large trailer. Just as they picked up a pipe that had been resting on the ground, one of them let out a loud yell and all the workers scrambled away from the pipe, running some distance before stopping to look back! Curious as to what had scared them, we walked over to where they were standing to ask what all the ruckus was about. Pointing in the direction of the pipe, several yelped out "snake!" Wow, we thought! Must've been a big one! So, we moved up very warily to see the creature.

It was a snake, all right, but it wasn't very large at all! It might've been 3-feet at the most. But the color was spectacular! It was a brilliant luminous green! Not ones to be intimidated by such a small snake, the other pilot and I armed ourselves with 4- or 5-foot sticks, then proceeded to get up close and personal with the snake, prepared for battle with our new-found weapons. Urged on by the onlookers, we poked at the snake and he took offense to this. Enough so, that he would strike, then recoil for another attempt to get us! However, on the last strike, we managed to get a good "whap" on his head, stunning him! We then scooped him up with our stick and left him draped on either side, much like a rope. Wanting to show off our snake hunting skills, we began walking the rather long distance to our quarters, straight through the middle of our camp. We were not prepared for everyone's response!

As we began walking toward our quarters and showing off our new "pet" along the way, everyone began parting to give us a wide berth! It was like Moses and the parting of the Red Sea! The longer we walked, the more we began to think maybe having a snake on

a stick wasn't such a good idea! Finally, we saw the camp doctor. He was Australian, and a very nice fellow. He was also one of our drinking buddies. We always knew when it was 5:00 o'clock somewhere, because we would hear his familiar tap on the door. When we opened it, he would be standing there with his personal shot glass and pronounce: "G'day mates! I just happened to be in the neighborhood. Mind if I step in for a drink?" Our response was to always throw the door open and welcome him in! As educated as he was, he would surely know what manner of snake we were lugging around the camp!

He was looking in another direction when we walked by, but a "Hey Doc!" from me got his attention. Turning around, he first looked at us, then dropped his gaze to our new friend draped over the stick, still a bit woozy, but starting to squirm. The Doc jumped back about three feet and said "Bloody Hell! That's a bloody two-stepper!" All of a sudden, that snake took on the appearance of a fire-breathing dragon! The Doc didn't want him, the guys gathered around us didn't want him and we sure as hell didn't want him! Nobody in the entire camp wanted him! Except one.

Just as we started to get panicky, an Australian miner who had just come down from a mountain camp, strolled by and calmly said: "Wot's on the stick, Mates?"

"We have a bloody two-stepper!"

"AHH, let me see", he said.

We poked the stick closer to him. The snake looked like he was beginning to wake up, and he was pissed! I might also add the miner had probably been in the mountains for months without a bath, and he reeked to high heaven! Perhaps it was the smell that jarred our snake out of his nap! Without any sign of fear whatsoever, the miner simply reached over to the snake, grabbed him behind the head, and dragged him off the stick! He then held him up very close to his face.

"Yeah! It's a bloody two-stepper all right! I'll play with him a bit, then turn him loose. You blokes have a g'day!"

He then turned and walked away with his newfound friend. Everyone just stood there dumfounded until someone finally someone spoke up.

"If that snake bites him, he'll die for sure! And I'm talking about the snake! Whew! I've never met a bloke who smelled that bad!"

I think our friend was a bamboo viper, but I never found out. The Doc had said that if he bites you, "you might take two steps, then die." Calling it a two-stepper was good enough for me. My snake hunting days were over from that moment on.

No More Peanut Butter & Jelly Sandwiches

June 23, 1971 was a spectacular day, but I don't remember what day of the week it was. When you're flying literally every day of your 20-day work schedule, you tend to lose count. I remember it was a beautiful day because I logged 8.5 flight hours in my logbook, and flying all day long while conducting sling load operations didn't happen but three times the entire time I was there. A typical day began while it was still dark at the jungle camp we called home, even though you could see the rising sun begin kissing the peaks of the distant mountains. Sometimes you could still see lights of various camps on the distant mountains, and I was always in awe, because I knew I would soon be unloading cargo up there. What had taken Forbes Wilson 17 days to get there, might take me an hour in my helicopter.

As clear and beautiful as it generally was in the morning, we all knew the mountains would begin to fill with clouds and rainstorms by noon, thus shutting down access throughout the valleys that were our highways through the mountains! On that particular day, it was different! The clouds never arrived, and the mountains remained crystal clear all day long. We were exuberant! So much so, we elected to continue flying as long as we could that day. Everyone was happily doing their thing when someone called over the radio to say they were hungry. Not a problem, the dispatcher said! We'll get some food ready and send it up to the staging area where all the sling loads were being prepared.

Soon, all our stomachs were growling but not a problem; the sandwiches were ready! Knowing how good the food was at our mess hall, we could hardly wait for those sandwiches! After the helicopters began settling into the staging area one-by-one, we rolled our throttles back, tightened the controls, then stepped out to stretch and enjoy a couple of deluxe sandwiches. Unwrapping them with great anticipation, we were stunned when we saw what they were! Peanut butter and jelly sandwiches! Are you kidding me?! We were so hungry, we ate them! But that evening, we had a little chat with our dispatcher and camp foreman. If we ever have another day like that one, do not bring peanut butter and jelly sandwiches, or we'll shut down and eat at the camp mess hall like everyone else!

Believe it or not, the next day was a mirror image of the previous day; beautiful clear skies. In fact, it was the only two-day stretch of beautiful weather the entire time I was there! Again, the decision was made to continue flying if, and it was a big if, we got good sandwiches! The cooks came through in a big way. We all had choice of roast beef, ham & cheese, potato salad, chicken salad, bologna, grilled cheese or whatever we wanted! I think the camp foreman about flipped out when someone ordered a peanut butter and jelly sandwich over the radio, but he was just kidding! The trouble was, after all that good food, everyone needed a nap!

Living Off the Fat of the Land

One day, after the helicopters were released from re-support operations in the mountains, a couple of us had been dispatched to port-site where we would begin re-stocking the forward staging areas. Everything seemed normal enough, but when we neared the port site, I noticed a very large contingency of natives had set up housekeeping around the entire perimeter of the port site camp. They weren't actually inside the perimeter, but were literally jammed up against it. Having never seen that before, I thought it was quite strange. After landing on the helipads inside the perimeter fence and shutting down, I went to check out what was going on.

After finding the camp manager, the other pilot and I asked what gives with all the natives camped around the perimeter! We thought maybe some really cool event was in store for us. "No", he said, "the Asmats (another tribe) were working their way down from the mountains to do some trading with us." Well, that didn't mean too much to dumb pilots like us. Seeing our "so what" looks, he went on to explain the Asmats were far more warrior-like and militant than most of the tribes in the area. Consequently, the more docile tribesmen, like the ones huddled up all around the perimeter, had come to them for safety since the camp did have a few armed guards for security. "The reason for that", he explained, "is because the Asmats travel light. They live off the fat of the land, so to speak". The "fat of the land" being any poor soul, human or animal, unlucky enough to be caught, because they could very well end up in the soup pot!

"So", I asked. "Should we be concerned"?

"Not as long as you don't have a forced landing and have to land in the middle of them when they're hunting."

That wasn't exactly the response I was hoping for!

As a side note, ASMAT was a word that had long scared people! It's a word that became a synonym for cannibals. ASMAT was a tribe whose members in time of war would eat brains of their enemies mixed with a few delicacies like worms from the halved skulls of their victims. Some of them even used human skulls to sleep on instead of pillows. The young son of the famous Rockefellers disappeared in the territory of this tribe, not far from where we had been standing. It is a word that evokes much more than respect in that wild country. As far as I was concerned, extreme fear might have been more appropriate!

July 4, 1971

July fourth didn't mean much to us in the jungles of Irian Jaya. It was just another day that I flew a total of 4.6 hours, except for a passenger I picked up on the last flight before heading into the

mountains. He was a very nice man who was pleasant to chat with. He also seemed to know more about the area than the average guy you might just happen to pick up in a place like that. He turned out to be Forbes Wilson, the same gentleman who had started it all! He certainly looked the part, too! A very handsome, silver-haired man who still looked very fit, even rugged! He had been the man who had trekked inland for more than 17 days through the literally impenetrable jungle just to get to get his first glimpse of the Ertsberg.

He still remembered every step of his journey and was glad to share his stories, all of which I thoroughly enjoyed! So much so, when I dropped of my last load at the base of the Ertsberg itself, I asked if he would like to see something special! His response was "absolutely!" My response was to pull up on the collective, thus pumping more power into the main rotor blades, and began climbing even higher than the 10,000-feet MSL (Mean Sea Level) we were at when we took off! We were flying even higher so Mr. Wilson could get a bird's-eye view of one of the world's few glaciers in tropical Asia. As soon as we topped the last ridge line to see the glacier for the first time, I knew it had been worth it! It was a spectacular sight, and the sheer joy I saw in his face made it all worthwhile! I have to admit, the way the sun glinted across the ice that day made even special for me as well! As we were flying over the glacier and taking it all in, Mr. Wilson turned to me and said, "you've made this one of the most memorable July 4^{ths} I've ever known!" When we landed back at base camp, he thanked me again and went on his way. It was a great day!

No Easy Day in the Mountains

All PHI employees worked a 2 for 1 work schedule, meaning every two days at the camp site earned you one day off. Typically, we worked 20 days, then returned to Darwin for 10 days of "rest & recuperation". There were 11 pilots and 10 mechanics, so that meant you were literally flying every day for 20 straight days. Mountains and weather presented formidable speed bumps that always tested us, both mentally and physically. The strain on pilots was brutal at times! Some would become dizzy after flying multiple round trips from sea level, land on flimsy helipads built with logs, rocks, twigs or whatever else could be found at such altitudes, then fly back down to sea level and do it all over again with the next load!

I still recall landing on the one such pad located on a peak some 10,000-feet above sea level. It was built of small logs that literally encapsulated the very tip of the pointed peak. Large enough to hold just the helicopters skid gear (landing gear), it was God-awful landing there! Even while sitting on the flimsy pad, you were still literally flying! You didn't dare lower the collective all the way to the down stop for fear the pad would give way and you would find yourself in a horrifying freefall for thousands of feet! It was a real test of skill and mental toughness.

During the entire time I was there, I only lost two sling loads. The first occurred while I was sling loading a 2,200-pound load of what we called tunnel steel. It was massive curved steel arches utilized to strengthen tunnel walls. The loads were prepared so you would pick it up from cables attached to either end, and it would resemble an upside-down C, except one leg was longer than the other. I had picked it up and was chugging up the mountainside when all of a sudden, the helicopter began an un-commanded right roll to such an extent I thought I was going to roll upside down! I put in instant cyclic stick correction and then, just as quickly, the helicopter rolled level. Someone on the radio yelled: "There goes your load!" Just like that, it was over and done with, but it could have been catastrophic! The cable had snapped free on one end, thus allowing the entire load to swing pendulum-like! Just as the helicopter was at the point of going upside down, the remaining cable snapped, dropping the entire load into the jungle thousands of feet below! I flew back down, picked up another sling load and finished the day like nothing had happened. I have to confess, though! I drank more beer than usual that night!

The other load I lost was one of the lightest loads we carried,

even though it was one of the largest. It was a load made up of Styrofoam mattresses to be utilized at various mountain camps. I never counted them, but they probably wrapped a hundred mattresses in a huge mesh bag! Again, I was making another vertical flight heading up into the mountains when I must have hit a gust of wind that set the load into a resonant vibration it didn't like, because it jerked free! I immediately rolled into a bank to try to mark some geographical spot in the event someone wanted to look for them. It looked like multi-colored snowflakes dancing in every direction as they slowly drifted down into the jungle! Wow! I was thinking they sure stuffed a lot of mattresses in those mesh bags! When I got back for another load, I asked if we needed to look for them. The response was: "Nope! We got hundreds more. The natives probably already picked them up!" I bet they did, too! Probably thought it was manna from heaven when they first looked up! I also bet they slept pretty well that night, or maybe they even used them as walls for their huts!

Wrapping Up

The month of August 1971 was representative of helicopter operations at their peak. Although the rainy season was in full swing, pilots were able to carry record amounts of cargo by moving their bases further inland as the road crept up the mountains. Although we lost three weather days, and had minimal operations on two other days, six helicopters flew 636 flight hours carrying 1,495 loads into the mountains. My contribution to that effort was 83 hours. From the time I flew my first flight on May 17, 1971, until my last flight on October 27, 1971, I flew a total of 409 hours; 350 of those were carrying sling loads. It remains to this day one of the most challenging, and satisfying, jobs involving helicopters I've ever done!

I'm standing beside a faded, well-used Bell Model 204 helicopter in Irian Jaya, Indonesia (1971).

Two Bell Model 204 helicopters conducting sling-load operations (1971).

Local native Pieter was PHI's assigned helicopter refueler in Irian Jaya (1971).

Camp worker holding the bamboo viper snake we "captured"; it was more commonly referred to as a "two-stepper" due to its fast-acting venom (1971).

CHAPTER 9

Back Home in Louisiana

Mysterious Disappearance of 206 Helicopters

When I returned to Lafayette after completing my six-month tour in Irian Jaya, I was immediately transitioned into a Bell 206 Jet Ranger. Even though the Bell 206 Jet Ranger had not been in the field that long, every pilot at PHI was drooling at the mouth just to get qualified in one. I had been flying the antique Bell Model 47 and by comparison, the new Bell 206 Jet Ranger was beautiful, fast and comfortable! However, all that changed in late 1972 and early 1973 when Bell 206 Jet Rangers began mysteriously disappearing! We finally located a missing one when one of our pilots took off early one morning from PHI's Lake Parlourde base and thought he saw a glint in the swamp about 5 or 6 miles from base. It just so happened the sun's rays had hit the crashed helicopter's plexiglass windshield perfectly, or he would have never seen it! In the meantime, while PHI and Bell Helicopter were frantically searching for the first clue regarding what was happening, our passengers and pilots alike were getting spooked! So much so, a few customers refused to fly in a Bell Jet Ranger! That created a problem because there weren't too many alternatives for a Bell 206 aside from a couple of Alouette III's and a few Hughes 500 helicopters.

It turned out the entire transmission and main rotor blades were separating from the helicopter, thus allowing the fuselage to drop in much the same manner as a brick, taking both pilots and passengers with it! It was not good times! I had been flying a BH 206 for Shell Oil, and my mission was to fly from platform to platform, delivering personnel and parts on a regularly scheduled basis. My call sign was Sierra-41, and I shared that particular job with a pilot by the name of Stan Bozek. We had worked shifts opposite to each other and would often brief each other, just as we had the previous week before I left on my break. When I showed up early the next Friday morning, the lead pilot asked if I had heard about Stan. "No", I replied, "what about Stan?" On February 19, 1973, he had been killed when Sierra-41 had mysteriously crashed, killing Stan and four passengers flying with him. Suddenly, I had an overpowering urge to vomit! It was surreal! Not just losing Stan and all those passengers, but also knowing it could have been me! I had been flying the same helicopter the entire week before Stan's shift! After much searching, PHI finally located the wreckage, and just like the other crashes, the transmission had separated from the fuselage.

Meanwhile, both Bell Helicopter and PHI were desperately working hand-in-hand, trying to solve the mystery. It took a while, but they finally discovered the culprit. It had been the material utilized in the construction of the transmission mounts. Unfortunately, they couldn't replace everything overnight. In the meantime, every pilot was issued a dentist's mirror and flashlight. During every preflight, each and every pilot could be seen diligently peering into the transmission compartment, armed with mirror and flashlight to inspect the V-shaped mount. I went a little overboard because I checked mine after every flight!

Bell eventually resolved the problem and the BH 206 Jet Ranger eventually became one of the most popular and safest helicopters ever built. I don't remember exactly how many Bell 206 helicopters were lost during that perilous time, but I know it was several. Most were eventually recovered, but I think one or two, plus their passengers were never recovered. They're still resting in the depths of the Gulf of Mexico. It wasn't a fun time, but believe it or not, the Bell 206 Jet Ranger is still one of my favorite helicopters to fly.

My Alouette III Check-Out
August 8, 1972

While the Bell 206 Jet Ranger saga was ongoing, my customer at the time was one of the more vocal ones, always talking about not wanting to fly in the Bell 206. So much so, I was called into the office one day and told I would be transitioned into an Aerospatiale Alouette III to accommodate my unhappy customer. It certainly looked different than the 206, but a helicopter is a helicopter, so I couldn't complain too loudly. A fine pilot by the name of Loren Foster was my instructor pilot and it was a very pleasant experience. After a full day of training that went well, Loren told me to go out early the next morning and fly the Alouette III for a couple of hours solo flying (as in, only me aboard) just to get a "feel" for it, plus build flight time in it. No problem!

The next morning I was out flying the Alouette III when I got a call from the PHI control tower asking me to return before my scheduled two hours was up. Being in no position to argue, I flew back into PHI's Lake Palourde heliport and landed on one of the small helipads, then began going through my shutdown procedures. That's when I got another call from the tower requesting my fuel status, then asking if there was enough fuel for a 30-minute flight to Lafayette. I advised them there was enough fuel remaining for the trip, and their response was to "keep it running".

Out the door came three PHI pilots, headed in my direction. Opening the door, one of them said to "keep it running and they would swap out with me". They were all going to Lafayette so two of them could pick up two Bell 206's just coming out of an inspection cycle, then ferry them back to Lake Palourde. One of them slid in, grabbed the controls, and I exited the aircraft on the other side. The other two pilots jumped in and, after I cleared the heliport, away they went. The thing that was different about the Alouette III was the fact it was configured with a long bench-type seat with enough room for the pilot and two passengers. Normally, it only had one set of controls since it was a single-pilot helicopter. What turned out to be a stroke of luck was the fact that particular helicopter had been outfitted with two sets of controls since we had been doing flight training.

Having gotten up early to get my solo flying in, I was quite

pleased with that latest turn of events, since I could have a cup of coffee a little earlier than I had planned! Life was good! I had been sitting, enjoying my second cup of coffee, when a pilot walked by and said we had a helicopter missing. "Which one?", I asked. It was the Alouette III! They had filed a flight plan for the 30-minute flight and now, some 45 minutes later, it was overdue! I was stunned! I walked quickly up into the control tower to find out the latest details, plus offer my assistance to help with the search.

When I entered the control tower, it was evident things weren't going well. The base manager was there, and they were busy making plans to launch a couple of helicopters to fly the same route they had flown to search for them. Things weren't looking good at all. The thought of losing three pilots, all good friends, was simply too overwhelming to think about! I'm sure everyone there had the same thoughts. That's about the time the phone rang.

The base manager grabbed it and the gentleman on the phone was a local farmer wanting to know "if we were missing one of them helicopters". "Yes!", screamed the base manager. "Well," he said, "there's one of them laying on its side buried in the mud!"

"What about the crew?"

"I don't know; I just saw it crash!"

Time stood still! The base manager pitched the phone to one of the tower operators, then both he and I raced down the stairs and sprinted toward a Bell 206 sitting at the first helipad. We quickly started, spooled up and departed with the tower providing instructions as to where to go. We arrived over the crash site in about twenty minutes and, sure enough, there was the Alouette III, laying on its side, buried in the mud! On short final, we could see several people standing around, and three of them wore the familiar forest green uniforms. Both of us began breathing again! All three pilots were fine except for a few bruises, scratches and frayed nerves! A couple of days later, all pilots were back at the Lake Palourde base "holding court" with other pilots to explain what had happened.

Everything had been fine until the helicopter suddenly

developed a vibration so severe, it literally shook their headsets off before they could make a mayday call! The helicopter then began lurching and rolling, completely out of control! Both pilots grabbed their prospective controls trying to control the cyclic but were unable to control it! Then the pilot sitting in the middle braced himself between the other two pilots, then wrapped his arms around both controls to assist! The three of them working together managed to control the helicopter just enough to crash in a semi-controlled condition! On one point they all emphatically agreed! Had there only been one pilot aboard flying as a solo pilot, the aircraft would have crashed with no survivors. It was another lucky day for me!

I had scheduled a two-hour flight that day, when the tower called for my return to base after only one-hour and 15-minutes! Why? I sat and thought about that incident a long time before I came to a conclusion. Someone up above had simply decided it wasn't my time to go. Somewhere, there had been a plan. There were still bigger and better things left for me to do in my life and I have strived to honor that gift of life presented to me that day. It was made even better knowing three pilots weren't taken in my place.

Hey, Where's My Helicopter

In the summer of 1972, I was flying a BH 206 Jet Ranger helicopter on a job that required me to spend the night on an offshore platform in the Gulf of Mexico. I had already landed and was securing my helicopter to the deck when another Bell 206 circled once, then made an approach and landed at the other end of the helipad. After shutting down, the pilot jumped out, paused just long enough to say hello, then scurried down the stairs and disappeared into the living quarters. I finished securing my helicopter to the deck, then followed him down the stairs.

After dinner that evening, I walked back up to the heliport just to check everything one last time before going to bed. It was absolutely beautiful, even if it was a hot summer evening. The winds

were calm and there were no waves; just a gentle swell. I also noticed the other pilot had left his helicopter exactly as it had been when he walked away from it earlier in the day. Not even the blades were tied down. Later on, as I was walking down the hallway, I happened to meet up with the pilot again and told him he really should secure his helicopter before he went to bed. "Ah", he said, "it's OK. The wind's not blowing". OK, I thought to myself, but PHI doesn't see it that way. But, whatever, dude! Without another thought about the unsecured helicopter, I went to bed.

The next morning, not long after sunup, I went up to the heliport and there sat one helicopter: mine. It was still secured to the deck. That's odd, I thought. I hadn't even heard him start up that morning, but maybe I had still been asleep when he departed. I had begun the process of untying all the tie-downs when I heard someone bouncing up the stairway. As soon as the pilot stepped up onto the helipad, I saw a surprised look on his face. Somewhat startled, he stopped short, looked at me, then looked over at the vacant spot where he had parked his helicopter, then looked back at me and yelled: "Hey! Where's your helicopter?"

"Right here", I said.

"Where's mine?" he asked.

"Gone" I said.

"Where?"

That conversation was going nowhere fast! I finally told him his helicopter was gone when I first got to the helipad, and I had no idea where it was! His brain simply could not compute the fact his helicopter had disappeared!

After checking with the night shift, they confirmed a strong summer thunderstorm had blown through about 3:00 AM with 70 knots of wind! Aha! Just like that, the pilot had an answer to all his questions. I have to say, it was a very distraught pilot who called in to report that his aircraft was missing. They proceeded to ask him all the questions he had been asking me. They also asked if he had tied it to the deck properly, as per PHI's strict tie-down policy. His response of, "No", sealed his fate. It wasn't long before another helicopter flew in to pick him up and I never saw him again. I guess it was just as well, though. They never found his helicopter either.

Escaping A Hurricane by Boat

Hurricanes are not welcome in the Gulf! Or anywhere, for that matter! The thing is, Mother Nature doesn't care! She listens to no one. She does exactly what she wants, when she wants and how she wants! There's not much we mere mortals can do about it! The best we can do is lay out extensive plans on how best to cope with such nuisances, and when to evacuate. All platforms in the Gulf had such a plan, as did PHI, and still does. It always works best if they "dove-tail" with each other to keep things going smoothly. Most of the time, it works. Sometimes, though, it doesn't. On one dark night, I was involved in one instance when it didn't.

The thing about it, no one likes to lose money, and when you shut oil platforms down in the Gulf of Mexico, or anywhere for that matter, oil companies lose money. Sometimes, lots of it! Consequently, they like to pump oil until they absolutely have to shut down. Like they did that time. The problem was, when they finally called for help, all of PHI's helicopters not assigned to contracts had already been dispatched to safer areas, far from the ravages of the hurricane coming ashore! No problem! They simply called for a boat to head out to SMI 58 and save some poor, stranded souls, of which I was one!

The boat didn't get there until sometime after dark, and the leading edge of the hurricane was already bringing in sheets of rain, wind and high waves! It stopped by one of the other platforms to pick up the personnel there before coming over to rescue us. Due to the desire to keep working, it just so happened another one of our helicopters had also been forced to stay there for the evening as well. Good, I thought. At least the other pilot and I can tell war stories on the ride to the beach!

Soon enough, we could see the lights of the boat as it made

its way over to pick us up. The bad news was the lights seemed to be blinking. Or maybe, just appeared to be blinking since the boat was pitching up and down in those monster waves! When the boat finally pulled up alongside our platform, it was really pitching up and down! So much so, I was getting a little nervous about being lowered onto the boat deck by a "Billy Pugh Net", a big mesh, personnel net used to transfer personnel from the boat to rig, or vice-versa. Plus, the rain was really coming across in sheets, propelled by high winds! It was not a good time to be outside in the elements!

When the boat finally pulled into position below the crane, the operator switched on powerful lights directed onto the pitching boat. As I watched, there appeared to be a rag-doll-like figure flopping around, sort of lurching from one side to the other in rhythm with the pitching boat. I thought that was rather odd, especially since it was a crew boat with a big, cozy warm cabin to rest in. The dreaded moment came when I, too, was being lowered down on the personnel net. Just before the net set down on the ship's deck, the "rag-doll" went lurching by and oh, my God! It was my fellow PHI pilot!

Finally stepping on board, it was hard keeping my balance, but I managed to get over close to the cabin in a small sheltered area. When my pilot buddy came lurching by, I grabbed him and asked what was going on! Why wasn't he inside the cabin? He was already sick! Very sick! But he managed to say, "Don't go in the cabin! It's full of puke, poop, piss, vomit and God knows what else! Everyone is sick!" A peek inside the cabin window was all it took. An inch or so of swill with everything he had mentioned was sloshing from side to side in rhythm with the boat's movements! Most of us never went inside the cabin. We just huddled up wherever we could to stay out of the wind and rain. It was a long boat ride to Morgan City! The next time a hurricane paid a visit, I tried to be long gone from any offshore platform! I had no desire whatsoever for another boat ride!

Rescuing My Buddy

Wayne Brown and I have been close friends for many years, dating back to our PHI days. Even though we had both worked at Fort Wolters, Texas as flight instructors, he had been in a civilian flight, while I was in a military flight; consequently, our paths never crossed. One day, many years later, he asked if I remembered when we had first met. I told him I'd never really thought about it, but no, I didn't remember where, or when, we met.

"Well," he said, "it was on Vermilion 39". "A couple of us had been stranded there with a hurricane moving in. We had stayed too long, and there had been no way off the platform in my Bell 47-G4 helicopter. In fact, PHI had even said there were no helicopters in the area to evacuate us. Then, in spite of all that wind and rain, you came racing inbound, flying a BH 206 Jet Ranger to pick us up! It was a beautiful sight, and I've never forgotten it!"

"Well,", I said. "I've been caught in those same circumstances, and I made myself a promise! I would never abandon anybody out there in a hurricane"!

I was glad to hear I kept my promise.

Disaster in New Orleans

One of the worst things to experience is getting hit by something that's spinning really fast! Unfortunately, that happened at PHI's small New Orleans base where another pilot and I were stationed to fly special missions in Hughes 500 helicopters. On that day, the other pilot had already finished flying his mission and returned early to take care of his paperwork, then wait for me, since we were sharing a car. I had heard nothing after he had landed so I was totally unprepared for what awaited me!

When I flew overhead prior to circling back in to land, I saw his helicopter sitting there, and an ambulance, lights flashing, was just pulling away, leaving the pilot and several others standing there. I landed, then glanced over at the other helicopter and noticed a lot of blood on the heliport directly under the tail rotor! I knew it was bad! Really bad!

    Before I could even shut down, the other pilot rushed up to the door, opened it and said: "I think I killed her!" Still stunned, he quickly called out her name. Oh no! It was the young lady with the beautiful smile who was always so friendly and helpful; such as bringing us doughnuts to brighten our day! Even after I shut down the engine, I just sat there! The pilot was in shock as well! One of the nicest guys I'd ever worked with was on the verge of a mental breakdown! Sensing that, I got out and grabbed him around the shoulders, then led him away from the helicopters, and blood, into the office area. I also passed a discrete look toward the mechanic and he knew what to do. Clean the helipad!

    The young lady had previously flown with us, so when the other pilot landed a little early, the mechanic mentioned a couple of other young ladies would also like a quick ride around the heliport. They had insisted the secretary go with them to "hold their hand" since she had flown a couple of times in the past. Reluctantly, she went with them. The flight went great and when they returned, the mechanic walked up to the pilot's door, opened it, then stood there to ensure no one got out of the helicopter until the pilot had shut off the engine, and the blades had quit turning. He hadn't seen the young lady sitting in the rear, on the far side, open the door and get out. Instead of going forward, she had ducked her head and ran under the tail boom, directly into the whirling tail rotor! The pilot felt the pedals jerk just as the mechanic looked back and let out an unearthly scream! The pilot immediately shut everything down, but it was too late! The damage had been done.

    Her injuries were horrific, but somehow, she had clung to life. Her will-to-live had kept her alive! As soon as we finished taking care of business at the heliport, we rushed across town to the hospital to meet up with her family to play the waiting game, and pray! Unfortunately, her injuries had been too great! Sometime in the early morning hours, the young lady with the beautiful smile, and personality to match, left us. It was an extremely difficult time for everyone!

Tying a Bridle on a Helicopter

I had just taken off from SMI 58 for my first flight of the day when I heard a desperate mayday call coming from a BH 47-G4 helicopter that had been stationed offshore on SMI 23, a platform located about 20 miles away. I immediately called him on the radio to find out the problem and asked what I could do to assist him! The winds were fairly calm, and the sea state was about 2 or 3, which has only one to two-foot waves. By way of explanation, "sea state" is the general condition of the free surface of any large body of water with respect to wind and waves at a certain location and moment. Measuring from 1 to 8, the higher the number, the worse the waves are.

My model 47-G4 was no speed demon, so it took a while to finally arrive over the helicopter sitting in the water, slowly pitching up and down with the waves. The pilot was able to talk to me since the radio was still operational. Sitting on the ocean, he had limited range, but with me overhead, I could relay his position to PHI's various radio stations.

In addition to the pilot, there were two passengers on board who were already seasick! The helicopter was afloat, and a rescue boat had been dispatched from the nearest platform. Everything was looking good! Before long, a helicopter coming out of Lafayette contacted me to advise me he was PHI's safety director. He began asking questions about what was going on and began laying out a plan of action. To tell the truth, I thought we already had a pretty good plan already activated. I mean, the three guys on the helicopter were a little green around the gills, but nobody was injured and a rescue boat was on the way. I responded with that bit of news, but he recommended we start preparing for the arrival of the rescue boat.

"Roger that! What would you suggest?", I asked.

"Advise the pilot to pull the rope out of the basket, jump into the water and begin tying a "bridle" onto the skids so they'll be ready to start towing him!"

Hmmm! I didn't know what the pilot would say about that piece of advice, but I asked.

The pilot immediately came back with "What? Say again?" I repeated what the safety director advised him to do. He responded pretty much how I thought he would respond!

"Tell that f-ing idiot to go f-himself! I ain't jumping in the water to tie a bridle! I don't even know what that is, plus I'm getting seasick!"

"Roger that!"

Contacting the safety director, I simply said: "The pilot says negative! He doesn't have any rope!" That answer didn't go over too well, but the boat arrived, and the three seasick "sailors" were soon aboard the boat.

As soon as the men were on board, I saw the rescue boat maneuver close to the helicopter, trying to tie a rope onto it. Unfortunately, one of the helicopter's fixed floats hit a sharp edge on the boat and, quicker than you can spit, the helicopter turned upside down in the water. The weight of the helicopter was too much for the remaining float and it, too, popped off! Then, bloop! Just like that, the helicopter quickly descended into the depths of Davey Jones's locker! I was still circling overhead staring at the boat, trying to resurrect the helicopter, when the safety director sailed overhead!

"Where's the helicopter?", he asked.

"It sank!", I responded. The safety director was not happy!

"Why did it sink?"

"All I know is, the boat crew took your advice and tried to tie a bridle on it, then it sank!"

Silence!

"Do you have another plan of action?", I asked. I took the silence as a subtle hint I had best depart the area!

Shark Fishing with Sledgehammers, or Harpooning Barracuda

One has a lot of spare time to kill when flying offshore. You pick up passengers, take them to their destination, then normally wait for them, sometimes for hours! You could read books, watch snowy TV, eat doughnuts washed down by coffee, or you could fish! I tried my hand at fishing, but never really had much luck. I also watched others and came to the conclusion fishing in the Gulf could be dangerous!

In the hot summer months, often-times, the hands working on oil rigs or platforms would hang a stainless-steel hook, baited with whatever type of bait the rig's cook would dole out, then suspend it deep into the depths of the ocean on a long, stout rope, or cable. They just wanted to see if they could catch a big fish, and on one particular night, they did. A really big one!

Since it was about dusk, I was doing my usual thing of watching the sunset when all of a sudden, that rope went tight, then started yanking around like crazy! Several other hands saw it and launched into action! They started cranking that rope up and sure enough, caught firmly on the hook was what appeared to be a six- or seven-foot shark! Unable to crank it up out of the water, two big Cajun boys jumped onto a personnel net while the crane operator manning the controls picked them up and lowered them down to the shark. Our two fishermen used a huge gaff-like hook to drag the shark onto the net where they had been standing. I say: "had been", because they had already climbed half-way up the net just to stay out of the grasp of the shark's huge, sharp teeth!

There was a lot of screaming and yelling going on when the crane operator began pulling the two Cajun boys and their pissed-off shark, up to the deck! When the net swung over the deck, now void of any other living human being, the operator plunked it onto the deck! A couple of other older, wiser deckhands and I had decided to observe all the action from the safety of the helipad where we had joined my helicopter. When the crane operator dropped the net onto the deck, the shark flopped out of it! He was then free to catch and eat whomever he pleased!

The two shark fishermen finally decided they would allow the shark to calm down, sneak up behind him, then slam his head with

a sledgehammer in hopes of getting his attention! All of us onlookers encouraged them by agreeing it was a good plan. Sure enough, after several minutes, the shark lay silent! Ominously so! Grabbing a 5-pound sledgehammer, one of the fishermen silently slipped up behind the shark, straddling him. Then, with all the strength he could muster, he swung that sledgehammer in a huge, powerful arc and it landed directly on top of the shark's head! In fact, everyone winced from the impact!

That 5-pound sledgehammer recoiled off that shark's head in much the same manner as a 5-pound shot being fired out of cannon! That hammer left the grip of the startled fisherman and arched upwards well above the level of our helipad as it rotated slowly, end-over-end, past the other side of the platform, then disappeared into the sea below! The young Cajun lad was also rotating ass-over-teakettle across the deck but was saved when he hit the railing! His mighty blow had awakened the napping shark and he now owned the entire deck! In fact, he owned the entire platform! My older, wiser buddies and I suddenly had company up on the helipad! It was getting crowded! Even the second fisherman joined us! He had abandoned his buddy who was now on his feet, but not totally aware of where he was, or what peril he was in! Since the shark was wide awake and hungry-looking, everyone started yelling he should either (a) join us up on the helipad, or (b) jump overboard!

Unfortunately, all the screaming had gotten the attention of the supervisor who poked his head out the office door to see what was going on! Seeing the angry shark, it didn't take long before he figured it out! His orders were very short and sweet: "Get that SOB off my platform!" He then went back inside and slammed the door. It took some doing, but finally, the crane operator picked up the personnel net and swung it over to where the two fishermen were standing. Armed with two long gaffs, they stepped on the net and the operator swung them over just above the shark, where they managed to hook the shark with a gaff. The crane operator then swung the net over to the side and the shark belly-flopped

back into the ocean! Every vestige of shark-fishing apparatus disappeared the next day, never to be seen again!

I also decided I didn't need to take up harpooning barracuda either. One day, I walked down stairs onto the lower deck located just about 10-feet above the water level. Some of the hands had crafted some really neat harpoons and left them for anyone to use in their spare time. I was watching a couple of new roustabouts trying their luck at harpooning a few barracuda when one of them let fly and made a perfect strike! The next thing we knew, the young man was being dragged across the narrow walk area before he grabbed hold of the railing. Puzzled, we ran to help him and soon figured out what had happened!

Each harpoon had a coil of strong cord attached so the loose end could be attached to the railing. It was important to do so, because a four or five-foot barracuda can swim with the speed of light, thus generating a lot of energy as the cord ran out! Unfortunately, our intrepid young harpooner had tied the end of the cord to his wrist, and except for the hand-railing that stopped him, he had almost taken a swim with a barracuda! That's when I decided reading books might be the safest way to go!

Besides, I could always get shrimp if I wanted tasty seafood, at least sometimes. I say that because on another day, I was flying a couple of deck hands to an unmanned platform when we noticed a shrimp boat tied up to it. It was still early in the morning, so I just figured they had tied up to take a break from a hard night of shrimping. My two passengers seemed to be pretty excited about seeing the boat, too! As soon as I landed and shut down, I heard one of the deck hands over at the side of the platform shouting down to the boat crew. "Hey", he said, "catching anything?"

"Yes", they said. They were catching the "biggest, fattest, juiciest shrimps", the size of which they had never seen before! Just to prove it, the captain went below and returned on deck shortly with a nice basket of shrimp! Even from where we were standing, they certainly looked the size of turkey legs, as in drumsticks!

"Wow!" said the deck hand, "need any diesel fuel?"

"Dat would be good!" said the captain in his Cajun accent.

A basket of those shrimp in exchange for diesel fuel! A great horse trade made even better since they shared their bounty of fresh shrimp with me! The old boat Captain had been telling the truth. I still recall with great relish how large and tasty they were!

Where are my Playmates?

A Gulf Oil Company platform, designated Ship Shoal 154D, was a large production platform that produced a lot of oil for that well-known company. It was also home to a PHI radio operator and mechanic. Early on, PHI had negotiated a contract with Gulf, whereby they would have access to a room designated as a PHI radio room, as well as a large refueling tank. It functioned in much the same way a full-service gas station did, since a helicopter could swoop in and land on the heliport, the mechanic would pump the fuel, and the helicopter would continue on its way! The mechanic could also address minor maintenance problems on occasion.

Meanwhile, PHI's radio operator sat in the designated radio room from daylight until dark, all the while providing flight-following services to every helicopter operating in that area. It was an invaluable service, since it kept track of every helicopter movement throughout the Gulf of Mexico. In the event a helicopter didn't show up at their destination, a search and rescue operation would be launched.

One day, I landed on the platform to refuel, plus I also had a minor maintenance problem for the mechanic to check out. Since it would take about an hour or so to repair, I wandered up to the radio room to visit the operator, as well as to kill time by listening to all the radio traffic. I had never been in the room before, so I was quite surprised to see foldouts of Playboy's famous Playmates whose photos were published in the monthly issues. The foldout of every Playmate-of-the-Month from day one had been taped to three walls, and had just started on the fourth wall! I was most

impressed! So much so, I started with the first and was working my way around the walls when the mechanic entered the room to tell me I was ready to go! Rats! Oh well, I would resume where I left off on my next visit!

Sure enough, about a month later, I was again parked on the heliport at Ship Shoal 154D. Again, I went up to check on "the girls" but, much to my surprise, the walls were bare! There wasn't a single Playmate foldout to be seen! The radio operator began laughing at the surprised look on my face. He then explained PHI's Chief Pilot paid a visit to the platform and had been appalled when he saw the photos taped on all the walls! He ordered them to be taken down so as not to ruin the good relationship PHI had developed and nurtured with the Gulf Oil folks throughout the years! So, down they came!

As luck would have it, it wasn't long afterwards that I found myself on Ship Shoal 154D again, so I went up to visit with the radio operator. After entering the radio room, I was pleasantly surprised by the re-appearance of all the Playmate foldouts, including the latest ones! The radio operator was happy to explain that shortly after the foldouts had been removed, the Gulf Oil Superintendent, who was in charge of all offshore activities in the Gulf of Mexico, had paid a visit. He had been based in New Orleans, so he paid regular visits to Gulf's offshore platforms. The radio operator explained the Superintendent would also stroll through to "check on the girls". He had also been shocked by their absence, so the radio operator and mechanic were all too happy to tell him about the demise of the bunnies! After hearing the story, he went to his office and called PHI's management to explain how Ship Shoal 154D was the property of Gulf Oil, not the other way around! And no one was to ever mess with the Playmates who adorned the walls of Ship Shoal 154D!

That all happened many years ago, so I have no idea if Playmates are even allowed on offshore platforms today. They certainly were in those days, and the radio room on Ship Shoal 154D had a lot of folks dropping in just to say hello to all those Playmates!

Motorboating up the Atchafalaya River

My last week at PHI was spent at their Intercoastal City (ICY) base, a pleasant little base just south of Lafayette, LA. Virgil "Jerry" Luttrell, one of my closest friends, and I had already turned in our two-week's notice to resign from PHI to join Bell Helicopter International. I was flying a Bell 206 out of my assigned base at ICY, and Jerry's Bell 206 was based at Morgan City; however, he had been assigned to an offshore platform. Simply put, I worked onshore and he worked offshore. Normally it wasn't a problem getting ashore for break day. That time, it was.

It was December 1973, and fog can be very prevalent that time of year. It is also unpredictable as to where it appears. I had seen fog so thick you could barely fly onshore, yet when you punched through the coastal area and got offshore, it was often Ceiling and Visibility Unlimited (CAVU)! We learned through years of flying offshore how to deal with it.

I had finished my preflight and was sitting in the cockpit getting my paperwork done when I heard someone shouting, "Hey!" Looking up, I was shocked to see fog! Fog so thick, I could barely see past the nose of my helicopter! The yells were from pilots who had parked out behind me in a pasture-like area with no concrete pads. And they couldn't find their way back to the office. I have never seen fog like that, before or since! We finally got everyone safely back into the safe confines of PHI's operational building and began drinking coffee. With fog like that, we drank a lot of coffee during my last week at PHI!

Meanwhile, my buddy who was stationed offshore was flying non-stop all week! In fact, he never missed a day flying in those clear blue skies! The same sky that was totally wrapped in heavy fog up and down the entire coast of Louisiana, was clear and blue offshore! Even though it was an impenetrable wall of fog stretching from the beaches all the way up to Lafayette, home of PHI's main base, it wasn't a problem for Jerry as he steadily flew from platform to platform during the week. However, it sure became a problem when it came time for his break. After all, it was his final week working in the Gulf of Mexico for PHI! He had to get home!

Our workweek at PHI went from Friday morning through the next Thursday. And it was Thursday! Break day! Jerry was bound and determined to get home even though the fog was as thick as ever! Truth is, it was still an impenetrable wall, at least for most pilots! But not so for Jerry! About mid-afternoon, Jerry and his passengers set out to find their way to PHI's base on Lake Palourde located just outside Morgan City. All went well until they got close to the beach when, sure enough, there was that wall of fog, just like it had been all week! But my friend Jerry had an ace up his sleeve that fine day!

Carrying his load of passengers, he continued inbound, descending lower and lower until he finally spotted the Atchafalaya River that flowed past Morgan City into what is called the Atchafalaya Bay, then empties into the Gulf of Mexico. And Jerry knew exactly where that river was! The dense fog forced Jerry lower and lower toward the river until finally, with no other choice, he popped his floats! From that moment on, he was no longer flying a helicopter; he had turned it into a motor boat! Unperturbed, he began motor-boating up the Atchafalaya River just like the big boys!

The boat Captains and crewmen were stunned when a black and gold helicopter, masquerading as a boat, putt-putted alongside them in his quest to find Morgan City. At long last, a bedraggled-looking helicopter with moss and swamp grass hanging from its floats called in for landing. He had made it. I heard his passengers had all fainted somewhere on the journey! The only thing Jerry ever said was he sure could've used one of those big fog horns like the boats and barges cruising up and down the river have, so he could let them know he was coming through! As it was, he said he had never seen so many big eyes and open mouths on the faces of all the boat crews he passed! He became famous, at least for a while. But who knows, some of those old boat Captains may still be telling the story of Jerry putt-putting up the Atchafalaya River.

On the same night of Jerry's valiant run up the river, I was walking into the office when I heard the voice of Morgan City's lead pilot on the phone screaming: "I'm firing that damned pilot who had the audacity to motorboat up the river!" The area supervisor was patiently explaining to him that he couldn't fire Jerry since he had already quit! The response to that was: "Hell no, he ain't quitting! I won't let him quit! I'm going to fire him!"

Even though my close friend Jerry Luttrell never worked for PHI again after we returned from Iran, he took a job with another company flying in the Gulf of Mexico and his Bell 206 Jet Ranger disappeared without a trace one stormy day on January 27, 1991. Although the Coast Guard and civilian helicopters searched for days, they never found a trace of Jerry, his young passenger or his helicopter. It was a tremendous loss of a wonderful friend. I still miss him.

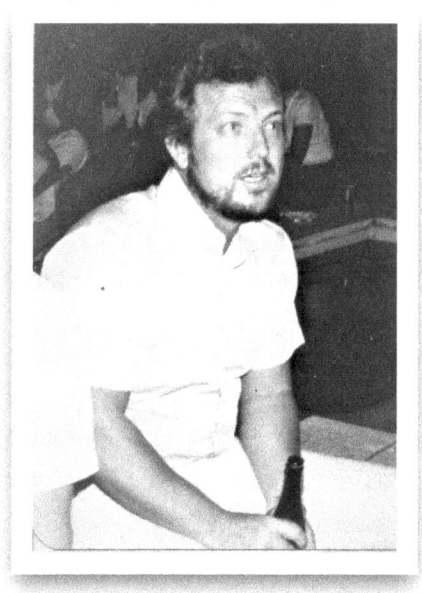

My good friend Virgil "Jerry" Luttrell who was lost in the Gulf of Mexico during a winter storm on 01/27/1991.

CHAPTER 10
Bell Helicopter International (BHI)

After a long period of flight demonstrations and negotiation between the U.S. government, Iran and Bell Helicopter, a contract had been signed in December 1972, whereby Iran agreed to purchase 202 twin-engine AH-1J model Cobras and 287 Model 214 helicopters. It was a "turn-key" type of contract that called for training Iranian military personnel in all aspects of maintaining and flying their newly purchased helicopters. Simply put, it essentially called for creating the Imperial Iranian Army Aviation Training Command (IIAATC). The ultimate plan was for Iran's army helicopter force to become a mirror image of the vaunted United States Army Aviation.

It was a massive foreign military sale so Bell Helicopter International (BHI), a totally owned subsidiary of Textron Inc., was formed for the explicit purpose of conducting business in Iran. It was responsible for the performance of a training program and any subsequent in-country services rendered to the Government of Iran. Its first president was Maj. Gen. Delk Oden, U.S. Army, (ret)., one of General George Patton's premier tank commanders in the famed Third Armored Division in World War II. On April 1, 1973, the first 43 employees boarded a Flying Tigers "stretched DC-8" and departed Love Field for Iran. They were the vanguard for hundreds of adventurous employees that followed to fulfill the ultra-large contract. My small family of four would be among them.

Fortune-Telling Flowers

In late spring, 1973, I was on the small Petroleum Helicopter International (PHI) flight ramp in New Orleans with a couple of pilots enjoying a beautiful day, when a Bell 206 Jet Ranger helicopter

circled overhead, then landed at the heliport for refueling. After shutting down, the pilot of the 206 ambled over to get a cup of coffee and visit. It wasn't long before he casually mentioned he had heard about some new company by the name of Bell Helicopter International that was looking for pilots to fly in some country by the name of Iran. Our first question was: "Do you mean Iraann, as in West Texas?" It was an odd-sounding name, but Iraann (pronounced Ira-Ann) is a small town in West Texas named after Ira and Ann Yates who were owners of the ranchland where the town was built. His response was: "Nope; it's Iran. As in the Middle East! At least I think so!"

I was familiar with Iran only because of the severe oil shortage, subsequent price increases and how the Shah of Iran had played a large part in breaking OPEC, thus reducing the price of oil. But geographically speaking, I had no idea where Iran was located, nor did any of the other pilots I spoke to. My limited view at that time consisted of nothing but the Gulf of Mexico where helicopters flew like gnats while servicing drilling rigs and oil production platforms! Even though I spent six months in Irian Jaya, my family had been living in Lafayette, Louisiana since I had left the military on October 10, 1969. Consequently, I knew all about Cajuns, crawfish, beans and rice, but absolutely nothing about Iran. However, it certainly sounded interesting!

I had been working at PHI four years and had come to the conclusion I didn't want to spend the rest of my aviation career flying in the Gulf of Mexico! It wasn't so much that I didn't like working for PHI, because it was a great company. I really enjoyed working for them, plus I made many life-long friends. The primary reason for my restlessness was the simple fact I had grown tired of the monotony of flying offshore. It was like enduring a never-ending groundhog day that kept repeating itself. A new company might provide an avenue of escape. The problem was, I had no idea where to begin researching the background of that new company I just heard about. But I was determined to do so!

Virgil "Jerry" Luttrell, a fellow pilot and good friend, and I began checking further into the new company and before long we both had our applications in for future employment. As soon as we thought they had our applications on file, Jerry and I pooled what little money we had and drove nonstop through the night from Lafayette, Louisiana to Bedford, Texas to visit the gentleman in charge of hiring. Not having money for a motel, we stopped at a local truck stop located just outside Fort Worth to shave and change into our suits for an interview at BHI. It was easy to find, and soon enough we found ourselves seated in BHI's lobby where we waited patiently for a window of opportunity, since we had arrived with no scheduled appointment. Fortunately, we were able to meet the pilot hiring rep and were basically told we met their qualifications. "Go home", we had been told. Go home and wait for them to call or send updates to our applications in the mail. So, with basically no sleep in the past twenty-four hours, Jerry and I ate another quick to-go sandwich and headed back to our homes in Lafayette, Louisiana.

Even though both of us were tired from our long trip to and from Fort Worth, Texas, neither of us were sleepy. We had been too excited about the prospects of new employment in a place we had never heard of! A place by the name of Isfahan, Iran. The headquarters of BHI was in the capital city of Tehran, Iran. We had never really heard of either one of them, but it really didn't matter! We were going to Iran if given the chance! All of those things were being discussed and before we knew it, we were pulling into the driveway of my home in Lafayette.

It was just after sunup so my wife was already in the front yard watering her flower beds. After hugging her hello, I began telling her about what we had learned on our trip to BHI, and how I hoped we would be presented with a job offer.

"Oh, don't worry" she said, "you will. We're going to Iran."

"So", I asked, "how do you know that?"

Half turning around, she pointed to a little bush that had several beautiful flowers in full bloom.

"That's a tulip bush and there's an old saying about how it only blooms when the homeowners are leaving", she replied. "I've been nurturing that plant ever since we moved into this house and have never seen so much as a bud! But when I came out of the house this morning, it had all those beautiful flowers! So, we're going to Iran, or wherever that place is."

I told her it was sort of foolish to put too much faith in something that dumb. One month later I received a letter from BHI offering me a job as a flight instructor in Isfahan, Iran. There's an old saying about how "words should be made of chocolate so if you ever had to eat them, they wouldn't taste so bad." It would have certainly been tastier than the crow I ate!

My First Day at BHI

My journey with BHI began on Monday morning, February 11, 1974, when I reported in at the Bell Helicopter Flight Training Academy in Fort Worth, Texas for Bell 206 Jet Ranger flight training. It was a full house because BHI had also hired another 20 flight instructors about the same time I'd been hired; consequently, Bell's flight training staff was literally overwhelmed with so many pilots to train! As luck would have it, out of all those new pilots, Jerry Luttrell and I were the only pilots who had previous flight time in a Bell 206 helicopter. Consequently, as soon as I completed my own training, I had been assigned the duty of training the other new-hire BHI pilots in the Bell 206. All things considered, it was pretty fortunate because it had been several years since I had been a flight instructor at Fort Wolters, Texas. Training BHI pilots allowed me to get back into my "instructor mode", plus, teaching experienced American pilots also prepared me for the more difficult task of instructing Iranian students in the Bell 206!

Welcome to Tehran

Accompanied by my wife, young son and daughter, and in the company of twenty other new hires and family members, we departed DFW International Airport for New York's JFK International Airport on February 28, 1974. The next leg of our journey out of JFK had been on Air Alitalia that had taken us to Rome, Italy, where we got to enjoy an eventful day touring the beautiful city of Rome. One of the things I still remember about that stopover was the total absence of any vehicles moving other than emergency vehicles and buses! Everything else had been prohibited because of the fuel shortage! I think it was the only time that had ever happened! However, it didn't deter us from arriving at Tehran's Mehrabad International Airport on Sunday evening, March 2, 1974. Although everyone was tired and sleepy from the three-day trip, the introduction to Iranian driving habits soon woke us up!

The "finer art" of driving in Iran was introduced to us by a "Mario Andretti" want-to-be driver, who had become infatuated by my lovely blonde-headed wife! After seating everyone else in the rear of the small bus, he gallantly invited my wife to sit up front with him. He then started the bus, revved it up and, with a quick shifting of gears, he proceeded to entertain everyone with his driving skills! With my wife seated beside him, he roared through signal lights at intersections, passing cars on either side, while continuously honking his horn to warn everyone in his path! He actually brushed the clothing of some poor, hapless pedestrian who had been either deaf or simply not nimble enough to leap out of the way! In addition to being our transportation to our hotel, it had also been a fast, scary introduction to driving in Iran! My wife said she had survived by grabbing a good handhold on the seat arms, then closing her eyes during the really close calls! She was a real trooper!

We spent the next night in the Commodore Hotel located in the heart of Tehran. It was a welcome respite from the tiring journey from Texas for all of us, especially my two young children. We finally arrived in Isfahan on Monday evening, March 4, 1974. It was hard to believe our new home was half-way around the world from Texas! Our adventure in Isfahan had officially begun and one of the first things we had to adjust to was weekends!

In Iran, Friday was the holy day, not Sunday! Consequently, our work week was Saturday through Wednesday. Our weekend was Thursday and Friday. It was rather strange at first, but it soon became the norm for all of us.

Isfahan, Iran
(Now spelled Esfahan)

Isfahan, Iran was one of the most beautiful cities in the world. Known for its exquisite arches, tiled mosques, minarets and tree lined boulevards, it also has some of the largest, most beautiful roses in the world. In fact, the locals explained to us roses had been originated in the beautiful gardens of ancient Shiraz before being exported to European gardens. Still remembering those beautiful roses, I have no reason to doubt that. It was also where the rugged game of Polo began, played by the various sheiks who, centuries ago, rode such magnificent Arabian horses! Even the world's finest caviar comes from sturgeon that inhabit the Caspian Sea.

Isfahan dates back over 1,000 years, and somewhere amid that time span, it was one of the largest cities in the world. Even now, it is referred to as "half the world!" Just being there seemed to whisk you back into another time, and it was as if ghosts of the past were still present. In spite of all the world's ongoing turmoil, I have been told that Isfahan is still one of the most beautiful cities in the world, especially the gardens that we still remember. I hope so, because my family and I still harbor wonderful memories of the four exciting years we lived there.

Orientation Flight
March 7, 1974

On the morning of March 7, 1974, my good friend Jerry Luttrell and I were taken on an orientation flight prior to our assignment to a training flight where we would both become flight instructors in a Bell 206. I cannot recall the name of the flight instructor we flew with, but I do remember he was an older gentleman who

smoked; in fact, he smoked a lot! Every time he finished a cigarette, he would immediately light up another one! I was very glad we had windows to open up!

After starting up, we made our takeoff and were soon airborne on my first flight in Iran. Our plan was to fly throughout the entire Area of Operations (AO) we would be flying, so we would have a good geographical idea of where we were when we started flying students. I was at the controls and, having logged more than 1,000 flight hours in a Bell 206 model, I was feeling pretty comfortable! I had really been looking forward to our orientation flight since it was an area I had never even seen before, much less flown over. I was prepared for an interesting flight, but I wasn't prepared for how strange and eerie it would be.

In just a matter of minutes, we crossed over a ridgeline, then flew directly into a valley that immediately became another world! Like the famous "Shangri La", it had been a valley totally void of anything that might have been interpreted as "modern civilization". There were no poles, no towers, no roads, no nothing! Absolutely nothing! It was absent of anything newer than 1,000 years. We passed over villages where you could see camels, donkeys and other livestock, but no automobiles, trucks or anything modern to which we were accustomed. We even dipped low over a large wheat field where workers were harvesting wheat with scythes, while others were bundling the wheat with cord! Small donkeys had been standing nearby, seemingly patiently waiting for whatever should happen next. I did not see one single thing, be it alive or otherwise, that appeared to be of this century! It was unnerving!

Even though the helicopter was still responsive to my controls, I felt as though I had simply been a passenger in a machine that had penetrated a time warp and had flown into another dimension! Like back in time a thousand years or so! It was very disconcerting! I even cast a glance toward the instructor pilot, hoping to catch just a glimpse of some uncertainty on his face, but he just sat there staring out the window, chain-smoking his cigarettes, acting as though

he didn't have a care in the world! He was totally nonchalant. I even looked back at Jerry, but he seemed to be perfectly fine as well.

We flew for almost two hours, and I cannot recall seeing anything that remotely looked modern. At one point, I even became a bit dizzy and nauseous because of the increased stress and uncertainty of where I was flying, and what I was seeing! Finally, after flying over every bit of terrain imaginable, we topped the same small ridgeline we had flown over going outbound, and there sat Isfahan and the airport! I cannot begin to tell you how ecstatic I was! Finally, there was something I could recognize! Something new! Something modern! I had never been so glad to complete a flight in my entire career. It was like I had flown through an invisible time warp before returning to the present.

I still recall that flight with great clarity and I'm glad to say, I have never had that feeling again! I don't even know what really caused it, but memories of that single flight into another dimension remains with me to this day.

JP-4 for Heating

When I first arrived at Isfahan, the BHI training facility had been located at the main airport located just south of the city. It only had one runway that lay east-west, and I was shocked to see Iran Air operating there at the same airport where we would be flying, and training, in helicopters! The buildings we were housed in had been constructed with cement and stone, and were very primitive. They were cold, dank and musty, plus they only had one toilet at one end of the long building where the instructor pilots worked. Making matters worse, the lone toilet was the kind you had to squat over to use, a difficult task, especially when wearing flight suits! Needless to say, women refused to use it, and I didn't blame them! Plus, all the buildings were cold. Even though they had a couple of black, kerosene-burning stoves in a couple of the rooms, our weekly "allowance" of kerosene (or "naft"), was a mere 5 gallons, which might have been enough for one day, but us pilots had been pretty clever. We all knew where we could get all the kerosene we needed! Actually, it wasn't kerosene, it was JP-4, the same kind we burned in our turbine-powered Hueys and Bell 206's!

Every morning, the first pilot who arrived had the duty to take our pilfered 5-gallon bucket, then go out and "milk" 5 gallons of JP-4 out of one of the nearby Hueys. The next step was to fill the decrepit looking stoves with JP-4 and get the heat going! It worked really well, too! What we had forgotten was JP-4 burned a wee bit hotter than normal kerosene, plus it had a lower flash point that could create a little volatility. As in, explode! But ignorance was bliss, and all of the pilots were happy campers as they sat and briefed students in our warm, cozy room made of concrete. Fortunately, the black stoves had all been vented by means of a 5-inch stove pipe that ran vertically from the stove, up the wall for about 6 feet before making a 90-degree turn that allowed the stovepipe to disappear out through the wall where the stove's thick, black smoke could be mixed with the earth's atmosphere. It worked great, until one morning it didn't!

On that morning, about a dozen instructor pilots had been briefing their students when a small explosion disrupted everything. I had been sitting at the opposite end of the building from the black stove and when I looked, all I could see was a huge ball of black smoke stretching from wall to wall, rolling down the room engulfing everyone! My students and I jumped up and bolted for the doorway, barely making it through just as every pilot and student behind us came barreling out the same way. It was quite a sight! Everyone was totally black! We couldn't help ourselves and began laughing at one of the funniest things we had seen, at least there in those dank, musty buildings! Everyone who had witnessed it started laughing with us! I don't think we even flew that day because everyone had to go clean up.

In the end, we had to confess to what we had done. The managers "huffed and puffed" about how dumb, stupid and every other constructive bit of criticism they could dole out! But they couldn't

fire us because we were all desperately needed instructor pilots! They finally calmed down and "forgave" us, providing we wouldn't use any more JP-4 for the stoves! We all agreed we wouldn't, provided they would increase our "allowance" of kerosene for the stove, which they did. Instead of 5 gallons per week, we got 5 gallons for two days! That was enough because spring hadn't been far off.

Dangerous Toilets

The cold, primitive stone buildings we first inhabited at Isfahan's International Airport had a single toilet located at the end of the hallway. Actually, it was very close to the entrance so we could appreciate its "ambiance" every time we entered, or departed, the building. It reeked mightily, plus it always seemed to have a black cloud of flies buzzing around, literally attacking anyone brave enough to venture inside. I might also add it was of the old French style, as in no commode! You simply dropped your flight suit down around your ankles, then squatted over a small hole. It was definitely for men only! Ladies would never enter such a place!

It could also be dangerous! I still get a chuckle when I think about the day, I was strolling down the hallway toward the toilet door when it suddenly burst open with enough force to slam against the wall, immediately followed by a half-clad, half-naked instructor pilot desperately clawing his way out of the toilet on his hands and knees! Stunned, I could only stand and watch as he proceeded to curse those flies between bouts of retching and gagging! After what seemed like forever, the poor, ashen-faced pilot finally looked up and seeing me, managed to scrunch up against the wall to fill me in on what had happened!

It seems while doing his thing with his flight suit at half-mast around his ankles, he had been keenly observing the ever-present cloud of flies buzzing around him, totally entranced by the whirling and buzzing of so many flies. He had been astutely observing them when an exceptionally large black fly had suddenly peeled off from the massive cloud, then proceeded to fly a straight trajectory directly towards his mouth which had been somewhat gaped open. Completely caught off guard by the fly's sneak attack, he had pitched backward, gasped, and, bloop – down his gullet went the black, nasty fly! That's about the time I saw the door fly open, followed by the pilot desperately trying to upchuck the fly. All of a sudden, I no longer had to use the toilet – I was laughing too hard! I'm glad to say, my friend survived the dive-bombing fly, and is still alive and well to this day. I still consider him to be a close friend.

Helicopter vs. Iran Air

As I stated earlier, when I first arrived we were based at Isfahan's main airport located just south of the city, the same airport serviced by Iran Air, the nation's flagship airline. It was difficult at first, but we quickly adapted to the coming and going of Iran Air's jets and made it work. In fact, it worked very well, except for those few times it didn't. I just happened to be very close to the action on one of those times it didn't work out so well!

All of BHI's training helicopters sat on the south side of the main east/west runway and the main terminal sat on the north side of the runway. It was a fairly simple situation whereby the helicopters would hover over a designated takeoff pad on our side of the runway, request permission to hover onto the runway, then takeoff. The only thing to remember was Iran Air had priority over everything. When an Iran Air Captain called the tower, everything else was put on hold until Iran Air had either landed or taken off, whatever the case might be. Surprisingly, many of the captains were British. Some of them also had nasty dispositions!

On that particular day, a BHI pilot who had been assigned the duty of flying the search/rescue helicopter, had started it up, then called for clearance to reposition to the takeoff pad. As luck would have it, I called for clearance to follow him and was cleared to hover behind the crash/rescue helicopter hovering toward the takeoff pad. About that same time, a British Iran Air captain called

in for landing clearance at Isfahan and was immediately granted permission to land. Meanwhile, the search/rescue helicopter had just reached the takeoff pad, then turned directly toward the runway.

Hovering behind the search/rescue helicopter, I had mentally prepared myself for a five or ten-minute wait for Iran Air to land, then clear the runway before being cleared to proceed. In fact, the Iran Air jet had already made the final turn onto final when the search/rescue pilot requested takeoff clearance. I was shocked by his request because the jet was already on final approach for landing! The tower immediately denied his clearance request! The pilot responded with "Roger", then proceeded to taxi out onto the runway! That was not good, and the tower began screaming at him to clear the runway! Thinking he couldn't hear the tower, I even called out for "helicopter on runway, clear immediately for landing traffic!" He just sat there! Meanwhile, the British captain began screaming to get that "bloody, f-ing helicopter off the runway!" The tower screamed to get that bloody helicopter off the runway! Everyone was screaming! Suddenly, I had an overpowering sense of urgency to get the hell as far away as possible from that runway. So, I fast-taxied over to the nearest landing pad and plunked the helicopter down in the middle of it, then proceeded to monitor the situation.

There was so much screaming on the radio that it was unreadable gibberish! I also saw the Iran Air jet pitch nose-up just above the runway as he began making a go-around! Not good! The landing gear was still tucking into its belly when it screamed past, barely skimming over the helicopter still sitting in the middle of the runway. I couldn't believe it! Then, smooth as you please, the helicopter took off advising "he was going to make a closed pattern". I didn't hear the tower's response! All I heard was the British captain informing everyone he was going to "slice, dice, chop and make mince-meat out of that bloody, f-ing poor f-ing excuse of a f-ing pilot in that bloody f-ing helicopter!" He was not a happy camper! I had to say I couldn't blame him. The crash/rescue pilot was told to clear the area until the tower called him back in to land.

In the end, the Iran Air jet came back around, landed and taxied in to discharge its passengers, some no doubt with dirty ditties! The British captain probably had to be restrained from making mincemeat out of BHI's poor dumb-ass helicopter pilot! And the helicopter pilot? When he landed, he was welcomed into the waiting arms of a small army of security guards! He was immediately taken to his home and advised he had been given 24-hours to get out of the country! His wife and family were given an extra week to sell their belongings, clear the post and get out of the country! As far as I know, no one from BHI was even allowed to talk to him. I still have no idea what his intentions had been that day when he had tried to mix it up with an Iran Air jet and lost! Fortunately, there was only one "fatality"! That was the job of that misguided pilot who had thought he had priority over Iran Air in the traffic pattern.

Mid-Air Collision
February 4, 1975

On the morning of February 4, 1975, I had been standing at Flight Operations to get aircraft assignments for my flight, when a frantic voice came over the radio telling us two of our Huey helicopters had collided and crashed! Since there had been a designated Huey crash-rescue helicopter parked just a short distance away, I raced to it and was quickly joined by another pilot. It didn't take long to start-up and we were soon flying at maximum cruise toward the crash site. We were fervently hoping it was the type of crash whereby both helicopters had somehow managed to land! Those hopes had been dashed when we made contact with the Flight Commander in the Huey circling over the crash site. It seemed like forever, but when we finally arrived overhead, we were stunned by what we saw.

I called back to base operations to report the location and request more assistance before spiraling down to land at the crash

site. There was no need to hurry because I had seen similar accidents while in Vietnam, and even at PHI. At that time, I had no clue who the instructor pilots were in the ruins of those two helicopters, but I knew they had been friends of mine! I say that because I was one of the first pilots to arrive in Isfahan, so I knew just about everyone. Suffice to say, it was a very terrible time for me, but I had a job to do. I began making a long approach towards the first of two crushed, smoking hulks that just a short time ago had been helicopters, each one crewed by one flight instructor and two student pilots.

While we sat idling for our engine to cool down, neither of us could take our eyes off what lay before us. Sitting on the left side of the aircraft, we could tell the cockpit and cabin was still sort of intact even though the tail boom had been badly burned. I had also seen what appeared to be a huge gash cut through the cockpit, just behind the copilot's seat. The instructor's seat! I could visualize the other aircraft's main rotor blade slicing through that cockpit! Whoever had been sitting there had never known what hit him! He died quickly, never knowing how the helicopter had free-fallen from 2,500 feet onto the hard, unforgiving desert floor of Iran. It had not been kind to the helicopter, or crew. There had been three pilots on board; one American flight instructor and two young Iranian students. It was hard to think about the young Iranian students who had probably survived the initial impact, then rode the mortally wounded Huey into eternity.

I slowly stepped out of the crash-rescue helicopter and began a slow walk toward the smoldering, smashed and crumpled hulk, pausing only when close enough to almost touch it. We could see one body inside, and I had to force myself to begin walking around the helicopter, knowing what I might find, yet not really wanting to accept it! But there they were. Lying about 50 feet in front of the nose of the helicopter were two bodies. I could readily identify the instructor pilot by his distinctive BHI-issued flight suit, name tag and boots. I also knew with great certainty he had been sitting in

the left seat; the same seat where the blade had sliced through so efficiently. Much like a guillotine, it had decapitated the pilot with that same, deadly efficiency. All I could do was walk off and squat down, trying to process what I had just seen. And we still had one more to go!

As I sat there, I could not help but wonder how such a tragedy could have occurred because it had been such a spectacular day! It had been clear, blue skies with no wind and even though it was still early February, it was a mild day. Absolutely perfect! Even though I didn't know the BHI pilots' full background, I knew both of them were combat veterans who had served honorably in Vietnam. Both were heroes who had served their country when called upon, only to end up there in the Iranian desert. I just couldn't grasp how in the world that mid-air collision could have happened.

The familiar "wop-wop-wop" of another Huey approaching shook me out of my bewilderment. Slowly, I stood up, took a deep breath, and began the long walk over to the other helicopter. It lay about 100 yards away, about the length of a football field. It was a long walk, but I needed it. The other helicopter landed alongside the crash site and began the shutdown process. Like me, the pilot was in no hurry to shut down. He had been in no hurry to find his friends he had shared that last cup of coffee with before departing on their training missions. It had been just another training mission. Another day in in that desert "paradise". Then, in the blink of an eye, everything had changed!

Soon enough, we were both standing alongside the torn and twisted pieces of metal that bore little resemblance to the beautiful Huey it had been just a short time ago. The pilots were still strapped in their seats even though the Huey had caved in around them. It had been their helicopter's blade that had sliced so neatly and efficiently through the cockpit of the second helicopter. Torn to pieces, the main rotor was no longer capable of providing lift. Once that happened, the helicopter had all the aerodynamic efficiency of a brick, as it dropped vertically from 2,500 feet!

It had probably only taken seconds to slam into the ground, but for the three pilots, it had been a lifetime! The last few seconds of their life! It had been best not to think about that. I had seen enough. Slowly, I turned and began the long walk back to the crash-rescue helicopter. Other helicopters had begun arriving, no doubt bringing specialists who would look at every aspect of that horrendous accident. They would digest and analyze every bit of data they could glean from that accident in hopes of arriving at a conclusion that would satisfy all. But for those six young men who had begun their day with so much hope, it really didn't matter about the who, what, when or where. All their tomorrows had ended there in the Iranian desert.

When I arrived back at our Isfahan training facility, it had taken a while before I could sit and write my own observations for the accident investigators. I kept playing it over and over in my brain, perhaps trying to re-wind the tape to start the day over by having both helicopters fly in different directions! The pilots were finally identified as Ed Thornton and John Romanski, and I desperately wanted to see both of them report for duty tomorrow, but I knew it would never be. I didn't sleep too well that night, nor did I sleep for several nights afterward. Even today, I can still see with great clarity what I observed on that terrible day. Just as I have so often in the past, I stored those memories alongside others, in a box somewhere in the far reaches of my mind. I often thank God for the wonderful long life he has given me when so many other close friends have already flown west on their final flight.

Motorcycles and Camel Caravans

Motorcycles seemed to be the best way to relax in Isfahan, so several good friends and I had taken an all-day trip into the desert on our newly purchased Yamaha motorcycles. After riding for what had seemed like hours, we pulled up to a small knoll where large boulders had created a shady outcrop where we could take a lunch break and cool off. Pulling out drinks and sandwiches, everyone had been thoroughly enjoying themselves when we began to hear something that sounded much like bells tinkling. It was certainly odd to hear such a sound out in the middle of a desert landscape, and it seemed to get louder! It also had a rhythm to it. Sort of tinkling in conjunction with someone walking. As our curiosity peaked, we scrambled atop one of the stone outcrops and peered down into an arid valley that stretched for miles back into the ancient mountains. We were literally stunned at what we saw.

Stretching as far as we could see was a camel caravan accompanied by what appeared to be thousands of sheep! The camels were loaded with the personal belongings of the nomads who were in various stages of walking, or riding camels or horses. Magnificent horses! Arabian stallions being ridden by fierce-looking men armed with rifles! Out-riders who rode on the outer fringes just alongside the massive caravan. It had been a scene directly from the long-told story of "Ali Baba and the Forty Thieves"! We were all mesmerized by a scene dating back thousands of years! We saw absolutely nothing we could tie to modern civilization. After what seemed like hours, the caravan finally passed, and the tinkling of the camel bells slowly dwindled into the distance. The silence of the vast desert again prevailed.

Pilot's Strike

August 1975 was undoubtedly the worst month for me in Iran. That was the month about 175 instructor pilots, out of approximately 200, went on a strike! Having arrived in Iran on March 1, 1974, I had known, and worked, with practically every one of them! We had all been friends! We had worked together, drank beer together, partied together and had been there for each other in times of need! Then, one day it had all ended. Unfortunately, it hadn't been a fast death, either. Much like a cancer, it had lasted for what seemed like months while both sides argued and negotiated until the "patient" finally died.

Of the 175 pilots who walked out on strike, 100 instructor pilots and their families eventually left Iran. I'm sure it was the beginning of tough times for them, but it was also the beginning of a very tough time for the 100 pilots who remained! Each of us were flying anywhere from four to six students a day, as we struggled to honor BHI's contract with the government of Iran. It was brutal! But everyone stepped up and did their jobs, and in so doing, saved the contract. It was the worst of times, yet the best of times! Everyone simply did what they had to do, and we worked together to make it happen. Soon enough, a new group of instructor pilots began to arrive. Like the Phoenix, Bell Helicopter International arose from the ashes, and flourished until February 1979.

The memories of those dark days have never left me, and one memory is especially haunting. Another pilot and I, both Flight Commanders, became good friends during that time. Having lost many friends during the strike, his friendship had been very important to me during a very difficult time. He was someone to drink beer with, ride motorcycles and share my feelings. Then, one day he announced he was joining the pilots still on strike! His announcement staggered me! His friendship had been a tremendous loss. It didn't get much better, but the one constant had been my wife! She was there beside me, just as she always has been. The only thing left for me to do was suck it up, move out, and move on!

About a year later, we had pretty much recovered from the strike and were moving forward with new instructor pilots in the fold, and still hiring more. One day the hiring rep came into our office with a list of new pilots for us to review. He also asked if anyone had known a former pilot by the name of John Doe. It had been my friend who had left at the very last moment to join the pilot strike! Casually, I asked the rep if the former pilot had re-applied for his job. "No", he said, "he robbed a bank and the FBI is looking for him!" I was so shocked I was speechless! A young man's future that had been so bright and promising had ended in a disastrous manner!

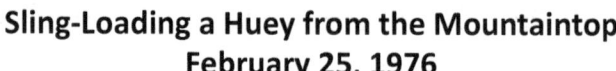

Sling-Loading a Huey from the Mountaintop
February 25, 1976

Early in the morning of February 23, 1976, one of my two tactical training flights began the day conducting cross-country training in the huge tactical training area that lay just south of Isfahan. The flight had required 12 helicopters, each one crewed by a BHI instructor pilot and two Iranian students. It was a typical training mission, whereby each helicopter would fly a pre-determined course utilizing navigational skills they had learned in ground school. Each helicopter would depart at a specific time to keep some distance between each helicopter. The flight commander would launch first, then climb to 10,000 feet MSL so he could provide flight-following for other helicopters to ensure each of them had been accounted for throughout the cross-country training period.

Everything had gone extremely well for the most part, at least it had until the next-to-last helicopter failed to check in at his RP (Reporting Point). After repeated calls from the command ship resulted in no response, word went out that a helicopter was overdue and presumed crashed in the training area! As Chief Pilot of Advanced Flight Training, the missing helicopter was under my control. I quickly responded by grabbing another pilot and launched in a standby Huey. It was déjà vu all over again for me! I had responded in much the same manner to another accident on February 4, 1975, a year earlier! My greatest hope was this accident would not end as tragically as the previous one had.

As we were flying out to the tactical training area, we received a call from the command ship that the missing aircraft had been located. It had crashed on top of a mountain peak that had been the final checkpoint for cross-country training! The helicopter involved in locating it was about to land to check on its crew. Everyone held their collective breathes as they anxiously waited for the update on the crew status. After what had seemed like an eternity, everyone got the word. The instructor pilot and two Iranian students had been injured, but everyone was alive! What a great message that

had been! We had gotten our miracle, now we just had to get them out of there and back to a hospital!

As it turned out, neither student had been badly injured and even though the BHI pilot had a couple of broken bones, he would eventually return as an instructor pilot as soon as he healed. It was great news! Having taken care of that, there was one last problem! How could we salvage the helicopter? I would soon have my answer.

Since the helicopter had crashed atop a mountain that was over 5,000-foot-high, there was only one helicopter that could lift it off the mountain, and that was the newly delivered Bell 214 model. Even then, we had to remove whatever we could from the crashed Huey to get the weight down. It had been a daunting task, but maintenance crews finally sent word the aircraft was ready to be sling-lifted off the mountaintop.

The Bell 214 was powerful enough, but it had just recently arrived in country and only a handful of pilots were qualified in it, myself included. Consequently, no experienced pilots were available to accomplish such a daunting task of lifting a Huey helicopter off a mountain top, but it had to be done! After much mental anguish, I made a decision. It would have been unfair to assign such a critical mission to a pilot with little time in the 214, so I decided to do it myself. A leader's job is to lead, and not one person argued or disagreed with me! It was a go for me. I was the pilot that would sling load the Huey off the mountaintop!

On the morning on February 25, 1976, I was walking out the main doors of the pilot training building, carrying my flight gear and heading toward the Bell 214 model I had selected for the mission. Much to my surprise, I met Howard Moore, BHI's Director of Operations at Isfahan, just as I was walking out. I greeted him as he asked: "Where are you going?" I explained the situation whereby I had taken the job myself, since I hadn't felt comfortable assigning another pilot to do it. He then asked: "Can you do it?"

"Yes sir", I said.

He then walked beside me and draped his arm across my shoulders.

"Dwayne", he said, "let me give you some good advice."

I allowed as I could use some good advice.

"Well," he said, "you drop that son-of-bitch and you're fired!"

He then turned and walked into the building.

It hadn't been exactly the words of encouragement I had wanted to hear, but I hoped he had just been kidding! It really didn't matter, though. I had far too many things to think about, such as how to sling load a crashed Huey helicopter off a mountaintop! My copilot for the mission had been a close friend and excellent pilot by the name of Billy Wayne Walker, a true Texan! He had to be with a name like that! Not only did I have total confidence in Bill, I also had confidence in myself. I had to.

We took off and after a 30-minute flight, I landed the Bell 214 on top of the mountain alongside the crashed Huey. The lead mechanic walked over and provided an update on what had been removed, and what the final weight of the damaged Huey should be. After making a few calculations, I lifted up to a high hover over the crashed helicopter so they could hitch the cable onto the cargo hook. As soon as that was completed, I lifted straight upward above the crashed Huey. The ground crew flashed me the signal that everything had been good to go! It was the time of truth!

Slowly, ever so slowly, I began increasing the collective upward with my left hand, thus steadily increasing the power to the main rotor blades to lift the Huey off the ground. It was a slow process, but finally, I felt the Huey lift off. The only problem was the fact it had taken almost maximum power just to lift it off the ground! I barely had the power to fly off the mountaintop! It was a struggle just to stay airborne with the dangling Huey, but I was able to slowly lift the Huey over to the edge of the mountain. Looking down, I saw a sheer drop that extended hundreds of feet down the side of the cliff! I had to maneuver the Huey off the side of the cliff and pray I had enough power to attain forward flight! It took just the slightest movement when suddenly, we were dropping like a rock off the side of the mountain while I was trying to maneuver the helicopter to start flying! After what seemed like an eternity, I could feel the 214 sucking clean air through the main rotor, and we were finally flying! Bill and I could start breathing again! I didn't know it, but later I had heard the maintenance folks had all held their breath as I struggled to get the 214 flying, then erupted into yells and cheers when they saw us level off, then head toward Isfahan!

By the time we arrived at Isfahan, I had burned off enough fuel that our landing hadn't been a problem at all. After dropping off the Huey, we then returned to the heliport and shut down. Bill and I also exchanged high-fives! It had been a great day! I also walked back over to the flight training facility and saw Howard Moore talking to my boss, Nick Psaki. I had a tremendous respect for both gentlemen and I had no problem interrupting their conversation. I walked up and said: "Well, Mr. Moore, your encouraging words certainly paid off!" He looked at me quizzically, so I said: "I didn't drop that son-of-a-bitch so I guess I still have a job!" All of us shared a good laugh about that. That is still one of my best memories of Iran!

Near-Miss with a Mountain

Training so many students was always a challenge, especially since we always seemed to be short on instructor pilots. As the Chief of Advanced Flight Training, I couldn't instruct on a regular basis, but there were times when I was able to fill in for a missing instructor pilot. I still recall when one of my flight commanders requested my assistance for a night flight. Readily agreeing to his request, I was assigned two students, each of whom I would fly a two-hour cross-county training flight with. It had been a bitterly cold night and one of the ten assigned Hueys had a heater that wouldn't work. Rather than pass it along to someone else, I took that helicopter myself. The students were woefully unprepared for such cold weather, so I managed to find an old, over-sized parka for the flying student, while the other student stayed behind in the warm classroom.

Even though I was wearing a heavy jacket and flight gloves, the student and I just about froze to death on that first two-hour flight! It had been so miserably cold, I took a little longer between flights just to warm up before going out on the second flight. I had finally fortified myself with enough hot coffee, and we were soon off for another cross-country flight across the desert surrounding Isfahan. I have to admit, flying cross-country flights to train students how to navigate had not been my favorite mission! But that was part of the job, and I was really looking forward to putting an end to it in such miserable conditions. In the meantime, sitting there in that frigid cockpit, the bitter cold and late hour made me sleepy. Very sleepy!

The flight had gone well, and our second leg of the flight had taken us far out to a small village located east of Isfahan. It was midnight when we finally turned onto our last leg and we could see the lights of Isfahan far in the distance! It was a most wonderful sight. I was almost frozen, and my student wasn't faring much better. I was so ready for that night to be over! The student was flying and even though we only had about forty-five minutes remaining, I began to nod off. At first, my head would drop, then I would snap awake! Finally, that no longer was the case, and I nodded off into a deep sleep.

I had no idea how long I had been sleeping when my eyes suddenly snapped open and I was wide awake! I glanced first at the instruments, then quickly looked out the windshield, and saw total darkness! Something wasn't right! I couldn't see the lights of Isfahan! I grabbed the controls and yanked the cyclic hard to the right, making an immediate 90-degree turn! I knew a large mountain was located just to the left of our final course as we headed for Isfahan and somehow, when I had seen nothing but darkness, I had instinctively known we had drifted off course due to a strong wind from the north.

The student, cold and inattentive, had allowed the helicopter to

drift south of our intended course, thus putting the mountain between us and Isfahan! Something, or someone, had awakened me just in time to grab the controls and turn away from the mountain! It seemed like we flew forever before I could finally see the lights of Isfahan appearing off the left side of the helicopter. Only then did I realize just how close we had come. Had I slept another minute or two, we would have smashed head-long into a sheer granite wall!

It was as if someone had thrown a bucket of hot water over me, because I was no longer cold! I was still flying the helicopter after taking the controls from the student and it had taken a minute or two before I could even speak. The student also began to grasp the enormity of what had nearly happened and began to shake. I certainly couldn't lash out at him for being so careless when I had nodded off myself. I told him we had both learned something that neither of us should ever forget! I know luck can always be a factor, but I don't think luck was involved that night. I had been in a deep sleep, yet when my eyes opened, I was instantly awake and, acted decisively and without hesitation! I truly believe someone directed me to take those actions! I've never dwelled on it, but I've been forever grateful! I have no idea how close we came to that mountain, but someone did, and they saved our lives.

Taking Watermelon Breaks

I still have wonderful memories of training young Iranian students in that distant land that had so much history! There had been such a wide variety of training environments that we could accomplish pinnacle and confined area training, low-level flying, and of course, desert training. The area itself was massive! We were limited only by how much fuel we could carry. It still stands out in my mind as one of the most interesting places I've ever flown.

A normal training day consisted of two separate flights, each one made up of ten or twelve Huey helicopters who could train at the same time. The tactical training area was so massive, each flight could conduct their scheduled training without conflicting with the other. Each helicopter had a BHI instructor pilot and two Iranian students who switched out at the end of each training period. The norm had been to fly each student one hour, then switch out with the other student. We even had tanker trucks sent out to the area for refueling in the desert. Most of the time, we also paid a visit to our favorite little area nestled high in the mountains that surrounded Isfahan.

Dubbed "Waterhole No. 3", our favorite little area had been found by some of our instructor pilots while flying in the outer reaches of our tactical training area and had become a popular place to visit during training operations. It appeared to be an oasis in the middle of nowhere, with no signs of civilization! The centerpiece had been a gurgling artesian spring of clear, cold water that seemed to flow out of the side of the mountain. I'm sure it had saved the lives of many nomads who had wandered through that high valley in times past, but whatever happened there has remained a secret. The oasis always kept such secrets, just as it has ours!

The thing that enhanced that special place of peace and tranquility was the discovery of a watermelon patch! Lying far down in the valley was a large watermelon patch, carefully tended by a dedicated farmer who had learned the technique of raising watermelons. In a land well known for fantastically tasty watermelons, his had always been best! Our students would always negotiate a fair price for the ripest ones available, then leaving the farmer with a handful of Rials, we would fly toward our spring, our "Shangri-La", to "store" the watermelons in that oasis of clear, cold water. That was important to ensure everyone had a cold slice of watermelon to enjoy as other helicopter training crews flew in to take a break while enjoying our impromptu party!

Drinking crystal clear, artesian spring water, and munching on cold watermelon, created wonderful memories for instructors and student pilots alike there in that ancient place where time seemed to stand still! Laughing, storytelling and dining on watermelon was an equalizer for all, and there had been no distinction between

instructor or students. Like the fabled Fountain of Youth, its magical waters turned each of us into young men, enjoying life's simple pleasures. Then, all too soon, we had to get back to the business of training students. Helicopter blades whirred to life, then one by one, we had departed, our helicopters quickly thundering off into the distance. Once again, only an empty silence remained in that tranquil oasis. The spell had been broken. I fervently hope it remains as we remember it. Hidden high in the mountains, silently waiting for the return of watermelons, helicopters and the young men who flew them. If only we could.

Air Show at the Gun Range

Every year, a branch of the Iranian military would host an annual event, whereby they would invite representatives from the other military branches to showcase their new equipment, personnel, capabilities, etc. In 1977, General Manouchehr Khosradad, Commanding General of the Imperial Iranian Army Aviation Training Command (IIAATC), had volunteered his new training facility to be the host for the prestigious military event. He also wanted to put on an airshow that would fully demonstrate the capabilities of the new arm of the military, to include a firing display put on by the AH-1J TOW Cobra. A pilot by the name of Dayle Courts and I, both of us Chief Pilots of Advanced Training, were tasked with planning an airshow worthy enough to present to every general in the entire Iranian military command!

Since Dayle and I had both served combat tours in Vietnam, it really wasn't too much of a stretch for us to simply replicate a combat assault-type airshow that would allow every helicopter in the military's inventory to be a participant. Since we would be incorporating live firing of the TOW missile, we decided to utilize the gun range for the airshow. After spending long days and nights preparing such an airshow, it was presented through the entire chain of upper BHI management, then to General Khosradad himself. Fortunately, he loved it! The only recommendation he had was for

us to allow the nearby Iranian National Guard to participate so they could present their new CH-47 Chinook helicopters. Not seeing a problem with that request, Dayle and I quickly agreed!

Once we had an approved airshow in hand, the next step was to start preparing and training for it. That posed a problem at first, because we had to utilize the training command's same helicopters utilized to train students. We overcame that speed bump by blending training into an actual combat assault-type experience that we intended to execute for the airshow. Since there were two separate training branches, A and B, that alternated morning and afternoon training on a weekly basis, we decided to conduct training only in the afternoon to ensure every instructor pilot and student would become proficient in flying in the airshow. The only problem we had was the Chinook pilots!

We briefed every day at 1:00 PM and we saw the same BHI instructor pilots and their students everyday. But, not so for the Chinook pilots! Each day we had three Chinooks show up and each day we had a completely different crew! When we requested that they send the same pilots every day to ensure continuity of training for the airshow, we were told they could not. It seemed they had far more pilots than Chinooks, and every pilot had been eager to participate in such a prestigious air show because every General in the Iranian military command would be in attendance! Dayle and I both realized there had been no way we could change the status quo, so we accepted the challenge and moved on with our training.

Since the airshow was planned to occur in June 1977, we decided to accomplish a dry run of the entire airshow to work out a myriad of details and develop split-second timing to ensure a successful show. We also decided to utilize live TOW firing for the final run-through!

TOW stands for "Tube-launched, Optically tracked, Wire guided". It was an anti-tank missile that had been designed just about the time Vietnam was winding down in 1970 and was actually used successfully against North Vietnamese tanks. It is an incredibly accurate weapon and, as far as I know, it's still being produced today. The only problem was the fact TOW missiles were expensive even then, costing approximately $65,000 dollars each!

Since we were allocated only three, we decided to fire only one missile during our last day of training. Everything else went as planned, all moving like clockwork, even with the three Chinooks. As it turned out, those "hook" pilots had been pretty sharp, plus we practiced so much we had cycled through every Chinook-rated pilot. The only problem we encountered was when we fired the TOW missile!

Since we had utilized the gun range for the airshow, there had already been numerous targets such as old tanks, trucks, etc. to shoot at, plus we had added a few more for the big show. The pilot who fired the missile had also done a fantastic job of center-punching the old tank hull he aimed at! The only problem was it had been anti-climactic. There had been no big bang! No explosion at all! Had we not been watching it through binoculars, we wouldn't have known the missile hit the tank.

That was a problem, but we came up with a solution. We would fill two or three 55-gallon barrels with both JP-4 and high-octane fuel, then place them just inside, or behind, the tank hull. Being short of TOW missiles, we chose not to fire another missile during training because we wanted to have a "spare" in the event the pilot missed with the first shot during the airshow. We had no clue as to whether or not our "Big Bang" plan of fuel exploding when smacked with a missile would even work!

The day of the big airshow finally arrived, and the gun range had been turned into a showplace worthy of the Oscars! Bleacher-type seating had been hauled in and erected in such a manner that every General in attendance would have a ringside seat for the airshow. There were many Generals, Colonels and Majors from all military branches in attendance. It lived up to its hype of being a very well-attended event. After everyone was seated, it was time to let the games begin!

Working together, Dayle and I had shared duties and each of us manned a radio to talk to each and every aircraft element to ensure good timing for the combat assault, complete with the live-firing TOW Cobra. It went off without a hitch! Every element of the airshow had been on time, and had hit their marks, including the Chinooks! The grand finale was the live TOW missile firing and it exceeded all expectations. When the missile hit the tanks, along with barrels of JP-4 and high-octane fuel, there had been a huge explosion, followed by a thick plume of black smoke! It had been spectacular! However, when I looked down toward the bleachers so I could bask in all the clapping, cheering and glory being heaped upon General Khosradad, what I saw were several Generals arguing mightily, one of them being General Khosradad himself! He was an excellent horseman and always seemed to have a riding crop in hand which he often used to "emphasize" a point, such as he was doing at that moment!

Since the airshow had gone off without a hitch, we were confused about what all the commotion was about. We were still trying to figure it out when one of the General's aides got our attention and told us to bring one of the helicopters back for the General. We wasted no time in complying with his request and very shortly a helicopter arrived. After the Huey landed, General Khosradad and several other Generals climbed on board and the BHI instructor pilot immediately asked for clearance to fly down the gun range where the targets had been. "Clearance granted!" With that, the helicopter flew down range and began flying all around the target area as if they were looking for something. Finally, after several minutes, the pilot advised us he was returning back to the pad to drop off his passengers.

As soon as the Generals got off the helicopter, they all walked back over to the bleacher area and everyone began shaking hands and laughing like they were the best of friends! Seeing all that, I cleared the pilot for takeoff, then asked what in the world was going on. "Well, he said, "you're not going to believe what we were

doing down range. We were looking for little men in foxholes who had been throwing grenades into those barrels of fuel to make such an explosion!" Too stunned to respond, Dayle and I just shook heads and proceeded to shut down for the day. As far as we had been concerned, it had been a successful day.

Later on, we found out some of the Generals had accused General Khosradad of "staging" the TOW missile strike! "Nothing could have been that accurate", they said. Even worse, they had also accused him of having soldiers hidden in foxholes and throwing grenades on cue to create the huge explosive fireball! Enraged, the fiery little General had taken offense to such an insult and the verbal fight had escalated to the point that he wanted a helicopter to take both him and his "doubting Thomas" buddies down range to look for those "little men in foxholes armed with grenades!" Finally satisfied the TOW missile strike had been legit, the Generals offered their apologies and they had all become best friends again. They also had good things to say about the quality of students who could have accomplished such an awesome airshow!

Dayle and I were happy just to have it over with! The fact it had been successful made it even that much better. We had several beers that evening to celebrate our success. We also brought several beers for our pilots who were celebrating with us at one of our favorite watering holes, The Long Branch Saloon. We also got a letter of appreciation from BHI's Director. I think General Khosradad also thanked us because we had used only two TOW missiles which had saved him $65,000 dollars! He might've used it to buy himself another horse!

Weather Forecasting

Some folks might think it was hot all time, but we actually had four distinct seasons in Isfahan, much the same as we have in Texas. The summers were hot, but like Arizona, they had low humidity. The old "swamp" coolers we used in West Texas worked fantastically in Iran! The winters were cold, and we far more snowstorms than I can ever remember in Texas. As in the U.S., Fall and Spring were always beautiful, and much appreciated by everyone, especially Spring. What we were missing was weather forecasters! Weather is always important in aviation but in Iran, we really didn't have a "FAA" type of weather forecasting in Iran. Like good soldiers, we came up with another way to forecast the weather, thanks to a foresightful pilot.

When we first arrived in Isfahan, one of the pilots, with much more foresight than anyone else had, remembered to bring a large National Oceanic and Atmospheric Administration (NOAA) chart of cloud formations and a synopsis of weather normally associated with a specific cloud formation. It worked so well most of us referred to that chart on a regular basis to prepare our weather briefings! Typically, the only time we really had to worry about weather was in the winter.

Tragically, we did lose an Iranian instructor pilot and two students when they got lost in a snowstorm late one evening and flew into a small valley with no way out in December 1977. It was a terrible loss since the flight instructor had been one of our best Iranian students who had completed flight school, been assigned to a helicopter unit to build flight time, then returned to be integrated in as our first instructor pilot. It was a tough loss of such a wonderful young man who had a very bright future in aviation.

Going Home

By the time January 1, 1978, rolled around, it had become increasingly clear things were changing and storm clouds were starting to gather in regard to the political climate in Iran. Unrest had begun in some of the distant areas of Iran and was slowly working its way into larger towns and cities. A former cleric by the name of Khomeini who had been exiled to Iraq was making increasingly loud noises about returning to Iran. All things considered, there were enough signs to warrant my departure; consequently, Wayne Brown, BHI instructor pilot and close friend, and I

departed Isfahan on March 1, 1978, almost four years to the day after arriving on March 2, 1974. When I returned to Arlington, Texas, I began working at BHI's corporate office located in Hurst, Texas.

As it turned out, my intuition had proved prophetic because things really started getting scary for BHI's employees still remaining in Iran, especially in December 1978. In fact, so much so, a relocation team was formed to call every employee that was home on vacation and advise them we thought it would be unwise to return to Iran. Many took our word and remained stateside, never to return. Others had been indignant that we would do such a thing and returned, only to come back in January all upset that we hadn't made them stay home! Go figure!

It all ended on January 16, 1979 when the Shah capitulated and was exiled. The Shah's reign had officially lasted from September 16, 1941 until February 11, 1979. Shortly afterward, four of the Shah's supporters had been captured, and General Manoucherh Khosradad had been one of those four. General Khosradad and I had not been close friends, but I was often in his presence in the performance of my duties as Chief of Advanced Helicopter Training at the IIAATC facility. He was rather short in stature, very handsome and was a dedicated father to his only daughter. I heard he had not been a man of great wealth, despite his rank, and that he had genuinely loved his country. Not only had he been a helicopter pilot, he was also an expert horseman and had been the founder of the Iranian Special Forces.

The KAYHAN (or, Cosmos) INTERNATIONAL, an English-speaking newspaper in Isfahan, published photos of four men just prior to their execution. The story also explained how the men had been given a ten-hour trial, during which three of them had begged for their lives. Their photos did not do them justice! On the other hand, General Khosradad basically told his captors he had always been loyal not only to his long-time friend the Shah, but also to his country. He had done nothing wrong and would remain a friend of the

Shah until his death! His photo showed a very nattily dressed man wearing a three-piece suit and French beret. He was looking directly at his captors with a smile. He had been executed with automatic weapons immediately after the photo had been taken. Regardless of what anyone had thought of him, in the end he had died like a soldier, true to his beliefs. He had been executed on February 16, 1979, and the paper published graphic photos of the aftermath of the executions. For me, it served as a graphic reality of what lay ahead regarding the future of Iran.

My time with Bell Helicopter International, Inc. ended on March 1979, a little more than five years since I had become an employee on February 4, 1974. It had been a wonderful experience! As for Iran, it was a massive change from what everyone experienced since Khomeini returned to power, and everyone still remembers when a mob of Iranian students stormed the walls of the U.S. Embassy in Tehran and took 60 hostages who were held captive for 444 days! The Iran where we lived and enjoyed so much, no longer exists. But the memories of that wild, wonderful and exciting time still remain.

My wife Lynnette, son Keith and daughter Nicole at our motel just before departing for DFW International Airport for our trip to Iran (02/28/1974).

BHI instructor pilots are standing beside a Bell AH-1J Twin-Cobra helicopter at Flight Training Center near Isfahan, Iran; I'm kneeling at far left in front row (1977).

I'm flying a Bell 214 helicopter that's sling loading a Huey that had crashed in the mountains (02/25/1976).

A Bell 214 helicopter is flying over a herd of camels near Isfahan, Iran (1976).

Bell AH-1J Cobras are lined up on training base flight-line near Isfahan, Iran.

CHAPTER 11

Goodbye BHI, Hello Bell Helicopter March, 1979

In early March 1979, I was sitting forlornly at my desk located on the seventh floor of the First State Bank building in Hurst, Texas, silently contemplating what I was going to do for the rest of my life, at least in regards to work. It was not a good position to be in. My sad demeanor had begun just a short time earlier, when my supervisor walked by my desk and quietly asked me to join him in his office. You never want to join your boss in his office, but I had no choice in the matter, even though I was pretty sure I already knew what it was all about! Taking a deep breath, I walked into his office, closed the door and sat down. There wasn't any friendly chit-chat either. It had been all business. My boss cleared his throat, then told me straight out my time at Bell Helicopter International would soon be over. Like at the end of March! Just like that, my five years with BHI was gone down the tube! Everything would end, and I would be without a job! It had been a great job, and I harbored no hard feelings against my boss, who was a very nice man and great to work for. But he had just turned a pleasant spring morning into a very bad day. He had a job to do, but I had a wife and two young children who were depending on me to take care of them. That's what husbands and fathers do. That's all I could think about on that Monday morning in March 1979. Then the phone rang.

    I had been so deep in thought, it startled me! Picking it up, a voice at the other end asked if I were Dwayne Williams. When I responded with "yes", he quickly introduced himself as Roger Huffaker. Roger was a member of Bell's pilot staff and I had met him

about two years previously when he was part of the New Equipment Training Team (NETT), conducting training in the Bell Model 214 in Isfahan, Iran. I was very pleased to hear from him and he got right to the point by asking me if I had a job, since BHI was winding down. Rather than say "how the hell did you know", my response was I did not, since I'd just been informed it would end in about three weeks, at the end of March. His next question surprised me: "Would you consider coming to work for Bell Helicopter as a test pilot?" I was stunned! Trying hard not to sound too excited, nor too eager, I responded by saying I had no experience as a test pilot. His response was: "Good; you don't have any bad habits and we can train you our way." I accepted his offer before he had even finished talking! Just like that, my day had gone from OK, to God-awful bad, then back to fantastic! Had I been able to crawl through the phone line I would have hugged Roger's neck! He remained a class act up until his death in 2019.

I won't go through all the details of what happened in between Roger's call and March 26, 1979. But suffice to say, March 26, 1979 was my official start date at Bell Helicopter Textron. I remember driving to work and, while sitting at a stop light, I actually pinched myself to confirm everything was real, not just some dream I was having. The pinch hurt, so I wasn't dreaming. I was actually starting work at the largest helicopter manufacturer in the world! It was exciting then, and it remained so until the day I retired. I never regretted my decision to join Bell Helicopter. It was quite a journey.

Bell Helicopter

At the time I joined Bell Helicopter Textron on March 26, 1979, it was one of the most famous helicopter companies in the world with a legendary history. A gentleman by the name of Lawrence, or Larry, Bell had formed Bell Aircraft Corporation (BAC) located at Niagara Falls, New York, on June 18, 1935. The company had grown to become one of the better-known aircraft manufacturers in the country, especially during World War II. It also became famous for

helicopters, thanks to a gentleman by the name of Arthur Young, who had received an invitation to visit BAC and demonstrate his "helicopter model that could fly out his barn door under remote control". Larry Bell, often referred to as "aviation's most seasoned dreamer", liked what he saw and offered Mr. Young a contract on November 1, 1941 to build and test two full-scale helicopters, just in case one of them crashed. The result of that agreement was the iconic Bell Model 47 that was awarded Helicopter Type Certificate No. 1 on May 8, 1946. It was the first granted by the Civil Aeronautics Administration (CAA).

It had taken a while to find a customer for the new helicopter, but the Korean War provided the perfect springboard into the military when the U.S. Army purchased a large number of them to perform a variety of missions on the battlefield. The medevac missions became the most widely known, and if you've ever watched MASH, television's hit show about the Army's forward medical units in Korea, you've seen the iconic Bell 47 model in the introductory scene. That new-found popularity prompted Larry Bell to relocate his helicopter manufacturing facility to Fort Worth, Texas in 1950. He wanted to get out of the extreme winter conditions of New York and into a more hospitable flying environment, plus he also liked the "can-do" spirit of the south. It proved to be a great choice!

On October 20, 1956, a helicopter referred to as the XH-40 made its first flight at Bell's main plant in Fort Worth, Texas. That historic flight was made before members of the U.S. Army's Research and Development Board and upon landing, Chief Test Pilot Floyd Carlson proclaimed it to be the "smoothest helicopter he had ever climbed into!" Everyone's enthusiasm was somewhat dampened when they received the word the iconic Larry Bell had died at almost the exact time Floyd was making the initial flight of the helicopter that would become famous in the Vietnam War. It was the U.S. Army's first turbine-powered helicopter, the Bell HU-1, named for Helicopter, Utility-1. It wasn't long before Army flight crews began referring to

it as Huey. Even though it would later be changed to UH-1, it is still referred to as the Huey, and it has become a legendary helicopter.

Reporting Into Bell Helicopter

My first day on the job had me reporting in at Plant 1, Bell Helicopter's main plant, located in Fort Worth, Texas. Chief Production Test Pilot at that time was Clem Bailey, a very pleasant gentleman who was also a very experienced pilot. Being the new guy on the block, it was quite intimidating just being in the same office with so many Bell pilots, all of them tremendously talented and widely known as being the most experienced in the industry. My first assignment was basically to shut up, sit quietly and pay attention to everything happening around me, plus attend all orientation meetings, classes, etc. I needed to make sure I was listening and taking notes because the very next week, Clem Bailey reassigned me to the old Globe facility located in Saginaw, Texas. Globe was Bell's 206 Jet Ranger assembly plant, and it also had quite a history.

Bell's Globe Facility

Located due north of Fort Worth near the little town of Saginaw, Texas, Globe first saw the light of day in 1938 when Fort Worth's first aircraft manufacturing firm, Bennett Aircraft Corporation, announced plans to establish a new facility. Built on an old farm, the horse barn actually became the firm's first office after all the hay and manure had been shoveled out of it. Their first aircraft was a bi-motor, mid-wing aircraft that made its first flight in 1940. Named the BTC-1, it would be made of plywood with an advertised speed of 196 miles-per-hour. It was a promise never kept and, facing bankruptcy, the new firm was soon refinanced and became known as the Globe Aircraft Company.

On opening day, a former War I pilot made a speech extolling the virtues of the newly formed company, but even though the new company didn't fare much better than the previous one, the outbreak of World War II saved the fledgling company from the same

demise as the first one. It became known as the U.S. Army Aircraft Plant and began producing large numbers of AT-10 multi-engine aircraft trainers for the USAAF. The contract lasted from 1941 through 1944, during which time the company built the largest all-wood building in Texas that would become synonymous with Bell Helicopter Corporation.

Built in 1942, the building was constructed using all-wood because of a critical shortage of metal due to the need for tanks, planes, ammunition and other war-time equipment. But in the post-war United States, there was no longer any need for military aircraft, or the huge facilities that housed them. Post-war plans to build an iconic little aircraft called the Globe "Swift" didn't work out and in the late 1940's, the facility became property of the government's General Services Administration (GSA) where it would languish until 1950. That's the year it was transferred to the control of the U.S. Navy and renamed the Industrial Reserve Plant in anticipation of Bell Aircraft Corporation relocating its helicopter division to Saginaw, Texas.

On January 6, 1951, Larry Bell inspected the facilities and signed a leasing agreement in early February so his helicopter division could begin relocating from New York to Fort Worth. On February 12, Bell was announced as winner of the XHSL-1 program, and the new Texas Division began operations on February 15, 1951 when five employees from Buffalo, New York, joined thirty Texas employees. Their first task had nothing at all to do with building helicopters inside the "big, dirty building". Their first task had been referred to as "burying the dead". That came about as the result of large amounts of rat poison being distributed throughout the building! Each morning when employees reported to work there was always a new selection of dead rats, mice or other "varmints" that had to be scooped up and buried. Only then could the new employees start building cribs, sweeping floors or doing a wide variety of odd jobs to prepare for helicopter operations.

Globe was a historic facility, and I regret not taking more time to really "explore" every facet of it. However, I was much too busy flight-testing new Bell 206 Jet Ranger helicopters. That kept everyone busy because Bell was producing 50 to 100 helicopters a month at that time. Even though the number of test pilots varied, there were always at least 7 or 8 production test pilots assigned there. All Bell 206 models were built in the large wooden building before being relocated to a large metal building that was utilized as a flight hangar, then they were prepared for flight test. With so many helicopters rolling off the assembly line, there was always an endless supply of helicopters to flight test, and I remained at Globe for a year before being transferred back to Bell's Hurst facility as a production flight test pilot. I might also add I loved every minute of my time spent at Globe and it's quite sad to pass by the now-vacant field where so much aviation history occurred.

Production Flight Test at Bell's Hurst Facility (Plant 1)

Even though a silver spade had been used to break ground for Bell's new Hurst facility on the morning of May 21, 1951, the Globe facility continued to represent Bell Helicopter's Texas Division until the 55-acre plot of land in Hurst became home to the first all-helicopter manufacturing facility in America. The administration building and factory layout were completed in just 60 working days, and complete occupancy of the new facility occurred in early 1952. From that time forward, Bell's new Texas division would continue growing until it became one of the largest helicopter manufacturing facilities in the world!

That facility was like hallowed ground to me since it was the home of the iconic Hueys and Cobras that had filled the skies of Vietnam, and were instrumental in coining the phrase "helicopter war". Consequently, when I returned to Bell's main manufacturing facility in Hurst, Texas, I felt very much like the child who had found himself in the candy store! I say that because at that particular time, Bell was still building the famous Huey and Cobra helicopters for the U.S. military, plus a wide variety of other commercial models to include the brand-new BH 222 model. In fact, just the thought of being a Bell staff pilot was exciting for me! While at Hurst, I was kept busy performing test flights on every helicopter Bell was building, not to mention helping out the Customer Training Center as an instructor pilot. I was quite happy as a "do-it-all" production test pilot, but one day I was assigned to Bell Helicopter's Flight Research Center located at the Arlington, Texas Municipal Airport to assist with "ground run" duty.

Bell's Flight Research Center (Plant 6)
Arlington, Texas Municipal Airport

Bell was just completing the long, laborious task of developing a new helicopter, and they needed pilots to participate in a long ground run test deemed necessary by the FAA. A "ground run" is when a helicopter is taken to a prepared cement area where chains are used to strap the helicopter to the ground. That would allow us to start the helicopter and increase power to what would normally be flight power (without flying), thus enabling us to complete a thorough series of tests without ever leaving the ground. It's very safe and efficient that way! Consequently, in the summer of 1980, I was one of several pilots assigned to accomplish a 100-hour ground run on what would become known as the BH 412 Model, a medium-sized, twin-engine helicopter. Ground run duty was rather monotonous, and definitely not on my wish-list, but it did provide me with the first glimpse of all the flight test activity that was ongoing at Bell's Flight Research Center located at the Arlington, Texas (Plant 6) municipal airport.

Not long after completing the Bell 412 model helicopter ground run activities, all production test pilots who had assisted in the 412 ground runs were invited by Lou Hartwig, Bell's Chief Pilot, to join him for lunch in Bell's executive dining room. He said it was in appreciation for our hard work, and since I was still relatively new at Bell, that was definitely an honor for me. I can't say the executive dining room was all that fancy, but I was surprised when Lou asked me to

sit next to him during the meal. Lou was a very pleasant gentleman, and I was enjoying our conversation when he suddenly asked me if I would be interested in being reassigned to the Flight Research Center (Plant 6). He went on to say how they could always use a good pilot, and he thought I could become a very good experimental test pilot.

Assuming he was simply making conversation, I responded by telling him how much I enjoyed working at Bell's main facility (Plant 1) and perhaps I would be more receptive to such an idea at a later date. I mean, I was thoroughly enjoying myself flying literally everything Bell was producing at that time, not to mention Hueys and Cobras, two of my favorite helicopters! When I explained that to Lou, he simply nodded, then continued eating his lunch. Thinking it was a dead issue, I resumed eating my lunch as well. Soon enough, everyone had finished their lunch, then began standing up and preparing to return to the pilot's office. I stood, then reached over to shake Lou's hand to thank him for the wonderful meal and his response was a total surprise! Nonchalantly, he simply said he had enjoyed it, too, and he would see me Monday morning when I reported in to Arlington! What did he mean, "see you Monday morning"?

His statement prompted me to hurry back to the pilot's office and find Clem Bailey to discuss Lou's comment, and what it meant! Clem's first comment was in regard to whether or not I wanted to go to Plant 6, to which I quickly responded no, at least at that time. "Well then", he said, "don't worry. I'll take care of it." Breathing a sigh of relief, I felt much better, at least until the next morning! That's when Clem walked up to my desk and said "Clean out your desk. You're going to Arlington." So, on a cool November morning in 1980, I moved all my flight gear to Bell's Plant 6 to become an experimental test pilot, and would remain there until I retired.

Huey Training for Royal Korean Navy (ROK) Navy
March 23, 1981

Even though I had been involved in a few projects following my transfer to Bell's Flight Research Center in November, 1980, my first assignment had nothing at all to do with experimental flight testing. Instead, I was sent to South Korea on March 23, 1981, to conduct flight training in a Bell 214 Model helicopter. Everything had gone well, but when I had finished the mission and was packing to go home, I received a call from Lou Hartwig, Bell's Chief Pilot. "Since you're already in Korea", he said, "can you remain a few more days to assist the Republic of Korea (ROK) Navy in training a few of their pilots in their new Huey helicopters?" I certainly wasn't thrilled about it, but I couldn't refuse since it was the Chief Pilot asking me to do it. Unpacking my bag, I called Bell's local dealer requesting him to make necessary arrangements for my new departure date home, plus arrange for a meeting with the ROK Navy. It certainly sounded simple enough, so I planned to visit the Navy's facilities the next day, on Tuesday, March 30, 1981.

Early the next morning, I walked outside my hotel to wait for Bell's local dealer to pick me up as planned; however, when I walked out the door, I was stunned! What had been a beautiful downtown area just last evening, had transitioned overnight into a war zone! Tanks were everywhere, as were machine gun emplacements; plus, some of the buildings had sand bags stacked around them! Jumping into the dealer's car, I asked "what in the hell was going on?" "Your President Reagan has been shot!", he said. My God! I couldn't believe what I was hearing! Immediate memories of President John F. Kennedy's assassination swirled in my head! I had many questions, but limited them to one; "Is he dead?", I asked. "No.", was his reply. It was a terrific shock, especially after seeing South Korea's response to such an overwhelming event, but I had been very relieved to hear he was still alive!

As we were driving toward the Naval Base, Bell's dealer explained how South Korea reacts to any type of world event such as the one when President Reagan was shot. It seems they were always concerned about North Korea attempting to use any type of incident to invade South Korea, just as they had done back in the 1950's when they had invaded South Korea, thus kicking off the

Korean Conflict. Since that time, the South Korean government had sworn they would never be caught off-guard again, hence the swift reaction of placing tanks in town squares, machine gun emplacements, etc. I have to say, I was very glad to hear that! I wanted no part in becoming a POW in North Korea, or to be assigned to their military as a helicopter pilot! I was still in the process of digesting everything that had happened in the last 24 hours when we pulled up to the large ROK Navy base. My concerns would have to go on the back burner. We were back in the business of training South Korea's Naval pilots.

We were greeted by several young ROK Naval officers and after a few minutes of trying to shout over so many folks trying to talk, we decided to go to the hangar where the Hueys were kept. As soon as we entered the hangar, I immediately saw six Huey helicopters, all without floats. The floats themselves were stored against the rear wall of the hangar. All of them were the large "banana-shaped" floats, and aside from the dust and cob webs, they appeared to be new. Taking note of that, I thought it rather strange they would be flying Navy helicopters without floats, especially since they would almost surely be flying over really deep water! I mean, flying over deep water is what they did in the Navy!

Seeing me casually looking at the floats, one of the young Naval officers spoke up and asked if I could fly with them if they installed the floats onto their Hueys. Taken aback by his question, I responded by telling him I thought I was there to fly with them in their Hueys, not check them out on floats. His reply startled me even more! It seemed they were already checked out in the Hueys. The problem was that they had never flown them on floats! "So", I said. "You're telling me you know how to fly the Huey, but only without floats installed." "Except once", he responded. "We tried to fly with floats only one time, but no more after autorotation!"

Aha! Everything finally started to make a little more sense! If they had flown the Huey with those huge fixed "banana-shaped" floats, and attempted to conduct an autorotation without an experienced instructor pilot, I knew they had more than likely scared themselves, big time! I say that because those huge floats were like airfoils and when the helicopter entered an autorotation, they actually tried to fly themselves, resulting in the nose pitching up, sometimes violently. I knew from experience it was a scary position to be in! I very politely asked the young officer if he had ever flown it with floats. "No sir", he said. I then asked if other pilots had attempted an autorotation, and maybe scared themselves. "Yes! Very bad", he confessed. Now that I knew what happened, I asked if we could fly the Huey the next day with floats installed. He got excited, then translated to everyone we were going to fly with floats tomorrow. Everyone got excited then! It was game on! We were going to fly a Huey with floats installed!

When I arrived the next morning, I was somewhat surprised to see all six Hueys sitting out on the flight ramp, but only one had floats installed. It looked rather ominous sitting there all alone, and I had a quick flashback about how long it had been since I had last flown a Huey with floats! I fervently wished it hadn't been so long ago, but it was a moot point! Today, we were going to fly with floats installed! I was also going to demonstrate, and instruct, how to do autorotations! Plus, I was going to do this in front of a crowd of at least 40 or 50 Navy pilots and crewmen, because that's how many I observed heading out to the other five helicopters for loading! Oh, by the way; we would do all autorotations in a large, fast-flowing river not far from Seoul!

A sharp-looking ROK Navy instructor pilot and I made up the crew for the Huey with floats installed. The other five Hueys had as many passengers as they could cram on board! Judging by the young pilot's face, I think he might have been the pilot who had attempted the autorotation with floats installed. He probably had also drawn the short straw, thus winning the dubious distinction of having to fly with me, and even worse, conduct autorotations! Oh well, if things went wrong, we'd have a lot of witnesses, plus someone available to fish us out of the river!

As we flew towards the river, all of the Hueys took turns flying alongside us to allow everyone on board each helicopter to take hundreds of photos of our float aircraft! Everyone on board those helicopters had multiple cameras hanging from their necks! I would like to say I had been cool, with no pressure involved; but I would be lying! As we flew towards the river, I just kept thinking about how long it had been since I had done autorotations on floats, and in the middle of a river, for God's sake! Plus, there was a large audience on hand, some of them high-ranking officers! All things considered, I would've much preferred being back in Texas rather than doing autorotations in a river I had never seen before! We finally made it to the river, and it was just as they said: really big, with a fast-flowing current! We had a lot of time to look it over because we had to allow for each Huey to shut down, plus let the passengers find a good observation point from where they could take photos. It reminded me of a Korean version of the popular old hymn, "Shall We Gather at the River", except for a different reason.

We made several approaches to a hover just above the river to check the wind before taking off, then circled back around to get established on a final approach over the river. Satisfied we were ready to proceed, I began telling the pilot what I would be doing, then when we were on final, I rolled the throttle off and entered autorotation. Muscle-memory did its thing on the descent by automatically reacting to every movement the helicopter/float combination tried to make! It wasn't unlike a cowboy riding a bull that bucks, rolls and spins, while the cowboy automatically counteracts every movement the bull makes. We continued our descent with engine at idle and main rotor producing zero lift. It was kept turning solely by air rushing upward through it, much like a child holding a small fan out a car window. It was our parachute! As we got closer to the river, I initiated a deceleration flare to stop forward airspeed, then leveled the helicopter and dropped vertically into the river with hardly a splash. It had been a text-book autorotation! I don't know who was more excited, me, the young Naval officer or our audience who was lined up alongside the river going wild with their cameras. Kodak was making a fortune from all the cameras and flashbulbs working in overdrive!

I turned the controls over to the Navy pilot and as it turned out, he was a damned fine pilot who quickly mastered the art of autorotations on floats. Satisfied with our efforts, we took off and flew low over our audience standing on the bank to let them know we were finished, then headed back to base where the young pilot was welcomed as a hero! The day ended with me accepting an invitation for drinks and dinner the next evening, my last night in Korea. I do recall the party was attended by several young Naval officers, plus their lovely wives, and a great time was had by all! But I have zero memory of how I got to the airport the next morning, or even who took me! I only remember waking up when the flight attendant announced we were landing at Honolulu International Airport. The best news of that trip was President Reagan surviving the assassination attempt and serving two full terms as President of the United States.

CHAPTER 12

Interesting Assignments at Bell Helicopter

Flight Testing in the Rockies

In March 1981, I was one of two experimental test pilots assigned to Bell Helicopter's new Bell 206L-3 project, the third model of a production line of helicopters. Since Bell had already produced the 206L series of helicopters, as well as a 206L-2, the new model was simply an extension of that very popular helicopter. The other pilot, Don Bloom, was a senior test pilot and, since he was considered to be one Bell's more experienced pilot at that time, he was to be the lead pilot for the project. It was a decision I certainly agreed with, since I was definitely the new guy at Bell Helicopter's Flight Test Center at the Arlington, Texas airport. In fact, I had only been on the experimental test pilot staff since November 1980, so I was both excited and anxious to be working with Don. Making things even better, the flight test engineer assigned to the project was a much-respected individual by the name of Tom Gardner. Knowing how experienced they both were, I was definitely looking forward to working on this particular project!

    The project started on schedule and, since the 206L-3 was really just a growth program of the existing 206L-2 series product line, we flew standard type of flight tests that were more oriented toward performance. It really didn't matter to me what type of tests we conducted; I was just happy to be involved in anything considered to be experimental! I didn't mind the long days at all, and another plus was the fact Tom and I worked so well together! Meanwhile, Don was quite happy simply sitting on the ground and being my mentor.

Flight testing went so well that by August 1981, we were ready to depart Texas for the historical town of Leadville, Colorado, an old mining town located in the heart of the Rocky Mountains. Not only is Leadville located in one of the most beautiful spots in the United States, but it also makes the claim of being home to the highest hard-surfaced runway in North America! Situated at 10,000-feet above Mean Sea Level (MSL), no one dared to challenge that claim! That is also what makes it a popular destination for every helicopter manufacturer in the world. The reason is quite simple. It you manufacture helicopters for a living, you want your products to be able to take off and land at the highest possible altitude. At least, you do if you want to be competitive with other manufacturers. Summertime is the busiest time of the year for helicopter testing, because that's always the hottest time. High and hot! That's what they all want. Manufacturers want helicopters to be able to operate in high, hot environments. Since that is so well known, it sometimes gets crowded at the Leadville airport with several manufacturers jockeying for a place to conduct their own flight tests. Fortunately, that wasn't the case for our project.

There's never a shortage of volunteers for offsite trips, especially if it happens to be in Colorado in August! Everyone wants to get out of Texas during those terribly hot, dog days of summer. Usually, it's a group of 10 to 12 specialists who either travel via a Personally Owned Vehicle (POV) or they fly commercial. Everyone except the pilot, that is. He always had to ferry the test aircraft across the country! I never complained, because that's always been the enjoyable part of working in the helicopter business and I must admit, I was really looking forward to flying the test aircraft into the heart of the Rocky Mountains! However, things can change sometimes. In that instance, I was taken off the trip in favor of Don. It had been determined he would be the logical one to accomplish the high-altitude flight tests, since he was more experienced in that sort of thing. I could see the wisdom of their decision, but saying I was disappointed would be an understatement. It didn't make me feel

any better when I watched Don and Tom depart for their "working vacation" high in the Rockies while I was left to work on my suntan in the Texas heat!

The Bell test folks had a well-laid out plan, but even the best of plans can change. Sometimes things just have way of working out and this seemed to be one of those times. The Bell 206L-3 test aircraft and crew had made it to Leadville, but flight testing didn't go as planned since the engine had to be changed due to some unknown problem. That changed the scope of things, at least for me. Since the flight tests would take longer to complete than planned, Don had requested to return home to honor a previous commitment back in Texas. So, my services were needed after all. I was soon on my way to Leadville!

Upon my arrival at Leadville, I was warmly greeted by the entire test team and they quickly brought me up to speed with regard to what already had been accomplished. It really didn't take long, because they had made very little progress, since they had just started flight testing when the engine failed. The good news was the engine had been changed, but that had been the extent of it! There was a lot to do and soon, I was in the swing of things and enjoying every minute of my time there high in the Rockies. We were now ready to get started on our first set of maneuvers in the flight schedule.

The first flight tests on the schedule were what we at Bell called "Run-Up-Down-Runway" or, R-U-D-R for short. The term had come from one of the more experienced test pilots when an engineer asked him one day what he was going to do, and he replied "R-U-D-R!" When the confused engineer responded with a "what do you mean", the pilot glibly replied, "Oh hell, you know – run up & down the runway!" It was meant as a joke, but that term had stuck because we still used it for the entire time I was at Bell; in fact, maybe they still do. In actuality, you really do go up and down the runway in an attempt to replicate the wind force you need. For instance, 5 knots, 10 knots, 15 knots, etc., etc.

The testing conditions are very strict, in that the prevailing winds must be no more than 2 knots, and the aircraft is loaded to maximum gross weight prior to initiation of flight tests. Once those conditions were met, we conducted the flight tests by hovering down the runway replicating wind conditions to establish such things as critical azimuth, limiting wind speeds for hovering downwind, sideways, etc. The interesting part was the utilization of a pace vehicle, driven by one of the more experienced team members. It's a difficult task that's not for the "faint of heart". That vehicle is always on the runway, literally underneath the main rotor blades while both pilot and driver accelerate to a pre-determined speed, then continue to hold that speed while the test pilot keeps the helicopter positioned alongside him to ensure the correct speed for the test point.

When those tests were completed, we moved on to autorotations, which were a whole different matter, especially with regard to being gut-wrenching and dangerous! For the uneducated, autorotations could best be described as an emergency descent and landing in the event you have a total engine failure. As long as the engine is running properly in a helicopter, it is driving the rotor blades and all is well! Enjoy the ride! However, should the engine quit for whatever reason, the helicopter starts to descend and the up-flow of the air through the rotor blade keeps the rotor turning. Think of a small child holding a small fan mounted on a stick out the window as the car goes down the highway. It spins like crazy as long as the car is traveling. However, when the car slows to a stop, so does the small fan. So, as long as the rotor blades keep turning, it's like a parachute, except instead of jumping out of the helicopter, the pilot sits inside the cockpit and maintains complete control as the helicopter descends at a sink rate of 2 - 3,000 feet per minute as it rushes to meet the ground. I might add, such tests are not for wimps or novice pilots! They're definitely for the more experienced pilots. It's all driven by the fact the Federal Aviation Administration (FAA) feels very strongly that regardless of where you want to fly your helicopter, you must be able to autorotate safely to the

ground in the event of an engine failure. And I agree wholeheartedly! There are times I'd like to argue the point, especially while I'm the test pilot responsible for development flight tests. But even then, I would still have to agree with them.

At that point in my budding career as an experimental test pilot, I had already performed hundreds of autorotations as an instructor pilot. In fact, as a helicopter pilot, you're always practicing them since you always want to be prepared in the event of an honest-to-goodness engine failure! The only problem was the fact all of mine had been accomplished at much lower altitudes. I had never performed an autorotation at a high altitude – EVER! And helicopters react much differently at higher altitudes. But, there's a first time for everything. So, Tom and I set about preparing a test plan that would allow us to live a little bit longer if we would just do all the right things, and we were both driven by that same desire! He had dreams of a long, successful career as a flight test engineer and I wanted nothing other than to become an old, bold pilot who beat the odds! But first, we had to take care of our business at hand.

For starters, we both decided the first touchdown would be attempted at the lightest possible gross weight, so I could get a good feel for how the helicopter handled at those altitudes. My own thoughts were that if we could just make it around the traffic pattern on a thimble-full of fuel and no other ballast, that's exactly what we would do! Both of us were quite certain that would be the optimum way to proceed with those particular flight tests, just in case something happened to go wrong.

In the world of flight testing, we all know that a mythical man by the name of Murphy is always hanging out somewhere close by, just waiting to apply one of his "laws"; the law that's based on "when an opportunity presents itself for something to go wrong, it usually does." Normally, such philosophy always made your day a bit more interesting; however, we were pleasantly surprised all our planning had paid off when we accomplished the first autorotation with no problems. It was just like the ones I had performed at much lower altitudes, except everything had happened a bit faster! The mechanics of an autorotation don't really change much, regardless of altitude. The biggest difference was thinner air that prevents you from slowing down the forward airspeed prior to touchdown on the runway.

After the first few autorotations, Tom and I were comfortable enough to build up our testing weight and it wasn't long before we were at conducting autorotations at the maximum gross weight for that altitude. Now that I was in my "comfort zone", we were ready for the next step of our flight tests. We were now ready for the FAA test team to arrive and begin what we call "certification tests". The FAA test team normally consists of two people. One member is a flight test engineer, the other is a pilot. Their job was to ensure we were adhering to all applicable FAA regulations during the course of our flight tests. I would like to add, throughout my career as a test pilot, it has always been my great pleasure to have worked with such professionals.

The FAA has established certain criteria that all aircraft, whether airplane or helicopter, must meet to ensure it can be safely flown and operated throughout its entire flight envelope. When testing determines the helicopter has met all the required criteria, the FAA will then send a flight test team to fly whichever test points they choose to fly, to ensure everyone is on the same sheet of music, so to speak. Overall, it's not a bad system and I always agreed with it; at least, most of the time.

The FAA flight test team appointed to participate in our flight test program responded quickly and arrived within a day or two of our requests. However, we had no idea which test points they would like to check. They always have the option of selecting whatever they so desire, but Tom and I were thinking they would want to look at some of the easier test points so they could build up to the more difficult tests. True to fashion, the FAA flight test team selected some of the easier tests to perform to get acclimated at such a high altitude. That had been the prudent thing to do, because a

helicopter does not perform the same way at altitude as it does at sea level.

Soon enough, we had gone through a list of tests for them to evaluate, then when they had been satisfied, we moved on to the more difficult tests to accomplish, like autorotations! Having gained confidence in flying easier flight tests, the FAA pilot announced he was ready to move on to begin the evaluation of autorotation characteristics. We all agreed that would be a good plan and suggested to prepare the helicopter for some light weight autorotations first. That would allow the pilot to develop a good "feel" for how to do them at lighter gross weights, then move on to accomplishing autorotations at the heavier weights, a more difficult task.

However, that's exactly what he chose not to do! We had fully expected them to check out the autorotations, but had also assumed they would utilize the build-up method that had served us so well. Flight testing can go awry very quickly, especially at altitude; consequently, we had chosen to conduct autorotations at the more benign weights to get a good feel for how the helicopter responds before tackling higher risk tests! But that's not what happened with our FAA pilot. He had chosen to conduct maximum gross weight autorotations first, certainly what none of us wanted to hear! I felt like we were on the brink of jumping feet first into the fire without benefit of the checking the temperature of the heat in the frying pan!

Tom and I both tried to explain the error of such a selection, and even attempted to convince him, rather tactfully, it might be better to accomplish the more benign tests before attempting what we considered a high-risk test. Unfortunately, neither of us were successful in our efforts. Accepting the inevitable, we began discussing what our recommendations would be so we could perform the flight tests in the safest manner. You can't imagine how shocked we had been when once again, the pilot announced we would forgo the buildup process that had worked so well for us. He would proceed by loading the helicopter to the maximum allowable gross

weight for that altitude! He said something about "that's how it would happen in the real world, when you least expect it." My God! Are you kidding me?! Somehow, I got the feeling our good friend Murphy was already beginning to chuckle to himself. After all, we were on the verge of presenting him with a whole truckload of opportunities to which he could apply his law.

One of the first things for us to accomplish when we traveled to an offsite facility was to make arrangements with the local officials to procure the services of a well-equipped fire truck and crew, ambulance and EMS personnel. We also had a well laid out accident plan, whereby everyone knows their specific job in case of an accident. It was company policy, and we always considered it essential in the event something should go awry, and we suddenly had a full-fledged emergency on our hands! In our case, all the dots had been connected and we were feeling pretty fortunate because we had savvy crews. They had quickly figured out on which tests they could relax and enjoy, versus the ones which warranted their unwavering attention. They all knew autorotation flight testing fit the latter. Consequently, since they had attended our FAA preflight briefing, they were most attentive when they heard what the FAA pilot's intentions were.

On the day of our tests, the entire flight test team had gathered in our "office", actually a dilapidated old trailer home, well before sunup for a preflight briefing prior to the initiation of flight testing. The pleasant smell of coffee permeated the small office and, needless to say, everyone had a cup of coffee in hand as we all tried to shake off the effects of not only an early morning get-up, but also trying to ward off the chill of the cool mountain air. Before starting our briefing, Tom and I had made one more attempt to convince our fellow pilot to abandon his thoughts of attempting an autorotation at maximum gross weight. Actually, it was more like a plea, but it was to no avail. His mind was made up! We were going for the end point. In one fell swoop, we were going for the proverbial brass ring!

Satisfied we had done our best, we ended our briefing. I stood and refilled my cup with coffee, then stepped outside to enjoy the best part of the day. Although the airport itself was still cloaked in the predawn darkness, I could look off to the west and see the sun's early morning rays just beginning to kiss the snow-covered peaks of Mt. Elbert and Mt. Massive. It seemed to bathe both mountains in a brilliant gold! Standing well over 14,000 feet above sea level, they commanded attention. Sensing that, I hesitated for just a moment so I could enjoy that spectacular gift Mother Nature had presented me. It was a sight I never tired of! It was also a sight I wanted to continue to enjoy for many days to come; however, today would definitely present a speedbump that could easily derail such pleasant thoughts.

There are times when you can do only so much, and I was satisfied we had done that. Committed to the task at hand, we completed last minute details and I began walking toward the helicopter that sat waiting for me on the flight ramp. There always seemed to be a lot of activity prior to the first flight of the day, and that morning was no exception. The flight test support crew had come in much earlier than I had, and were already hard at work preparing the helicopter for the day's flight schedule. They had already conducted preflight inspections, refueled it, cleaned the windows and warmed up the all-important instrumentation package. It captured all the test data from sensors located throughout the helicopter, each deemed critical for flight test operations. It was a go, no-go item. A quick thumbs-up from the engineer indicated all was well for the day's mission, a good thing; at least, most of the time. Today's events could prove differently!

The helicopter itself seemed to be awakening as a result of all the care and attention it was getting from a very focused crew. In fact, with just a little imagination, you could almost visualize the helicopter shaking off the doldrums of another cold night spent high in the Rocky Mountains! I've always found it rather amazing how an inanimate object like a helicopter can begin to take on human-like qualities. They seem to develop a personality all their own, and it almost becomes human to me. And its gender is female! Always! Beautiful aircraft and ships are always a "she"!

I resumed my steady walk toward the helicopter and began performing a mental preflight by taking note of such things as: how level is the helicopter setting on the ramp? Is the overall appearance normal? Are there any oily spots underneath the helicopter that would indicate a leak? Then, once I got to the helicopter itself there were the usual early morning greetings, or sometimes grumblings, as you exchange pleasantries with the support crew. Little things like: "please tell me again how much fun I'm having" or "man, what an ungodly hour to be up!" You know they're all good-natured complaints that you expect to hear from someone who enjoys their job. And you know each of them was glad to be there, high in the Rockies and not one of them would change places with anyone! It helps me to know that everyone one of them are specialists in their own field – and they're damned good at it! A pilot is always thankful for that because his life may very well hang in the balance. In regard to that particular crew, I had all the confidence in the world in them!

Opening the door and laying my flight gear inside the cockpit, I began the ritual of ensuring the aircraft was ready for the morning's planned flight test operations, starting with a thorough inspection of the aircraft logbook. I knew the flight test engineer had already made entries of everything that had been accomplished on the aircraft simply by reading that all-important document. I knew how much fuel had been pumped into the fuel tanks, how much ballast had been placed and secured on board, what the center of gravity was, and what changes or modification had been made since I had last flown the aircraft! Were there any open squawks I should be aware of? Suffice to say, there were numerous items of interest to me, so I had to read, and understand, every item. I read them as if my life depended on it, because it did.

Once I was satisfied with all that, I began pre-flighting the

aircraft itself and I paid close attention to all components, especially those painted universal orange! If something is painted orange, they're experimental parts. I also paid particular attention to all items noted in the logbook, because simply put, my life and that of the FAA test pilot flying with me were on the line. Finally satisfied with everything, I took the aircraft logbook and signed my name in the appropriate blank. Basically, I was saying the aircraft was now mine. It was my responsibility from that point forward.

After all preliminary rituals had been completed to my satisfaction, both the FAA pilot and I began to pull ourselves into the cockpit and settle into the seats. Even though I was the pilot-in-command (PIC), the FAA pilot settled into the right-hand seat (PIC seat) and I would sit in the left-hand seat. We quickly strapped in, put on our helmets, pulled on our flight gloves, then began going through the checklist to ensure all switches and gauges were precisely as they should be. I left nothing to chance because I know full well what has happened to other pilots who might have forgotten to have a switch in the right position, or maybe some other little oversight! Just some silly little something that might be significant enough to bring the helicopter crashing back to earth with a sudden jolt! Next, I took a preflight record by punching the orange button attached to the collective to check the instrumentation package. I verified it by observing the flashing amber light mounted on the instrument panel, which continued to flash until I pushed the button once again to extinguish it. That procedure established a record for all subsequent records we would be taking throughout the flight. It also verified it was working properly. Satisfied everything was functioning as it should, I started the engine, then began completing all checks in preparation for our flight.

The airport at Leadville was uncontrolled, but I always called the local Unicom to get prevailing winds, even though I already knew the winds were practically zero, since our strips of cloth, or flags, were hanging limply. It's always the prudent thing to do because I also wanted to check on any reported traffic in the area.

Actually, I would have been quite surprised if there had been any other aircraft in the air at that early hour, but it was always best to check. At that time of year, the winds are nearly always dead calm in the early morning hours. That's what the FAA decrees for our tests since it represents the worst-case scenario for helicopters.

The winds were calm, there was no reported traffic in the immediate area, and the flight test crew was standing by. All instruments were in the green and the FAA pilot signaled he was ready to go. Taking a last, quick look for traffic and, seeing none, I made the call on the local radio frequency to announce to everyone monitoring their radios that we were taking off to commence flight testing on the runway. As expected, there was no response from anyone in the immediate area. It was much too early for anyone else to be up, and I was very thankful for that small blessing!

You always take off into the wind because it not only provides better lift, it also assists in slowing the aircraft down in the event of an emergency during takeoff, and that's true for both helicopters and airplanes. However, when prevailing winds are practically zero, it's really irrelevant which way you take off since the wind is of no value to you. Our flight test criteria dictated less than two knots of wind so we could take off either way on Leadville's single runway that was laid out north and south, or more commonly known as 34/16. That simply means the cardinal headings are 340° for north takeoffs, and 160° for south takeoffs. Fortunately, the runway was some 6,400 feet long and 75 feet wide, and that was important! Longer and wider means you have a bit more room for error; always a good thing! It's also hard surfaced. In fact, I think it's made from some of the hardest material known to man! I say that because it takes on the consistency of sandpaper! The kind of sandpaper capable of grinding through the helicopter's landing gear!

After announcing our take-off, I turned control of the helicopter over to the FAA pilot since he would be doing the autorotations while I sat through them as safety pilot. We began our initial climb-out to the north, then turned left, thus setting ourselves up

for what we call a left-hand traffic pattern. Settling into our flight test routine, we made two touch-and-go type takeoffs and landings so the FAA pilot could get a "feel" for the helicopter prior to initiation of the autorotation flight tests. I have to admit, I still didn't have a good feeling about the test, at all! Truth be known, neither did anyone else!

Tom Gardner, Bell's on-site test director, was standing outside the test van and he informed us he had already alerted the fire truck and ambulance crew to ramp up their vigilance as we were flying on the downwind leg of our traffic pattern. He also asked again if the FAA test pilot wouldn't like to make some practice autos without actually touching down. Please! Anything to improve our chances! But my pilot was having none of that! He simply shook his head negatively as we began turning onto the base leg of the traffic pattern. We were one thousand feet above ground level, and after only one more left turn, we would be on final approach for runway 340°, ready for our unrehearsed autorotation. Oh boy! I fervently hoped Mr. Murphy was sleeping in! Or, maybe taking a coffee break!

The moment of truth had arrived! Taking my cues from the pilot, I informed Tom we were initiating a straight-in autorotation, just as the pilot rolled the throttle off to a ground idle setting. We were no longer in powered flight! We had the aerodynamic qualities of a stone, and dropping like one! Making matters worse, he had failed to lower the collective properly, thus allowing the main rotor RPM to droop significantly! That, in turn, set off an aural warning and an amber caution light! Damn! Not good! All my years as an instructor pilot served me well because I immediately knew he was already behind the power curve! We had barely started, and Murphy was beginning to chuckle!

Moving quickly, I grabbed the controls trying to get everything under control before we ran out of a pilot's best friends - airspeed, altitude and ideas! Slamming the collective further down against the stop to prevent the RPM from drooping further, I also tried to

roll the throttle on to restore engine power but couldn't. My pilot had a death grip on the throttle! In fact, he had a death grip on all the controls! Things had quickly ratcheted out of control, and Murphy wasn't chuckling anymore! He was laughing his ass off!

I was literally alone, at least mentally. I say that because the other pilot seemed to be having an out-of-body experience in addition to going deaf, because he couldn't hear me screaming over the microphone: "I have the controls! I have the controls! Get off the controls!" Thank God the rotor had returned to the green arc, but we were running out of altitude, and our airspeed seemed to be inching toward Star Trek's warp speed! I had just enough time to override his death grip and yank the cyclic control rearward to raise the nose way up in a flare, then shove it forward to level the aircraft! I also gave a simultaneous upward yank on the collective, hoping to cushion the impact with Leadville's unforgiving runway!

Out of the corner of my eye, I saw red flashing lights as both the fire truck and ambulance began rolling to intercept us. I also saw Tom hurl his clipboard away with all his precious paperwork tumbling about like snowflakes as he, too, began running toward the runway! I saw all that in SLO-MO just before we touched down – HARD! I think I also caught a glimpse of Murphy rolling around holding his sides from laughing so damned hard as we roared by! That SOB was having a field day and we hadn't even finished!

As soon as we made initial contact with the runway, the compression of the skid gear created a springing effect to launch us back into the air, then drop down to hit the ground for a second time. We finally stayed on the ground, but still had way too much speed! The aircraft began bouncing up and down as we skidded down the runway! Sparks, smoke and stench exploded from the aircraft during our mad dash! I was no longer a pilot; I was merely a passenger sitting alongside the other pilot! Fortunately, luck was with us and we finally ground to a stop, about a football field's length beyond our original touchdown spot. The nose also seemed to droop a bit, primarily because we had not only burned completely through our

tungsten skid shoes, we had also burned through a good portion of the skid gear! But what the hell! There wasn't any damage that couldn't be fixed right there in Leadville!

We had both been sitting there in sort of a daze before I noticed the bright red fire truck and ambulance that had pulled up and parked alongside us, red lights still flashing. Tom huffed and puffed up to the open window of the helicopter, then peeked through it a few seconds before quietly asking if we were OK. After what he had just observed, I think it was still hard for him to talk. That was okay, because I didn't have the energy to answer him. I think none of us could really grasp what had just happened aside from the fact we had been incredibly lucky! I did manage to nod my head to acknowledge all was well.

Then I asked the other pilot if he would like to accomplish another autorotation. I don't think he could speak just yet either; fortunately, he shook his head no. In fact, we didn't even have to repeat it! Hey, no sense in pushing our luck. After all, we had survived. He would accept the test data we'd already flown as sufficient. That was a good thing, because I wasn't up to performing another one, nor was he! At least not right away! I think old man Murphy had his fill of excitement, too, at least for that day. Maybe tomorrow. In the meantime, the other pilot and I decided a really cold beer might make us both feel better, even if it was still early in the morning. Heck, it was 5:00 o'clock somewhere! He and I became good friends before he made the trip west like so many of my pilot friends have.

The Summer Heat Wave of 1980

During my time at Bell Helicopter, I worked as a test pilot during two of Texas' worst weather periods, one of them being the hottest; the other was the coldest. The first weather phenomena of renown was referred to as the "summer heat wave of 1980." I had only been a production test pilot a little over a year when the summer of 1980 rolled around. I still recall the early days that summer as being hot, but summers are always hot in Texas! Then, along about June, three separate high-pressure areas moved in over Texas and much like a happy chicken settling onto her nest, they comfortably settled in to form a stable weather pattern that locked in searing heat that lasted throughout the summer!

The first triple-digit day had been June 7, but June 23 was the red-letter day that kicked off the longest string of triple-digit days in history! Forty-two days from June 23 until August 3, when a one-day break occurred, then another string of triple-digit days for a total of 69 days! June 26th and 27th were especially memorable for temperatures that ticked up to 113°, while Wichita Falls, a Texas town long noted for hot summers, saw 117°!

People, pets and thousands of chickens died in the suffocating heat, even as highways throughout Texas buckled in the heat! Local TV weathermen took their positions before cameras each night and tried to smile while forecasting "more of the same" to a suffering audience. It was going to be hot and dry, with no cool down or rain in the foreseeable future. Then, there were other predictions such as the twit who claimed Russians were bombarding us with laser beams from space stations, while a preacher spoke about how God was simply letting us know who was in charge. Whatever the reason, it was hot, dry and miserable!

With temperatures averaging 107° to 112° every day, folks simply tried to cope with a summer like none they had ever seen! Some of us sprayed our roofs and outside walls with water, hoping to increase the efficiency of struggling air conditioners; others simply stood in front of open refrigerator doors! Those lucky enough to have swimming pools chunked in blocks of ice before swimming just to cool the water down from steaming, to slightly warm and tepid. Others filled churches that conducted prayers, seeking some relief from what had become a living "hell on earth". However, nothing could compare to what it was like working on Bell Helicopter's flight line that summer! I can say that with some authority because I was flight testing U.S. Army AH-1S Cobra gunships on a black tarmac

ramp at Bell's main flight test facility in Hurst, Texas, where the temperature typically crested at least 120° each day.

Even though we began flying as early as we could each morning, by 9:00 AM the temperature was already sneaking into the 100's on the flight line. It is true the Cobras have air conditioning, but it always took at least 15 to 20 minutes for the cockpit to cool down; unfortunately, our flights typically lasted no more than 10 to 15 minutes. The hot sun beating down on the clear plexiglass covering the Cobra's small, 36"-wide cockpits, turned the cockpit into a sauna bath! At the end of each 15-minute flight, I was literally too weak to climb out of the Cobra cockpit! The mechanic graciously turned the main rotor blade to where its shadow covered me as I sat in the cockpit, unable to climb out. He would also hand me a bottle of water, plus a towel soaked in ice-cold water.

Sitting around the entire flight line were 35-gallon barrels filled with water, and Bell's flight-line supervisor would periodically stop by and drop large blocks of ice into each of the barrels to keep the water cold. We weren't drinking the water, either; we dipped large towels into the water, then draped them around our necks and upper body to combat the heat. Without a doubt, the summer of 1980 was one of the absolute worst times I spent as a test pilot at Bell! It is still the benchmark by which all other summers have been measured.

The Siberian Express

In December, 1983, every state in the continental United States was registering freezing temperatures, thanks to the onslaught of a cold front that brought frigid temperatures with it! Due to its origin in Siberia, it was quickly dubbed "The Siberian Express" and it brought freezing temperatures that lasted from December 18 through 30 - a total of 295 hours! It is still the longest consecutive number of days with freezing temperatures in Texas history. However, enduring such cold temperatures was good training for me, because I was planning a trip to an area famous for its cold

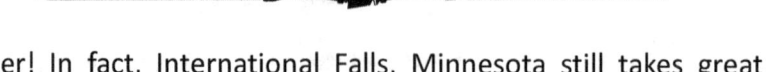

weather! In fact, International Falls, Minnesota still takes great pride in being referred to as "America's icebox"!

On the morning of January 3, 1984, I departed Bell's Flight Test Research Center, located at the Arlington, Texas municipal airport, in a Bell 222U Model helicopter heading for International Falls, Minnesota. We were going there to conduct cold weather operational tests on the helicopter I was flying. I had carefully planned my course to make a refueling/overnight stop at the Mason City, Iowa Municipal Airport. I had a special reason for going there, because a small spit of land just north of that small airport has become hallowed ground. It was there, in the middle of the "Winter Dance Party", on February 3, 1959, that musicians Buddy Holly, J.P. "Big Bopper" Richardson and Ritchie Valens had taken off into the teeth of a howling snow storm and crashed shortly after takeoff from the Mason City Airport. Millions of fans the world over still refer to the date of that accident as "the night the music died."

My hometown of Littlefield, Texas lies only 40 miles to the west of Lubbock, Texas, the hometown of Buddy Holly. Not only had I been a big fan of Buddy Holly, but one of the members of his band had been Waylon Jennings, a neighborhood friend of mine. As a pilot, I had read about the accident and probable cause with great interest, because the crash was eventually deemed to be the result of the young 21-year-old pilot's "unwise decision to embark on a flight in known inclement weather conditions without proper qualifications"; today we simply call it "pilot error". As a pilot, I can also say it might have been caused by something far simpler; their luck simply ran out. That has caused the demise of many good pilots.

We landed at Mason City late in the day on January 3, and spent the night there. After taking off from Mason City airport early on the morning of January 4, 1984, we departed along the same flight path young Roger Peterson (their pilot) had flown almost 25 years earlier. It was a solemn event, but I had at long last paid my respect to the young musicians who had filled our lives with such wonderful songs as young teenagers. It seemed rather ironic that all those

years later, I was heading north in the same wintry conditions that had cost them their lives.

Even though we were traveling on January 4, the Siberian Express had still made its presence known, thanks to the Arctic outbreak of frigid air that had rendered the continental United States colder than Alaska! In Denver, temperatures had remained below zero for 115 hours! Low temperatures in cities all over the U.S. had shattered 100-year-old records, and there we were, heading towards the "icebox of America". It was a cold flight, too! Even though the crew and I were both wrapped in all the clothes and socks we could wear, we were always cold! Maybe frozen would be a better word, because we both felt like popsicles when we finally landed at the International Falls airport! We were greeted by reporters from the local newspaper who had heard about the crazy Texans looking for cold weather in the midst of one of the coldest times in recorded history!

In spite of the fact that was my first trip to International Falls, we soon got into the swing of things and flight testing was going great! Then, one morning about mid-January, I was returning from flight testing when I noticed a TV van pulled up in front of the small airport terminal building. Always looking for ways to break up the monotony of long days away from home, I decided to go over and investigate the happenings at the terminal building. As I walked in, I spotted the TV crew interviewing an elderly gentleman, and something about him piqued my interest.

I quietly walked over and stood quietly while they finished interviewing the older gentleman, then profusely thanked Mr. Bronko Nagurski by name! Now there might be a lot of folks who don't have the first clue who Bronko Nagurski is, but I'm not one of them; I knew exactly who he was! Standing at 6-foot, 2-inches tall and weighing in at 230 pounds, he was the toughest, most feared football player in the early NFL days throughout the 1930's! He was a feared fullback, and equally good on defense. Money was tight in those days, so you played both ways if you were good enough. The old gentleman

had kept the NFL alive during some lean times, so the NFL was paying homage to him by inviting him to attend Super Bowl XVIII as part of the official coin-toss event. He was a long-time resident of International Falls, so the TV crew had come out to interview him prior to his departure for Tampa, Florida. I did manage to get a handshake from Mr. Nagurski and was properly impressed by how large his hands were as he engulfed mine in a crushing handshake! It was a great pleasure meeting the 76-year-old gentleman!

As soon as Mr. Nagurski departed for Tampa, the TV crew couldn't help but notice my Texas accent, and immediately inquired about what we were doing up there at that time of year. "Well," I explained, "we're looking for cold weather. Like 40-degrees below zero so our helicopter could be FAA certified at that temperature." That got their attention! So-much-so, they asked if they could join our crew in the hangar and visit for a while. "Sure; why not?" The three nice young men then accompanied me back into our hangar where our twelve team members were busy with the helicopter, plus trying to stay warm. They enjoyed their visit immensely, but had to get the interview tape back to the TV station. They then asked if they could come back out the next day to spend the entire day with us.

Sure enough, about mid-morning the next day, the TV crew showed up and spent most of the day with us, all the while interviewing our crew members and filming ongoing flight activities. We were accomplishing our flight tests, but also griping mightily about the temperature "only" averaging 20- to 25-degrees below zero, when our desired temperature was 40-degrees below zero! We also introduced them to Tex-Mex food! One of our team members loved to cook, so we had taken up a collection prior to departing Texas so he could purchase some of our more popular food in Texas, then loaded it onto a large semi-trailer truck, along with our equipment, for transit to International Falls. Needless to say, the TV crew enjoyed our spicy meal in the frigid weather, and we enjoyed their visit.

Early the next morning, we had a surprise visit from the supervisor of the TV crew. It seems their TV station in Duluth, Minnesota was an ABC affiliate and Peter Jennings, the renowned national news anchor, had picked up on the story about some "crazy Texans" who were looking for cold weather in International Falls, and was wanting to present our story on that evening's national news! I'm talking prime-time TV, and 15 minutes of fame! After calling home to advise all our family members, the entire test team gathered in the Holiday Inn bar to await the national news.

Sure enough, right on cue at 5:30 in the evening, Peter Jennings came on and his lead-in story was all about the frigid weather that had the entire United States in its deadly grip! He spoke for about 5 minutes, then went on to say "but in spite of the frigid weather that has the entire United States in its grip, it's not nearly cold enough for twelve tough Texans in International Falls, Minnesota! They're looking for 40-degrees below zero!" The news then flashed over to us doing our flight tests and the story actually ran for about 3 minutes or so. Everyone in our snug little bar thoroughly enjoyed it, especially since the manager set up a free round for his now-famous guests! It had been a great day!

International Fall's favorite son, Bronko Nagurski, also had a good day at Super Bowl XVIII in balmy Tampa, Florida on January 22, 1984. It was the Washington Redskins versus the Oakland Raiders, and their co-captains joined Bronko in the center of the field for the coin toss. Joe Theismann, the Redskins' quarterback, made the call, but lost; the Redskins also lost the game, 38 – 9. As far as we were concerned, Bronko was the star, because he tossed that coin in the air without a hitch! In fact, it was so good the bartender set up another round of free drinks for us to celebrate our hero! It was another great day!

Me and Bell flight test crew standing next to Bell 206L-3 test helicopter during high altitude flight tests at Leadville, CO (08/1981).

I'm flying the Bell 206L-3 flight test helicopter in Rocky Mountains near Leadville, Co (08/1981).

I'm flying alongside pace vehicle in a Bell 430 helicopter during flight tests in Rocky Mountains at Leadville, CO municipal airport (07/1995).

Me and FAA test pilot Eric Bries had just completed a cold weather test flight at International Falls, MN (01/25/1984)..

CHAPTER 13

Lost in The Bermuda Triangle

Having just completed a demonstration tour in the Dominican Republic, John Evans, a Bell mechanic, and I had planned an early morning departure out of Santo Domingo on the morning of October 29, 1982. We had hoped to make it all the way to Georgetown, a quaint Bahamian town with a small hotel on the beach. Planning to arrive there before sunset, it would have been a pleasant place to unwind after a long day of flying. Unfortunately, overzealous customs agents had delayed our departure and it had a domino effect, because it also delayed our arrival at the Caicos Islands, our last refueling stop before Georgetown. Making matters worse, after departing Caicos, we encountered stronger than forecasted headwinds and had flown almost two hours over nothing but an endless expanse of water. Thunderstorms popping up all around had forced me to fly a zigzag course, and our navigational equipment mysteriously stopped working at sunset, forcing me to revert to what is known as "dead-reckoning" navigation.

Dead reckoning is the most basic form of navigation dating back to Charles Lindbergh's historic flight across the Atlantic Ocean. Simply put, a pilot utilizes a magnetic compass to maintain a heading, a clock to keep time and an aeronautical chart to identify specific checkpoints such as bridges, railroads or, in our case, islands. If we had seen any islands to use as checkpoints, it wouldn't have been so bad! But there had been very few islands, and they had long since disappeared. For a long time, we saw nothing but water in all directions and a sky rapidly fading into darkness; a darkness that was intermittently illuminated by the feathery fingers of lightning flashing across the horizon!

John and I were intently searching for an island where we could land when the amber colored low-fuel caution light suddenly flashed

an ominous warning! Even though the light was small, it was as if a flashbulb had gone off directly into our eyes! Neither of us could pry our eyes from it because it had alerted us in stunning fashion, we had approximately 20 minutes of fuel left in our tank! It also meant our options had instantly diminished in an inhospitable bit of ocean located in what is widely known as the Bermuda Triangle! Folks who are more superstitious call it the Devil's Triangle! In either case, it's an eerie, triangular-shaped area of ocean that stretches from Bermuda, to Miami, then to the Bahama Island chain. It's also an area where numerous aircraft and ships have mysteriously disappeared without a trace!

After such a long day, my brain was having trouble grasping the situation we had suddenly found ourselves in. I began reviewing a mental checklist in preparation for an inevitable ditching into an ocean where so many others had disappeared! I also briefed John on emergency procedures for ditching, and he quietly responded by tugging at the waist strap of his flotation vest to adjust it. Running out of options, I then rolled the helicopter into a right-hand turn, intending to make a last-gasp 180° turn, desperately searching for some small island where I could land and escape the grasp of the Devil's Triangle!

Unfortunately, I was operating in such a stress-induced mental overload, that I forgot the heading I intended to roll out on! Trying to remember it had resulted in a kaleidoscope of numbers tumbling in my brain, none of which made any sense! I was so busy wrestling with the numbers, I failed to roll out on a new heading after completing the 180° turn! The spell was broken only when John began screaming: "Level off, level off; you're flying into the water!" Mentally wrestling with what should have been a simple task, I had failed to maintain altitude during the turn! The helicopter responded in much the same manner as a racehorse, when give a loose rein, or a whip! Instead of racing toward the finish line, we were racing toward impact in a spot of ocean where we would become just another mysterious statistic!

Responding to John's warning, I quickly rolled out of the turn and simultaneously pulled aft on the cyclic stick to level the helicopter. I fervently hoped my actions would prevent the belly of the helicopter from slapping the black surface of the ocean. Involved in mental gymnastics, I had involuntarily entrusted our well-being to luck rather than skill. After what seemed like an eternity, we both breathed a huge sigh of relief after realizing we weren't going to crash into the ocean! Our luck had somehow tipped the scales in our favor one more time! We had dodged another bullet, but both of us knew we would need more luck before our long night was over.

All hope rapidly diminishing, I began seriously thinking about a water landing and began peering intently through the windshield, desperate to determine wind direction and wave height, both critical to surviving a water landing. My intention was to make a controlled landing into the water with the engine still operating, rather than having our engine flame-out due to fuel exhaustion. The consequences of an engine-out landing in a dark ocean could be catastrophic! After a quick review of our emergency water landing checklist, I slowly began decreasing airspeed while switching to an international emergency frequency. Squeezing the microphone switch, I broadcast a "mayday" call, an emergency phrase advising anyone listening that an aircraft was in need of immediate assistance! The response was a thunderous silence that confirmed what I already knew. John and I were alone, in the grasp of the Bermuda Triangle!

Straining to visually penetrate the murky blackness, I reluctantly began a slow descent towards the black abyss that lay below. With all hope for a successful conclusion disappearing, I had a fleeting thought about the irony of that moment. After surviving a combat tour in Vietnam and flying thousands of hours in the Gulf of Mexico, it was hard to accept our disappearance in the Bermuda Triangle, a well-known graveyard for aircraft and ships that had mysteriously disappeared!

Surprisingly, I was filled with a calm resignation rather than panic. Minutes away from landing in the unhospitable ocean, I was lost in a myriad of thoughts, the biggest one being "what are the odds we would ever be found." That's when I thought I saw a quick flicker of light in the far distance. Directly in front of us, it diminished as quickly as it had appeared! Every fiber in my body became totally alert as I refocused my vision on that small spit of ocean, and uttered a quick prayer; "Please Lord; let that light be real!"

Instinctively, I pulled up on the collective to arrest our descent, then shouted to John about the light I had just seen! Quiet desperation propelled him to join my search for the mysterious light. Straining to find it, I tried to dispel the notion it might be a mirage. I knew from experience your brain can project an image your eyes can't see! It had occurred when I was desperately searching for an oil platform in poor weather conditions. Like apparitions, phantoms or ghost ships, a platform would appear out of the fog, then disappear. Being in a worst-case scenario and running out of options, that flickering light had to be real! Our survival depended on it!

I levelled off just above the ocean, and began skimming along just above the waves. My attention was split between flying and searching through the darkness for that flicker of light that could tip the scales of life and death in our favor! Knowing we were close to running out of fuel, our chances of survival were rapidly fading when John suddenly pointed through the windshield and shouted excitedly, "There! There it is! A light! Can you see it?"

I quickly turned my attention to where he was pointing. After a second, I finally saw it. Slowly emerging out of the darkness was a light! A small pinprick of light that seemed to be slowly bobbing up and down in the swells of a choppy ocean.

Midnight in the Islands

Now that we both saw it, we knew for certain the light was neither phantom, nor ghost ship! It was something tangible! Something real! Now we needed just one more miracle. Enough fuel to get us there! As we got closer, the lights slowly began to brighten, and John broke the silence by asking, "Do you think it might be a boat?" I reckoned as how it could be. His next question was "What if it's too small to land on?", to which I replied, "It doesn't matter if it's a canoe! They're having guests for dinner!"

After what seemed like yet another eternity, the light slowly brightened and began taking a distinct shape. One light seemed to slowly split into two lights! They appeared to be Coleman lantern-type lights; the kind so popular for camping. Just as we soared over them, a startled face turned upwards to look at us, probably not believing what he had just seen! John and I were both smiling. The boat appeared to be a small, catamaran-type fishing boat with two poles arching up from either end, each with a lantern suspended from it. Something else also caught my attention! A group of lights in the distance! Maybe a village! I quickly put aside the idea of landing in the water and turned my attention to the lights just ahead of us. I also asked for "Five more minutes of fuel, dear Lord; just five more minutes!"

I had no idea how long we had been flying since our 20-minute caution light first illuminated, but it had to be close to 20 minutes, maybe more! It's hard to judge distances at night, but the new lights couldn't have been more than five miles away. Concerned with our fuel status, I was totally focused on making it to those lights! The closer we got to them, the more certain we were it was small village along the coast of some unknown island! Now in sight of land, we needed to find a spot where we could set down, refuel, then be on our way to our destination. We needed a quiet, deserted spot. Maybe a beach!

As soon as we knew for certain we were looking at an island, I turned away from the lights of the small village, looking for some small spit of beach to set down on. It wasn't long before we saw exactly what we needed! Turning on my landing light, we could tell it was a smooth beach with no large obstacles. We had survived certain disaster in the Devil's Triangle! It was a wonderful feeling, but

I knew who ever lived in that village had surely seen our helicopter with its bright red anti-collision light flashing as we flew across the small bay. We had to refuel quickly!

After maneuvering the helicopter over the beach, then landing on solid ground, our euphoria was overwhelming! As soon as I shut down the engine and stopped the rotor, John and I jumped out, ran to the front of the aircraft where we high-fived each other, then sank to the ground on our knees! No matter what else happened that night, we had escaped the clutches of the Devil's Triangle! We were survivors! Plus, we had taken extra precautions and placed six 5-gallon jerry cans of JP-4 on board! We were feeling pretty smug about our foresight, but we had celebrated enough! We had to get our helicopter refueled and get the hell off that beach!

Without further ado, John and I quickly unloaded our jerry cans and he began pouring fuel into the fuel tank while I opened the next can for him. Now on land, I had time to enjoy what had become a beautiful evening, and how quiet it was. All I could hear was the fuel softly gurgling into an almost empty fuel tank, a wonderful sound! Then I thought I heard something else; some new sound! Sensing that, I turned to John and placed my finger to my lips, then motioned him to tilt the jerry-can down to stop the gurgling fuel. As soon as he did that, it was still silent except for what seemed to be bees buzzing in the distance, maybe in the direction of the village. It steadily grew louder, sounding more like people talking! A lot of people! That's when the hair began rising on the back of my neck!

Dropping onto my belly, I turned my face toward the village hoping to use the lights in the distance to discern whatever it was between our location and the lights. What I saw shocked me! It appeared to be a mob of people coming from the village! Maybe they were coming to greet their new visitors who had just arrived in a helicopter and landed on their beach! Why would they do that? I didn't like the way things began adding up! We were about to be

in a very serious situation, especially if the folks in our welcoming committee had taken offense to our impromptu visit. And might be armed!

I jumped up, turned to John and told him to keep pouring that fuel! Do not stop! We would need every drop if we had to leave in a hurry! I also told him to signal me when he'd finished refueling. After setting all the fuel cans within John's easy reach, I quickly grabbed a small moon-shaped bag of Bell Helicopter pins, pencils and whatever else was in it. I also grabbed my map and a flashlight before running about a hundred feet up the beach, away from the helicopter! My plan was to keep the un-welcomed guests away from the helicopter while John was refueling. I didn't even consider a Plan B. Plan A had better work!

I smiled, waved and welcomed each and every one of my honored guests to my "humble abode, albeit a helicopter, out there on their lovely beach! It was the most beautiful beach in the world! My friend and I really loved it! We might even build a permanent pad to park our helicopter on!" I dipped into my bag of goodies and began passing little doodads out as if I were at Mardi Gras! Here my good friends! Have some! By the way, what's the name of your lovely village? Ah, Rolletown! What a lovely name! Here, I happen to have a map! Shining my flashlight on it, I pointed out a small village on the map. Here? Yes, yes! They were excited to see their small village of Rolletown located on my map. I was pretty happy too! I was no longer lost!

My new friend with suspicious eyes suddenly asked in a very unfriendly manner, "Why are you here, and what are you doing?"

"We came to visit you and all your friends and neighbors! What else would we be doing in such a lovely place?"

"You are lying! You are flying drugs!"

I told him that's the dumbest thing I had ever heard! Even as I was saying that, his eyes narrowed, and he began to take on the appearance of a snake. I fully expected to see fangs and a forked tongue! Things were about to take a serious turn to the left! Just as

I was contemplating my next move, John appeared just outside the group of people. John! I had forgotten about him!

When I looked his way, he slightly nodded, then began walking briskly toward the helicopter. We were ready to go! Just as I distributed my last doodad, I abruptly turned toward the helicopter and offered my sincerest regrets about having to leave the party so soon, but we must be going! I had caught my suspicious friend entirely by surprise! Before he could even move, all the kids joyously ran along beside me, thus creating a barrier between him and me. I jumped in the seat and began starting the helicopter while John kept everyone away from the rotors. Soon as I was at 100% operating RPM, John jumped into the copilot's seat. We were airborne before he could even get his seat belt fastened!

I wasn't sure we were entirely out of harm's way, but at least I knew where we were, and what heading to fly to our destination! If my calculations were correct, Rolletown was twenty-minutes away from Georgetown, our original destination. Fortunately, I had flown into Georgetown's airport on a previous trip, so I was somewhat familiar with its layout. Before long, the lights of Georgetown appeared on the horizon, a most welcome sight. But when I searched for the airport, it began to get a little strange.

Strange, because I knew exactly where the airport was in relationship to the town, yet couldn't see any airport lights! Somewhat confused, I decreased my airspeed and flipped on the landing light, then began flying a series of S-turns over where I thought the airport was located. It wasn't long before the beam from our landing light flashed across a runway. Aha! We finally found it! Just as I was thinking happy thoughts, I saw barrels setting in the middle of the runway, stretching from one end to the other! Oh well, that was odd; but we weren't in an airplane! I could land my helicopter anywhere I chose to. As soon as I located the flight ramp, I did!

Immediately after lowering the collective and rolling the throttle to its flight idle position, John stepped out of the helicopter to look around. I took a quick glance at my watch when I shut down;

it was 10:45 PM. I was totally exhausted, and suddenly very hungry! I was also drained! My adrenalin had been on a roller-coaster ride for the last several hours, and like air escaping a balloon, my adrenalin high bottomed out one last time. At least, that's what I thought. What I saw next erased that notion!

When I landed, I had been so exhausted I forgot to turn off the landing light. As I looked over the instrument panel, I saw John backing into the nose of the helicopter with both arms held high over his head as if stretching for the moon! Emerging from the gloom was the silhouette of someone slowly walking toward John with what appeared to be a gun pointed directly at John's mid-section! My brain was trying to digest that bit of info when my own door was suddenly jerked open, and a gun was uncomfortably jammed into my side! We had already experienced more excitement in one day than most folks have in a year! Or, maybe two! My brain had been cycling from high, to low, to high and now this? A gun stuck into my rib cage? John and I were in trouble! How much trouble? That was to be determined.

I sat rigidly in the cockpit, not daring to breathe. John was spread-eagled across the nose of the helicopter while I was a captive in my own cockpit! Had we been in a movie, the cavalry would be galloping up about that time, just in the nick of time! Unfortunately, it was no movie! It was an unending nightmare, and the next move was up to whoever was holding that weapon jammed into my ribs! Suddenly, the weapon was removed, and someone grabbed my shirt and ordered me out of the helicopter. The voice was very British. I began to comply by sliding out of my seat, then gingerly stepping onto the ground. The very British voice then demanded to know where our drugs were! Oh, boy! I knew we were definitely in trouble!

The battery and landing light were still on, so I could see the fellow with the very British voice. First of all, he was black! But most of all, he was big! The kind of big that has bulging muscles! Hell, his muscles had muscles! I could tell that because he was clad only

in underwear! As in fruit-of-the-loom shorts! Oh my God! He had been in bed! We had screwed up his bedtime! He was joined by two other dudes in uniform who immediately opened the cabin doors and began ripping through everything inside. By that time, I was totally drained and asked permission to please set down while they destroyed my helicopter. Looking right at me, the muscle-bound gentleman who looked like a fugitive from the Chicago Bears motioned me over to a log. Very slowly, I walked over and sat down.

From inside the helicopter, I heard "Aha! I found it! Their moneybag!" Stepping out of the helicopter, he held up his prize! It was the little half-moon shaped bag that was a Bell giveaway! It even had Bell Helicopter Textron stenciled on it. It was the same bag I had distributed all the goodies from back on the beach. I began to curse the very day Bell had made that damned bag! Heck, I had to admit, it looked exactly like a money bag! Except it had no money in it. Or trinkets. Hell, I didn't even have anything to bribe them with. Just as I thought it couldn't possibly get any worse, it did! The second dude still digging inside the helicopter let out his own yelp of exaltation! "Here is their needle!" Needle? What needle? I snapped my head toward John, who had the look of a condemned man about to be shot!

The second dude stepped out of the helicopter and in his hand was John's shaving kit. In addition to John's toothpaste, toothbrush, razor and shaving cream, lay a syringe. A syringe with a needle attached. All I could do was stare at it, shift my head to look at John, then back to the shaving kit. John tried to explain he had a severe gum infection and the syringe was what he used to squirt an antibiotic solution into his gums. I don't think they believed him! Hell, I didn't believe him! Who knew what to believe anymore? Suddenly fed up with everything that seemed to be moving at warp-speed, I stood up, turned to the Chicago linebacker and began to talk.

I told him his heavy-handed way of destroying my helicopter had gone on long enough! I then explained how John and I were both U.S. citizens, Bell Helicopter employees, and had filed the proper flight plans from Dominican Republic to come there. We were completely legal! We weren't trying to lie, cheat or steal, and we damned sure weren't running drugs! If he was going to take us to jail, then take us! I'm good with that! My only request was to please take John and me to some restaurant where we could get something to eat, and drink! As I was talking, I happened to see John's face! He looked like he was about to pass out!

Surprisingly, our friend who had been dressed in fruit-of-the-loom shorts, lowered his weapon and identified himself as the local Sheriff. The other two gentlemen were customs agents. Someone had seen our helicopter buzzing up and down the runway, and had placed an emergency call to him. He then explained how he was in the process of taking a shower when he received the call. Then, looking around at our "stuff" scattered all over the ground, he told us to pick everything up, stow it in the helicopter and turn off the battery. Soon as we did that, he led us over to his jeep and said: "Get in". We did exactly as he said. The Sheriff was beginning to become a nice guy, but not so one of the customs agents. After discovering our "money bag", he never said another word. But there was something about his eyes I didn't trust! It's always the eyes that tell the story.

As we started driving toward wherever we were going, he actually became quite friendly and made idle chit-chat. Since he seemed to be getting somewhat friendly, I was quite surprised when we pulled up to a big brick and stone building that had the words SHERIFF'S OFFICE written in large letters across the door window. He then announced how he should be putting us in there, but he had to admit, we did look tired and thirsty. Consequently, he didn't think we would run away. After all, where would we go, hee-hee! I didn't think it was that funny! He then took us a couple of miles down the road, pulled up to a quaint motel, and dropped us off. As we were getting out, I casually said, "Why don't you come back and have a drink with us?" He responded with: "Thank you. Maybe I will". Then he drove off.

Walking inside, we walked up to the check-in counter where we were greeted by a very pleasant elderly gentleman. When I asked if the restaurant were open for dinner, he said: "No, it closed at 11:00". It was almost midnight. He then said if we were really hungry, he would go and "rustle" something up for us to eat. "Please", I said, "and if you don't mind, each of us would appreciate a couple of beers before you go!" We soon had two cold bottles of beer in our hands. Before drinking any, we hoisted the beer to toast all the adventures we had shared during that very long day! We shared another toast for surviving them!

Before long, the pleasant host returned with two dishes, each piled high with whatever it was he had rustled up! As it turned out, it was leftovers from the day, ready to be thrown out. A few more minutes and we would have been out of luck! It was a good sign that maybe, just maybe, our luck would continue waffling in our favor. As we were wolfing down the tasty leftovers and washing it down with beer, the good Samaritan sat down and began to visit with us. He certainly seemed pleasant enough, so I began a discussion about the good Sheriff who had deposited us there, and the circumstances about how we met him. He listened patiently, then began explaining a few facts about the Sheriff.

"First of all," he said, "he is no one to be trifled with!" He then went on to explain how illegal drugs had become such a problem throughout the Caribbean because of so many deserted islands and cays used to traffic cocaine from Colombia. The Sheriff had developed a reputation as being someone who was really tough on anyone even remotely connected to cocaine trafficking; but also, a fair one. It took a while, but he eventually touched on our situation. He went on to say if we were indeed who we claimed to be, we would have no problem. However, should we be even remotely involved in drugs, we would both in for a shock. He then explained how no one wants to be tossed into a prison anywhere in the islands!

He also said the Sheriff was no fool. In all probability, he had already taken drug-sniffing dogs to your helicopter and gone over it thoroughly. He probably had also checked out our claim about filing proper flight plans for transiting through the islands. But, if we were just two dumb-assed gringos trying to make it from one island to another, we would probably be OK. If not, well, the Sheriff would soon be paying us a visit! Then he excused himself to visit other customers still sitting in the bar to take their orders. His words had a very sobering effect on both of us. Sitting there nursing our beers, there really wasn't much we could say! After some small chit-chat and another beer or two, we both decided to call it a night.

Since we were both exhausted from our long ordeal, neither of us planned to get up early. However, it was a restless night for both of us, since our brains were too full of "what if's"! Consequently, we met each other in the small restaurant for breakfast earlier than either of us had planned. Being the only two early risers in that neat little place of paradise, we were nursing our coffee and discussing what we should do. Instead of the pleasant gentleman who had waited on us last evening, we now had a very pleasant older woman to serve us. We looked around, hoping to see our old friend from last night, but he was nowhere to be seen. We were disappointed, because we could have used his wise counsel. After finishing our breakfast and drinking our last cup of coffee, we decided it was time to go. We hoped we wouldn't be met at our helicopter by the Sherriff, or his two deputies.

After checking out of the hotel, our taxi arrived, and we were soon headed back out to our helicopter. I had a strong suspicion that having cash on hand might be beneficial; however, being unsure what to expect, I began sifting through a small stack of crisp, new bills of various denominations starting with ten-dollar bills, all the way up to one hundred-dollar bills. Which denomination I used was always based on how serious I thought a specific situation might be! I also kept each denomination in separate pockets so I could get to it quickly. I had just completed arranging my cash into

various pockets when the taxi driver announced our arrival at the airport. We were about to find out which denomination of money I would need for "expedition charges", if any.

After paying for our taxi, John and I collected our bags and began walking at a fast pace towards our helicopter. We were anxious to "get out of Dodge" before anything else happened! Surprisingly, nobody was waiting for us at the helicopter! Taking that as a good sign, I quickly dispatched John to order a fuel truck to refuel the helicopter, plus all the Jerry-cans. I didn't plan to use them again, but why take chances! They sure as hell saved our butts last night!

In order to expedite our departure, I grabbed John's passport and headed toward the customs office which was nothing more than an open-air type hootch with a long counter. Being so early in the morning, there was no one else in sight except an agent standing inside the office. Hoping everything would go quickly, I stepped up to the counter and was stunned when I got a good look at the customs agent! He was our "good friend" from last evening! The same gentleman who had ransacked our helicopter, then jumped out in great exaltation after finding our "money" bag! My God! My instincts had proven true, and my heart sank!

He was standing far behind the counter, leaning against a small desk in the corner. Keeping a poker face, I laid both passports on the counter and pleasantly said "good morning". No response! He simply stared at me! Immediately, I sensed he knew something. I just didn't know what. I also knew with great certainty; I was going to find out soon enough.

Never moving away from the desk, he looked squarely at me and said, "It seems you did not tell the whole story last night."

"What are you talking about?" was my response.

"You did not tell us about landing on the beach at Rolletown last night."

Aha! There was no doubt he had gotten that damning bit of information from my suspicious new friend on the beach! He had

probably spoken to the customs agent within minutes following our departure from the beach near Rolletown.

Taking a deep breath, my response was to the point. "I told the Sheriff and you would have heard it, too, if you hadn't been so busy throwing everything out of my helicopter!" Even as I was speaking, I was fishing around in my pocket for a one-hundred-dollar bill for "expedition services". There was no sense in going small and negotiating because he was exactly right! I had deliberately omitted the part about landing on the beach. I had no doubt that bit of information would have landed us in jail last night; maybe for many nights! It had already been a volatile situation, and telling the good sheriff we had landed on a beach near Rolletown would have been the last straw. I was convinced of that! Now I was betting it all on the gentleman standing between us and a clear path home. Standing there and waiting for his decision was an eternity.

Chances are, he had already suspected his early morning get-up would net him a nice "tip", and I silently agreed with him; in fact, I was betting on it! I deliberately kept eye contact with him as I picked up both passports, pulled out a crisp, new one hundred-dollar bill, inserted it into my passport, then said, "I hope this will cover all expenses for inconveniencing you." I then slid both passports across the counter towards him. He continued standing there for another 30-seconds or so, then walked across to the counter, picked up the passports, then stepped back to his desk, sat down and began stamping them. When he was finished, he stepped back to the counter, handed the passports to me, then smiled and said: "Have a nice flight."

It was early in the morning of October 30, 1984, and our next stop at Nassau had been uneventful. We made it all the way to Fort Lauderdale by noon where I dropped John off, then continued on to Tallahassee, Florida. Seeing the airport totally covered in airplanes, I asked the tower what was going on. The tower operator replied as if I had just arrived from Mars: "We have a big game! Miami's

playing Florida State tonight!" Of course, football! It felt so good to be back in the states!

Note: I found out later our navigational equipment had stopped functioning because all VOR transmitters had been shut down at sunset to discourage drug runners from using them for clandestine operations. The barrels stacked down the center of the runway were placed there for the same purpose. Neither of them deterred our overwhelming desire to survive the clutches of the Bermuda Triangle!

And John's hypodermic needle? As it turned out, he had recently undergone oral surgery, and squirting doses of liquid antibiotics directly into the gum cavity had been prescribed by his oral surgeon. It had been legit, but it sure was a shocker when I saw that needle glistening in the moonlight! I mean, what were the odds something as simple as that needle would almost land us in jail? We could only shake our heads. We had beaten the odds that night, and that was enough.

CHAPTER 14

Interesting Times in Colombia

The Medellin Cartel
January/February - 1983

On January 20, 1983, a Bell Helicopter tech-rep and I departed Bell's main plant in Fort Worth, Texas in a military camouflage-colored Bell 412 helicopter headed for Caracas, Venezuela, then later on, to Colombia. The 412 was a medium-sized, dual-engine helicopter, and since it was the first Model 412 to arrive in South America, it had drawn a lot of attention. That made it a memorable occasion, but even more memorable was a flight demonstration on February 14th for a gentleman who was well-known in the drug world. He was widely known as the leader of an alliance formed between leaders of other drug operations that had grown into what would become known as the Medellin Cartel. That group was extremely well organized, and controlled the manufacturing, distribution and marketing of cocaine. They were also well-armed and more than ready to protect their interests! They weren't anyone to trifle with, and the man I took flying was the leader of that group.

That particular demo occurred after several long days of flying demonstrations for companies and individuals who had the money to afford a large helicopter such as the 412. In fact, one of the flight demonstrations I had conducted was for a large bank that had dispatched us deep into the Columbian jungle to collect a load of gold bullion that had been mined from the river. I knew it was a working mine because when I flew over the campsite that had been chopped out of the jungle, the river was teeming with workers sifting for gold, similar to the movie about California's famed "forty-niners"! Soon after landing in the small, dusty airport, a decrepit, ancient-looking flat-bed truck belching black smoke pulled

up alongside us. Sitting in the cab was the driver and several workers. Atop the flatbed was a small wooden box being watched by an elderly, sleepy-eyed guard who seemed far more interested in continuing his siesta than the box itself.

After parking alongside the helicopter, a couple of workers spilled out of the cab, grabbed two long poles, ran them through leather straps secured to the box, then proceeded to hoist the small wooden box off the flatbed. It took both men to carry that little box in much the same manner you would carry a stretcher and, judging by their grunts, I suspected it was heavy. Struggling to the helicopter, they carefully loaded the box of gold onto a large piece of plywood we had placed inside to spread the load equally so as not to damage the cabin flooring. Then, much to my surprise, the old guard, shotgun and all, clambered aboard our helicopter before we took off for our return flight to Medellin to deliver the gold to the local bank.

The flight was rather uneventful except for my mechanic who was sitting in the back seat. He kept suggesting he could provide a "gentle" whack upon the elderly guard's head to ensure he would have a long nap. Then, according to him, we could head for the nearest port-site with our bullion of gold where we could buy a boat to begin sailing the "seven seas", and maybe even become pirates! "Just think of the good life we could have with all that gold!", he rattled. Fortunately, he was just kidding, or I think he was! At least he didn't whack the old man on the head, nor did we grab the gold and make a run for the coast!

Despite being known as the home of the infamous Medellin Cartel, Medellin is a beautiful old city located at the bottom of a large bowl-like valley. The airport is also located in the valley, and it's a demanding place to fly out of since you literally must begin climbing immediately after takeoff just to clear the sides of the valley. The airport itself was nice enough, at least one side of it was. That was the side where the Fixed Base Operator (FBO) had a nice clean facility with easy access for customers to come and go.

However, there was nothing on the other side of the airport. No FBO, no clean hangar – no nothing! Since most pilots like to take a break occasionally, plus make that all-important toilet run to drain the bladder prior to flying, I deemed the other side of the airport to be unacceptable for flight demonstrations. That promptly set off an argument between me, Bell's sales representative and the local distributor because they had a flight demo scheduled. It was two against one, but I had pretty much dug my heels into the dirt before relenting and told them if they could give me one good reason why I should fly a demo from the "ratty" side of the airport, I would consider it. That's when they somewhat reluctantly told me who the flight demo was for!

The requested flight demo was for a gentleman widely known as the head of the Medellin Cartel. Even I knew about the Medellin Cartel! My God, who on the face of the earth hadn't heard about it! I was caught totally off-guard by that new demo request, not to mention being dumbfounded for a moment. All I could do was stand there with my mouth gaped open before finally managing to let out with a "you're shitting me" comment! I mean, surely not! But yep, they surely were! It seems the gentleman had heard about the big helicopter with its camouflage paint scheme and had requested a flight demo! To be discrete, the decision was made to fly his demo from the other side of the runway. Across the tracks so to speak! The side with less people, and far fewer prying eyes!

After spending a restless night in one of Medellin's older hotels, we drove out to the airport early the next morning to begin preparations for the day's event, that of flying Senor Ochoa around the countryside, or wherever else he might request to visit. After conducting a pre-flight on the large model 412 helicopter, settling into the cockpit, then starting it up, there had been no questions asked when I requested clearance to relocate to the "other" side of the airport. In all probability, the local traffic controllers knew exactly what we were doing because it was far less populated, and "seedy" looking. The perfect place for what we intended to do, but I had

the feeling we were fooling no one. Everyone on the airport probably knew what our plans were, but were discreetly keeping their thoughts to themselves. Considering the circumstances and who we were flying, I would've done the same thing! It was much safer!

After relocating the helicopter to the "other side" and shutting down, we made ourselves as comfortable as possible, then began patiently waiting for our customer to arrive. After weeks of doing flight demonstrations in that part of the world, I had come to the conclusion that schedules meant very little, at least as far as customers were concerned. I'm not saying it was deliberate, but there always seemed to be some type of "speed-bump" or some other delay out of their control. Perhaps they just got busy at work, had to take their kids to school, or they got stuck in traffic! Knowing that, I was quite surprised when exactly on time, several large black "Jimmy" SUV's silently, and ominously, turned off the main road and made their way down the small pothole-filled road toward the helicopter.

With the arrival of the SUV's, everyone got to their feet in anticipation of meeting our newest potential customer. When the SUV's pulled up to a stop a short distance from the helicopter, the doors slowly opened and well-dressed gentlemen began to emerge, each of whom looked very business-like. Dressed in soft linen type of clothing, their appearance and smooth, deliberate movements suggested a high degree of professionalism in the armed body-guard business. Although there were no weapons in sight, small ripples in pants, shirts and jackets provided every indication they were there, and I had no doubt they knew how to use them!

After several of them emerged and took up positions around the entourage of vehicles, numerous young children of various sizes then began emerging from the SUV's! They seemed to vary in size from ten-years-old to teenagers, and they were all laughing, giggling and pointing toward the helicopter. As the entourage of bodyguards and children began walking toward the helicopter, I was rapidly looking at each of them trying to pick out the big, ugly monster who was the leader of the pack. After all, he was the head of the notorious Medellin Cartel! Surely, he had to be big, tough and ill-tempered with scars crisscrossing his face! All sorts of images flitted through my mind in regard to what he looked like. One version was of a large, broad-faced ogre with wild hair and piercing eyes who was dressed in camouflage, while another was that of a well-dressed, gold-laden sharply dressed gentleman right out of Hollywood central casting.

Stepping closer to our sales rep, I whispered: "Where is he?" and his response was a curt nod of the head towards the approaching group. Trying to whisper out of the side of his mouth he said: "There, see him?" Well, no. I didn't see him. Eyes bulging, again he nodded and said: "There"! "Right there!" Finally, I saw him. Are you kidding me? What I wasn't expecting was a slightly built, casually dressed young man who looked like a teenager! Are you kidding me? He was dressed in a white leisure suit with a gold chain hanging from his neck, and he wore a Panamanian hat. In fact, they all did. He was actually very sharp looking. But he still could've passed for one of the older teenagers. He also seemed to be as excited as they were as they continued talking, laughing and pointing excitedly as they headed towards the big camouflaged twin-engine 412 helicopter.

The entire group began to assemble around the helicopter and even the bodyguards seemed to be caught up in the moment, and very impressed! A really good thing I thought! The mechanic and I had already slid the rear cargo doors to the fully open positions but none of the new visitors attempted to climb on board; a very well-mannered group I thought. Bell's local salesman then began conversing with the young man and after a short conversation, he asked if we could take the children on a demo flight with us. My response was "absolutely" and when this was conveyed, the young children began laughing and clapping, then began loading into the helicopter.

The mechanic and I assisted and fastened everyone's seat belts. When I stepped down, Senor Ochoa began climbing into the

cabin area where the children and two bodyguards were already seated. Intending for him to ride up front in the cockpit with me, I instinctively reached up and grabbed his left shoulder to stop him and immediately knew I had made a mistake! Everything froze as the bodyguards instantly directed their full attention to me, and a couple of them even reached for whatever weapon lay just underneath their jackets! In much the same manner you would drop a glowing, red hot burning ember, I dropped my hand and stepped back away from the helicopter. Then, in what was probably a squeaky voice, I explained I was intending for Senor Ochoa to ride up front with me. His response was a big smile that lit up his entire face! Now everyone was happy and excited again, myself included. Thank God!

After the tense moment had passed, I began assisting my new copilot into the seat on the left side of the cockpit. After settling him into the seat, I adjusted his seat, strapped him in, and walked around to the right side of the helicopter to enter the cockpit. The engine startup took a bit longer than normal, since I slowly went through the checklist for the benefit of my copilot, but it wasn't long before I had the engines operating at 100% RPM with all instruments in the green. Turning my head towards my copilot, I gave a thumbs up indicating all was ready to go and his response was another big smile indicating he, too, was ready for a new adventure!

After a call to the control tower for takeoff clearance that was quickly granted, we began our takeoff, then initiated a climbing left turn to a heading that would take us to his small estancia located high up on the steep bluffs overlooking the valley. After stabilizing the helicopter's controls to maintain steady-state flight, I offered the controls to him. Somewhat apprehensive, he slowly placed his hands and feet on the controls, and after slowly taking my hands off all controls, he seemed to be very excited that he was actually the pilot now in control! It wasn't long before his estancia appeared, sitting very near the edge of the bluffs overlooking the entire valley. It was stunning! What a view he had!

Circling overhead so I could check out the landing area, plus ensure it was clear of animals, employees or family members, we turned onto a long final approach and continued our slow descent towards a landing in the middle of a massive cobble-stoned patio. After landing and conducting the standard two-minute engine cool down, I rolled both throttles off and continued shutdown procedures so he could give me a tour of his "little" ranch. He also wanted to show the helicopter to those unlucky family members who had been unable to make the trip into Medellin, and subsequent helicopter ride.

We were greeted by a large entourage of smiling, excited faces, all of whom wanted to touch or even better yet, sit inside the helicopter! Everyone was accommodated and it wasn't long before everyone was hustled off to a very elaborate meal the cooks had arranged for such a glorious occasion! What a wonderful time it was, with all the laughing and talking while several heavily laden bowls of excellent food were passed around, along with wine straight from his enormous wine cellar! I have to admit, it was quite difficult to bypass the generous offers of beer and wine, but I knew we would soon fly back down the hillside for the Medellin airport.

After a couple of hours of relaxing and enjoying the best the estancia had to offer, we were soon back in the air with my new copilot still at the controls (with my assistance, of course) as we slowly made our way back down to the Medellin airport. It was obvious he hadn't flown a helicopter before, but he was certainly an enthusiastic novice pilot who seemed to have had a good time. After landing and rolling both throttles to the flight idle position, I occupied the time during the two-minute cool down by pointing out and discussing the various instruments in my best Tex/Mex drawl before shutting down. I knew he liked it very much because as soon as we shut down and exited the helicopter, he motioned for Bell's sales rep to come over and join us.

It seems he really liked the paint scheme since it was already painted in military camouflage. I'm sure it was a color that would be

most beneficial! He also wanted to know how we would prefer to be paid, and was aptly prepared for providing either check, cash or credit card! As I said, he really liked the helicopter! Truth be known, I really enjoyed flying and visiting with him! Everyone in his family was very nice, courteous and pleasant to be around. I'm sure to some he was a bad man, and perhaps rightfully so. However, to me, he was a very pleasant young man who was really excited about having an opportunity to fly a helicopter. I also thought that was the end of my dealings with that young man, but I did manage to be involved in one more incident a few years later.

Be Careful Where You Land

Colombia is a beautiful country populated by wonderful people, but it was a dangerous place in the 1980's. The Colombian military was constantly doing battle with the FARC's, the Medellin Cartel was at its peak, and kidnapping wealthy citizens and holding them for king-sized ransoms seemed to be one of the biggest concerns. Astonishingly, helicopters played a large part in each of them! I happened to see that firsthand soon after my arrival at the Bell dealer's facility located just outside Bogota, Colombia's capital city.

My technical representative (or tech rep) and I were being shown around the facility when we noticed a damaged Bell 206B Jet Ranger helicopter sitting inside a hangar. The fuselage itself didn't look too bad but sitting next to it, completely detached from the fuselage, was the transmission and some pretty banged up rotor blades. Curious, I asked the gentleman escorting us if anyone had been injured in the crash. "Well," he said, "the two passengers survived the crash with some injuries, but the pilot was not so fortunate." I thought that rather odd since the fuselage itself looked like the crash had been very survivable. He then went on to say "actually, the pilot had very few injuries himself when the helicopter crashed into a large building, but when he crawled out of the damaged building and began approaching the homeowner, the homeowner shot him dead!"

As it turned out, helicopters had been utilized by terrorists to land and kidnap unsuspecting victims, then made their escape to some well-hidden destination to await their ransom demands. Making matters worse, the crash I described had happened at night and the homeowner, badly shaken up after hearing loud crashing noises of a helicopter plunging through one of his buildings, was in fear for his and his family's lives! Consequently, he took matters into his own hands and shot the pilot, even as the passengers scrambled to safety! It definitely wasn't the ending I had expected to hear, but I should've been paying more attention to what he was saying about using helicopters for kidnapping!

I say that because a couple of weeks later, I was advised by our dealer we had a very wealthy gentleman scheduled to show up for a flight demonstration in our Bell 412 helicopter. I hadn't been too concerned about it, since that's what I had been doing since departing Fort Worth a month earlier. In fact, I assumed the proposed demonstration might end in a sale, since our dealer had mentioned our guests were extremely wealthy! I mean, when someone was deemed to be extremely wealthy in Colombia, they were normally just that – extremely wealthy! They lived in huge mansions, had beautiful wives and drove top-dollar sport cars! But generally speaking, literally everyone I had met in Colombia had been very friendly and courteous. It's just that sometimes they asked you to do dumb things, much like our guest did during my flight demonstration for him. It was made even worse when his dumb-ass pilot went along with it!

The customer had arrived on time and surprisingly, he had invited five additional guests to join him on his demo flight on such a beautiful day. Fortunately, the number of guests was not a problem since the 412 was more than capable, plus we wanted the customer and his guests to enjoy the flight in a nice, uncrowded atmosphere. I didn't mind because it was much easier flying a helicopter when it wasn't at its maximum gross weight, especially when flying in higher altitudes, such as it was in Colombia!

After briefing everyone prior to boarding about the nuances of flying a helicopter to ensure their safety, the takeoff was just the way we liked it – uneventful! Flying in the skies of Colombia was always pleasant and beautiful, at least for me. After hearing all the bad things about Colombia, I was prepared for the worst but was absolutely blown away by how gorgeous it was, and that morning had been no different. I was thoroughly enjoying flying over beautiful ranches and lush jungle when our customer asked if it might be possible to land and visit one of his old friends who lived on a very large hacienda that was very close by. "Sure", I said, "no problem!" "Oh, by the way,", I asked, "you did tell him we might drop in for a visit, right?" "Si", he answered. Soon enough we arrived over his friend's big spread and it was definitely big enough to land in. In fact, it was large enough to land a fleet of 412 helicopters!

I circled overhead a couple of times to make a high reconnaissance check for any large, visible obstacles that lay in close proximity to the massive home, completely surrounded by barns, a swimming pool, gardens and a herd of cattle with few horses thrown in for good measure! Things like trees, bushes or rocks were easy to spot to ensure the area was clear enough for landing. Then, when I turned onto a long final approach into my chosen area, I began looking for small things like powerlines or antenna guy wires. Trees and bushes were easy to see, but it was the small things that could ruin your day if you weren't alert! Everything looked good and my passenger was helpfully pointing out an area very close to the mansion where he wanted me to land. The approach and landing were uneventful and the helicopter's skids were soon settling into the lush grassy courtyard where we would shut down and visit our customer's old friend. I rolled the throttle back to an idle position and had just settled back in the seat for a two-minute run prior to shutting the engine down when all hell broke loose!

Both pilots' doors were jerked open, scaring the heck out of the passenger sitting in the copilot's seat and I! We were both stunned! Before we could respond to that, screaming men with guns seemed

to be trying to climb into the cockpit with us! They were screaming in Spanish, and even though I couldn't understand them, I was pretty sure we were in trouble, so I shut down the helicopter. My passengers' eyes looked like saucers as they tried to comprehend what was going on! Making matters worse, more men opened the large cargo doors behind us and began pointing guns at the passengers sitting back there! Finally, after what seemed like an eternity, a gentleman appeared who seemed to be in control of the threatening gunmen. Fortunately, he recognized my passenger as being a friend of the owner of the large hacienda and ordered the security guards to back off and put away their weapons. Thank God!

After explaining we had landed so my passenger could visit the gentleman who owned the hacienda, the security captain quickly grasped what happened. He was a pretty cool fellow who quickly connected the dots soon as he recognized my passenger. He also spoke good English, so I could communicate with him as well. His calm demeaner resolved the scary moment, and soon everyone's eyes were back to normal, plus some coloring had returned to everyone's white, chalk-like faces, mine included! In fact, we were even laughing about what had happened, and how stupid it was for us to land in the middle of a rich politician's ranch! It could have easily been fatal because the guards thought we had landed to kidnap the resident gentleman! The captain didn't apologize for our rude welcoming, but he did tell us we would be welcomed back any time just as long as we informed them in advance! When the captain said that, I knew my passenger had not advised his friend we might stop in to visit him. Thinking about it, I thought it best to keep my thoughts to myself. Getting upset after such a scary moment wouldn't resolve anything, plus I knew he would never do that again, at least, not if he was flying with me!

Flying with Colombian Police Pilots

Bell Helicopter received an urgent telex from the American Embassy in Colombia on April 1, 1987. They were requesting pilot

training for Colombian Police pilots who had been assigned to fly several Bell 206 and 212 helicopters they had received from the U.S. State Department to combat the rampant drug problems in Colombia. Since it was a telex sent directly to Bell Helicopter's President, he was quick to provide a positive response to their request. I was called in and asked if I would volunteer for this particular task and my response was "yes", but only if another instructor pilot could accompany me. I knew there were a lot of pilots to be trained and two pilots are always better than one! I also wanted another pilot involved just for the sake of having another "gringo" with me. It's always much better when you have someone sharing the same hardships, or good times, whichever the case. A Bell Helicopter Canada pilot by the name of Leo Meslin agreed to accompany me. Since I had previously worked with Leo on numerous occasions, I knew he was a great pilot and enjoyable to work with. I was delighted he agreed to join me on such an adventure!

Prior to departing, we were given explicit instructions to travel incognito, meaning we would wear absolutely nothing to identify us as Bell employees, not even a ball cap with the Bell logo! We were to travel as tourists and would make no mention to anyone about what we would be doing in Colombia. Someone from the U.S. Embassy would meet us at Bogota's El Dorado International Airport to escort us through customs.

Sure enough, when Leo and I arrived at El Dorado International, we were picked up by U.S. Embassy personnel, and it was the easiest passage through customs that either of us had ever experienced! We went directly to the Embassy to be briefed, then proceed to an unnamed National Police base where we could coordinate our training. As it turned out, we eventually flew to the lovely little town of Santa Marta, located along the northern coastline of Colombia, to do the bulk of our training. It was popular as a tourist location due to its proximity to both beach and sun. Lots of sun, and very hot! Our first order of business had been to pick up our reserved rental car. As soon as we concluded the usual negotiations with the car rental agency, Leo and I strode outside and jumped into the rental car, eager to get some cold air going. Much to our surprise, we couldn't seem to find the controls for the air conditioner. After marching back inside to get instructions on how to operate the air conditioner, the young lady behind the desk informed us there was a good reason why we couldn't operate the air conditioner. It didn't have one!

It turned out they only had one car in the entire fleet that had an air conditioner! Since the rental price had been so expensive, she simply assumed we hadn't been interested in renting it. We proceeded to tell her we didn't care what the price was, we wanted it! It's wonderful being able to spend other people's money on such niceties, and that was absolutely one of the smartest things we could have done in Santa Marta! After a long day of sweating during training sessions, being able to travel about in air-conditioned comfort made it so much more pleasant!

After laying claim to our air-conditioned car, the next order of business was to check-in at the hotel they had arranged for us. There had been times when the folks who made our reservations screwed up and put us in a dump. But not so in that case! When we pulled into the parking lot, we saw an absolutely delightful set-up where a bunch of "cabanas" were sitting randomly beneath an enormous forest of trees that basically created an umbrella. In fact, when we first pulled into the parking area, we noticed several cars had parked in a sunbaked area some distance from those beautiful trees, but none under them. "How dumb was that?", we asked. We weren't stupid like those other dummies! We parked directly under those magnificent trees! They would really keep our rental car cool!

After checking in and eating a nice meal in the restaurant, washed down by a couple of cold beers, we decided to check it in for the evening. It had been a very long day and we were both exhausted. Early the next morning, we strolled out to our air-conditioned car and were stunned at what awaited us. Our car looked like someone had taken a gigantic egg, cracked it just

above our car, then let the whole mess dump on our car! It was covered in what looked like the clear, opaque color of egg white with red splotches all over it! Even worse, we couldn't even get it off the windshield! Fortunately, a young man appeared out of the shadows and told us in broken-English he could clean it up for us. It certainly sounded good to us, so we headed back inside for another cup of coffee for the next 30 minutes while he cleaned the car. When we came back out, we had a spotless car and happily paid him for his trouble. When we left, Leo and I began to wonder if he'd had some trained birds, or something else, to make such a mess!

We began the day by flying with various Colombian National Police pilots, and those guys were something else! They had all been dedicated to the task at hand and anxious to learn whatever they could from us. For the most part, we instructed in Bell 206 L-1 helicopters flying in the vicinity of Santa Marta airport and it had been a most pleasurable day! It's always a good day when you're instructing pilots who want to learn and respond well to instruction. Leo and I were both happy campers when we made it back to the motel later that afternoon. Not wanting a repeat of the "egg wash" we encountered the previous night, we parked the car alongside all the other dummies who parked their cars in the sun.

After we had returned to our respective cabanas, I took a quick shower and started walking toward the small bar when something hit the ground beside me with a very loud "splat"! Startled, I literally jumped sideways across the small stone walkway that meandered through the trees. I had been staring at the large, opaque mess when another "splat" occurred just beside me! Looking up into the trees, I was amazed to see gigantic lizards "roosting" in the trees! Aha! Now we have the culprits! The lizards appeared to be 2- or 3-feet long! They were huge! Leo and I could no longer blame the birds, or the "cleaning guy". It was just two dumb-ass gringos who made it a point to never park under the trees again!

Visiting the Lost City

While Leo and I were training the Colombian National Police pilots at Santa Marta, we had an opportunity to visit an archeological site of an ancient city located deep in Colombia's Nevada de Santa Marta mountains. It was known as Ciudad Perdida, or Lost City, and neither of us had ever even heard of such a place until one evening over dinner, when the pilots spoke of a "very interesting" place we might wish to visit. They explained the ruins had been built over 1,000 years ago but had remained remote and hidden deep in the jungle until sometime in 1972. That's when a gang of treasure looters just happened to stumble upon a series of steps in the mountains that led from the river up to the long-abandoned city. The looters tried to keep it a secret, but it hadn't been long before the word got out, sparking considerable interest throughout the world. Fortunately, it was still relatively unknown to most, since it was a very difficult place to access. Tourists could only get there after hiking many miles through the thick jungle, fording dangerous rivers and traversing steep climbs and equally steep descents.

Needless to say, both of us had been eager to pay a visit to such an interesting place, even though it sounded like a difficult journey; however, the pilots made an offer we couldn't refuse! They knew of a shortcut. Since we were training a group of National Police pilots, why not fly there in a Bell 206 L-3 during one of our training missions? In complete agreement with that idea, we took off the next day to visit the Lost City. I think the flight itself took no more than an hour as opposed to a five-day hike from Santa Marta through very difficult terrain. As we flew over mountain valleys toward the Lost City, it was easy to see why it was such a difficult journey, if one had chosen to hike. After landing in a small clearing and shutting down the engine, we emerged into an unnaturally silent, forgotten world! It was beautiful, yet eerie, and I was suddenly glad the two National Police pilots with us were both armed! It didn't look so much like the hidden paradise of Shangri La as much as it did the mysterious island that spawned King Kong!

The city had once consisted of a series of terraces carved into the mountainside, with tiled roads and small circular plazas that might have housed as many as 8,000 people before being abandoned hundreds of years ago. It was a beautiful, yet haunting place to be, and except for the absence of any sign of buildings, you had the feeling the inhabitants had just packed up and left yesterday. The tiled paths and elevated plazas appeared ready for use even though they were hundreds of years old!

The two police pilots had probably seen the ancient city many times before and were quite content to relax in the shade of a huge tree in close proximity to the helicopter. Eager to take advantage of such a once-in-a-lifetime opportunity, Leo and I began meandering throughout the old city and were serenaded by a wide variety of chirping birds, along with an occasional grunt of some unknown animal, perhaps in search for dinner. It was a surreal moment for us as we sat down on one of the tiled steps to enjoy the beauty with accompanying jungle sounds. We rarely spoke and, thinking we were totally alone, we were both quite startled when a young Indian tribesman appeared silently out of the jungle, striding purposely on the ancient tiled road, heading directly toward us!

He was rather small in stature, but also well-developed with muscular arms and legs formed by a hard life of survival in the jungle, not pumping iron in some local gym. He had long black hair cascading down his back, and was scantily clad in a mesh shirt and black shorts held in place by a hand-made leather belt. A well-used machete with a razor-sharp edge was securely attached to his belt, and I'm sure he could wield it most efficiently! He was bare-footed and moved with the silent, graceful stride of a cat as he approached us. You could tell by his body language he was unafraid of us and, as impressive as his overall appearance was, it was his eyes that stood out the most. He never took his dark eyes off us, yet remained alert and aware of everything around him. He was very much at home in the Lost City. Leo and I both were very impressed with our new visitor.

Stopping a short distance away from us, the young man seemed friendly enough, so Leo and I responded in a friendly manner by trying to communicate with him in broken English and Spanish. We tried for just a minute or two before finally giving up after realizing he didn't understand us. Our attempts weren't in vain, because his eyes remained friendly as we admired his machete, and even managed a slight smile before turning to look back towards the jungle trail from which he had emerged. With a quick flick of his wrist as if to motion someone, we were stunned to see a young woman emerge from the shadowy mists. It was a young woman dressed similarly to the young man, except she wasn't carrying a machete, and she was pregnant. She was also holding the hand of a young boy who appeared to be about four years old. She, too, seemed totally unafraid and completely at ease, even though her mesh shirt left very little to the imagination!
    Without the slightest hesitation, she began walking toward us in the same silent, graceful stride as the young man. She was also bare-footed, and very lovely! The young boy stood silently holding her hand while watching us with his huge brown eyes. I'm not sure he had ever seen "gringos" before. It was hard to believe such a perfect young couple actually existed in that ancient world! A world without TV, radio, baseball, football or any other niceties of life we always take for granted. But that environment was all they had known, and for them and others like them, it was enough. Life there was very simple. They just happened to be passing through when they saw us standing in the pathway toward their destination. It was obvious he had the young woman stay behind in the safety of the jungle while he came forward to check us out before exposing his wife and son to any danger. Sensing we presented no danger to either of them, our brief encounter was soon over. With a slight nod of his head, a fleeting smile from the young woman and a small wave from the young boy, they silently continued their journey and quickly disappeared into the dense jungle.
    I think we were both stunned by what we had just seen, because

neither of us spoke for a couple of minutes. I finally broke the silence by saying: "Can you believe that?" Leo's response was: "I'm not sure I do. Did that really happen?" I mean, they were there for just a couple of minutes, then they were gone. Neither of us had even thought of taking a photo, either! Perhaps we didn't want to break the spell of those few incredible moments when we were transported back in time, into another century!

Mysterious Drinking Buddies and Early Morning Phone Calls

After we returned to Santa Marta and had been there for a couple of weeks, Leo and I had come to the conclusion we were being followed. We would go to various restaurants around town and invariably we would see what seemed to be a familiar face, just different clothes. We hadn't thought much about it at first, but as time went by, we had begun playing a game by pulling up to one restaurant, walk inside and have a beer, then leave and go to another restaurant. Sure enough, we had finally narrowed it down to three different individuals who seemed to alternate on a regular basis. We hadn't been too alarmed, but our problem was not knowing whose side they worked on, the Colombian National Police or the Medellin cartel. They never seemed threatening, but not knowing what was going on, we made sure we never went to an out-of-the-way place. We always went to the more popular restaurants and always mixed with crowds wherever we went. The quaint little bar at our hotel sufficed quite nicely for our nightly cold beers to quench our thirst.

We only had about a week to go when I received an unexpected phone call about 4:00 o'clock in the morning. I had been sound asleep, and the loud ringing of the phone had startled me! I couldn't imagine who could be calling me at that time of the morning, but my first thought was that my wife was calling. It's never a good thing when you get a call in the wee hours of the morning! I was quite shocked when I picked up the phone, answered, and an unknown male voice greeted me with a "good morning"

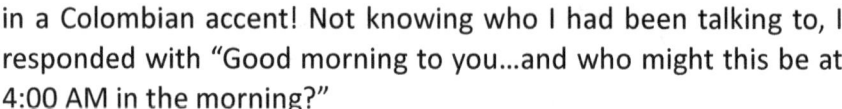

in a Colombian accent! Not knowing who I had been talking to, I responded with "Good morning to you...and who might this be at 4:00 AM in the morning?"

"Oh, I am senor so-and-so," said the voice.

Suddenly, I was no longer sleepy! The gentleman on the other end of the line worked for the Cartel. I had met him when I flew with Senor Ochoa in Medellin.

I was trying not to sound too disturbed when I pleasantly asked how he was doing at this time of the morning. He had been doing fine and "they" had heard I happened to be "in the neighborhood" and was wondering if I could pay them a visit in Medellin when I completed my "other" business in Colombia! It was a surreal conversation to say the least! My mind was racing as I was trying to carry on a normal sounding conversation while at the same time, think up a good excuse for not visiting "them" in Medellin. All things considered, it was a very pleasant conversation at 4:00 AM in the morning. I finally concluded it by telling him I would check with Bell and see if I could work their request into my schedule. He was certainly agreeable with that, and wished me a very pleasant rest of the night! Well, he could've saved himself the trouble with his last statement of "have a pleasant rest of the night"! I was wide awake and very curious how he had gotten not only a phone number, but also my room number.

I met Leo as soon as the restaurant opened and told him about the early morning phone call. He was as surprised as I had been! The question at hand was whether or not we tell the police pilots. Finally deciding I should be up front with the National Police pilots regarding my dilemma, we finished breakfast and headed toward the office.

When we arrived at the National Police facility, we had been surprised to find the U.S. Embassy's air attaché already there. He had decided to drop in for a visit and check on how things were going with our training schedule. After we assured him everything was going great and we should complete as scheduled, I mentioned

the call. "Oh, by the way, I received an unexpected, and unsolicited, phone call earlier this morning from Senor so-and-so who was just wondering if I could visit him in Medellin when I finish up here. What would you think about that?" There! I had told him!

Everyone in the office stared at me! It seemed to take a couple of minutes before the air attaché cleared his throat, then allowed as how he had taken certain steps to guarantee mine and Leo's safety and security. If I made the decision to stop by Medellin on my way home, he would no longer be responsible for my safety and well-being. I would be on my own! "But", he said, "you're free to make your own decision as to whether or not you stop by Medellin on your way home." That little speech did two things; first of all, it confirmed our observations about being tailed in Santa Marta, and secondly, I would be strictly on my own in Medellin! With that, the meeting was over, and Leo and I got on with our training assignments.

As my student and I walked toward the helicopter, we were talking about my phone call and he said, "Senior Williams, I heard what the embassy man told you about security. I promise you, if you decide to go to Medellin for the Cartel, you will be the safest man in all of Colombia, maybe even the world!" I thanked him profusely for his honest opinion, then we proceeded with our flight training. Even though I would have been the safest man in all of Colombia, I decided against it. There had been no sense in messing up what had been a great trip! Plus, I had been away from home for several weeks; it was time to go home.

Bell tech-rep and me are loading up Bell 412 helicopter we used for flight demos all over Venezuela and Colombia (02/1983).

Bell Canada test pilot Leo Meslin and me at the "lost city" in Colombia (06/1987).

Bell 206 L-3 helicopter lifting off from "lost city" in Colombia (06/1987).

A good aerial view of the well-preserved "lost city" hidden deep in Colombian jungle. (06/1987).

CHAPTER 15

Flying the 222/680 Helicopter

I've had the pleasure of flying many different helicopter models during my long career and invariably, the question always comes up about which helicopter is my favorite. My response has always been the same; I have two. The first is the legendary Bell UH-1, more commonly known as the Huey. Perhaps that's because I flew it during many hours of combat missions in Vietnam; and it always got me back to home base, safely. Because of that, I'm sure it will always be one of my favorite helicopters! The second is a helicopter most pilots have never even heard of. But for the sheer joy of flying, it was incomparable to any helicopter I've ever flown. Ironically, it was never a production model, nor was it ever meant to be. In fact, it was a model designation that had nothing whatsoever to do with a new helicopter.

It had all come about in the late 1970's when Bell Helicopter initiated a special project to develop an advanced rotor system for future applications. It wasn't launched with any great fanfare, because Bell was always seeking ways to improve rotor systems. Since rotor systems provide the propulsive forces and lift necessary for flight, any improvements to a rotor system can be critical to the success, or failure, of a helicopter. The goals of that new rotor system were quite simple. It had to have 50% fewer parts, 15% overall rotor system weight reduction and lower life-cycle costs. An additional goal was to achieve lower vibration levels with a minimum weight penalty. In keeping with their custom of assigning a numerical identification number for a new project, the next sequential number available was assigned to that new rotor system. I'm sure not much thought was given to it at the time, but the number 680 would become the defining namesake of what would become a remarkable helicopter.

Bell had gone through a number of designs and configurations before engineers finally decided on a configuration that featured a one-piece, four-armed rotor-head (called a yoke) as its basic component. The innovative new rotor-head assembly was formed and molded with a S-glass epoxy, a very high-strength fiberglass that was extremely strong! It had to be incredibly strong because the arms carried blade centrifugal and lifting forces, transmitted engine torque and accommodated both flapping and lead-lag motions. With that new design, the primary goals were met through the simplicity of design. But only if it all worked as advertised. The secondary goal, that of reduced vibration, could only be determined during flight tests.

A Bell Model 222 helicopter was chosen as the test aircraft for this project, not only because it was in the mid-sized range of Bell Helicopter's product line, but also because it provided the desired performance capability for proper evaluation of the new rotor system. Even though the Model 222 was selected as the test bed for the rotor system, Bell's engineers never intended to improve the performance of the production model 222. Flight testing was to be limited to a thorough evaluation of the rotor system by conducting a broad spectrum of tests. Any future expansion of the scope of flight tests depended entirely upon the successful completion of the original goals.

Before flight tests began, the new rotor system had to be adapted to a Bell 222, and that required a large number of modifications, one of which was the removal of the standard nodal beam assemblies with bipod transmission mounts. In a nutshell, the nodal beams were long metal straps that resembled the springs that connect the axles to the underside of a car. Just as their purpose is to reduce vibrations a passenger feels as the car goes down the road, the nodal beams accomplished the same thing. The only difference is the fact the helicopter flies through the air, rather than rumbling down a road.

The main rotor blades were standard Bell Model 412 blades that had been modified by removal of the inboard portion of the blade, then adding fiberglass, steel doublers, and grip plates so they could be attached to the yoke assembly. Another very important piece of equipment to be used with that system was a new vibration absorber designed specifically for use with the new rotor system. It was called a Liquid Inertial Vibration Eliminator, or LIVE system, for short. That important new system was mounted directly between the bipod mounts and transmission. In a nutshell, it was the direct link between the entire body of the aircraft and the transmission itself. Needless to say, it was critical to the success of the project.

The system had been previously tested on the much smaller Bell Model 206B Jet Ranger and it was rather unique in that it had a high density, low viscosity fluid commonly known as Mercury. That fluid acted as a damping mass inside a chamber, thus acting as a shock absorber to dampen vertical four-per-rev, high-frequency vibrations (since it had four rotor blades, the vibration in one complete revolution is referred to as four-per-rev vibrations). It was a very compact installation that provided a large reduction in weight, especially when compared to the Bell 222's nodal beam installation. In all aircraft, especially helicopters, weight is always the enemy of design engineers. In fact, during my entire career, every program I was involved with had a weight savings program! Always! So, if that compact vibration dampener worked, it would be a real plus for future projects.

After the 680-rotor system was installed on the Bell 222 (s/n 47004), it would forever more be referred to as the 222/680. Not long after the Bell 222 was modified with the 680-rotor system, a series of ground tests were successfully accomplished, thus clearing the way for its first flight on May 27, 1982. Bill McKinney, a senior experimental test pilot, was the assigned project pilot and he accomplished all flight tests during that period of time. All went well over the next couple of years, and it became increasingly apparent that Bell's engineers had achieved not only their goal of simplicity, but also such low vibration levels that pilots involved in

testing began referring to the ride quality as "jet plane smooth!" Other characteristics pilots liked was the fact it had no tendency for the nose to tuck under during pushovers, or to pitch up in turbulent air. In short, it had excellent gust response, a very important quality to have, especially in the unpredictable weather in Texas! That latter feature, combined with extremely low vibration levels, gave the feel of a much larger fixed-wing aircraft throughout the entire speed range. It felt very solid instead of being tossed around like a kite surfing in gusty air.

In mid-1984, I became the project pilot and during the course of flight testing, it became increasingly apparent the 222/680 project was one of those rare instances when all goals had been achieved, and everyone thought it might be capable of even greater achievements. I paid close attention to all test results, as well as rumors, because I knew I would be the project pilot for any envelope expansion program that Bell's Program Management team might dream up! Sure enough, I hadn't been the project pilot very long when Bell's Chief Experimental Test Pilot nonchalantly sat down at my desk one fine day and asked if I'd ever had any aerobatic training. Now I must confess, I'd never had any formal aerobatic training in my life! But I thought, "has any pilot on Bell's staff ever flown aerobatics, at least in helicopters?" Consequently, my response to the chief pilot was "affirmative!" That simple one-word response indelibly linked my future to that of the 222/680! For better, or for worse, I was going to attempt aerobatics in a helicopter! It wasn't long after that conversation that we had a fully-funded project for an envelope expansion of the 222/680. It was for the development of Split-S aerobatic maneuvers. I was greatly relieved when I found out it also provided for five hours of aerobatic training!

Knowing I had committed myself to the program with my simple one-word response, I wasted no time in searching for aerobatic training. Much to my surprise, there really weren't many aerobatic flight schools to choose from, at least back then. I finally settled on a well-known, highly respected aerobatic instructor by the name of Duane Cole, who conducted aerobatic training in a fixed-wing Decathlon. I wish I could you tell the training instilled great confidence in me, but that wasn't the case. Not at all! And the problem had nothing to do with the instructor. It was because I had only five hours to train for aerobatics!

I considered five hours to be inadequate for developing a skill level where I could transfer lessons learned in an airplane to the much more restrictive flight envelope of a helicopter! Plus, making matters worse, I hadn't been able to conduct any solo maneuvers during those five hours! Consequently, I must admit I wasn't looking forward to accomplishing split-S maneuvers flying solo for the first time in a helicopter when I hadn't even done them in an airplane! I think the saving grace was having to expand the 222/680 flight envelope prior to attempting split-S maneuvers. That allowed me adequate time to become well acquainted with the 222/680 flight characteristics, plus get a good feel for its overall handling qualities.

The goal of envelope expansion was to first establish a dive speed of 190 KIAS (Knots Indicating Air Speed), then develop a zero to three-G flight envelope. We hoped that envelope would allow us to attain our goal of performing split-S maneuvers. Establishing the dive speed of 190 KIAS was both stressful and exhilarating! It was also dangerous! But everything went very smoothly with minimal problems. It was also quite an achievement, since no other Bell helicopter had ever gone that fast. Once that was accomplished, we then turned our attention to the development of the zero to plus 3 G's envelope.

In preparation for that, we installed a cockpit camera in the center of the cockpit, located just behind the pilots' heads. The purpose of the camera was to capture the G-Meter, flight instruments and pilot actions during the flight test for post flight analysis. During our first preflight briefing for establishing the zero-G test point, there was some discussion about how we could best provide a visual presentation of achieving zero G. Yes, we did have a

G-Meter to register it, but I felt we needed a more visual presentation to satisfy everyone. I simply didn't think merely recording the G-Meter would be adequate, and I had no desire to repeat multiple zero-G test points just because "someone" couldn't properly see the G-Meter! The solution turned out to be quite simple; we attached a lime-green tennis ball I had brought from home to a short piece of string, which was then attached to the overhead structure of the cockpit. Mounted directly in front of the camera, the result of that last-minute bit of ingenuity was a picture-perfect shot of the tennis ball floating upwards, then "hovering" for just an instant precisely at the exact moment we attained zero-G. There was no further discussion about attaining zero-G, and the flight envelope for the next step of performing split-S's was complete!

Having successfully completed the envelope expansion, our thoughts were then directed toward accomplishing split-S maneuvers. We also discussed whether or not we wanted to utilize a two-man flight crew. Normally, a flight test program of that magnitude would require two pilots; however, the 222/680 was unique. It had only a small window instead of a door on the left side of the cockpit, and even though the window could be jettisoned, it was still too small. Parachutes would be required for the duration of the flight tests and I was hesitant to have an additional pilot assigned as copilot, because if we had to make an emergency egress, he would have to escape through that small window or follow me out through the right-hand door. Knowing how difficult that would be, especially while wearing a parachute, I was reluctant to place that risk upon a second pilot. Even though we had always conducted previous tests with a two-pilot crew, I made the decision to attempt the split-S maneuver alone. It was my decision, but I have to admit I felt rather lonely, especially on that first flight out to attempt a split-S!

October 17, 1984, was an absolutely gorgeous fall day. A perfect day for conducting flight tests! Mother Nature had presented us with a generous gift for our first attempt at accomplishing a split-S! Not long after our preflight, startup and completion of pre-takeoff checks, the aircraft was at full operating RPM and ready to go. We also had a chase aircraft for the day's mission. Not only would the flight crew assist by keeping me clear of other aircraft in the flight test area, it also had a video photographer on board. In fact, I think his job might have been tougher than mine, since the rear door had been removed and he was perched halfway outside the Bell 206, nestled into what is called a Tyler-mount. Plus, it was going to be cold at 4,000 feet!

Shifting my gaze back inside the cockpit, I got back to business and requested a takeoff clearance for a flight of two. Approved for takeoff, we made a takeoff to the north, then made a climbing left-hand turn and departed Bell's Arlington Airport Flight Test Center to the south, heading to our assigned flight test area. After takeoff, both aircraft began the climb to an altitude of 4,000 feet, since we had determined that to be high enough for me to accomplish split-S maneuvers. It was also high enough for me to exit the aircraft and parachute to safety in the event something went horribly wrong! At least I really hoped so! After putting all negative thoughts aside, I began accomplishing a series of turns, with sequentially increasing roll rates, and soon felt comfortable enough to get on with the task at hand, that of performing split-S maneuvers. As we often say in Texas, "it was time to cut bait or fish", and we would soon know whether or not our faith in the 222/680 abilities was justified!

Actually, it turned out to be a lot easier than I thought, because a slight aft and almost simultaneous right cyclic input resulted in an immediate right-hand roll that literally snapped the 222/680 upside down, even as the nose began dropping downward! The horizon disappeared above me as the aircraft began accelerating, but a slight aft input on the cyclic resulted in a return to level flight. It also kept the airspeed under control. After the aircraft was once again established in level flight, all I could think of was how exhilarating that had been! Totally exhilarating! Never having had an opportunity to roll a helicopter upside down before, I had an adrenalin rush of epic proportions! After completing the first split-S, it was almost

anticlimactic doing several others because each one was easier to accomplish than the preceding one. It went so fast, before long it was time to return to Bell's flight test facility.

Needless to say, there was an abundance of "high-fives" all around after I had shut down the engines and exited the aircraft! An added plus was the fact we had a photographer and cameraman on board the Bell 206-chase aircraft to document the achievement! It was also an excited flight test team that gathered in a large conference room to debrief and share the videos with Bell's Executive Vice President, and he was impressed. Well, almost! I say almost, because his first question surprised me since he wanted to know if there had been any particular reason why I hadn't accomplished any of the split-S maneuvers to the left. It was a "you've got to be kidding" moment, because I had no good answer for his question. So, early the next morning, off we go to accomplish a series of left-hand split-S maneuvers, plus a couple more to the right for good measure! Everyone was happy during the next "show & tell" briefing, including the VP.

After our successful entry into the aerobatic world, we returned to the business of developing sufficient airworthiness criteria, so we could begin flight demos for selected guests from the FAA, NASA and all branches of the military. Being able to demo split-S maneuvers in a helicopter intrigued most pilots, especially in a helicopter! Interestingly enough, the biggest problem we ever encountered during flight demos had nothing at all to do with aerobatics. In fact, it was a problem we had never even considered.

It all came about because helicopter pilots tend to fly by cues, be it visual, aural or, in that case, vibrations. Or I should say, the lack of vibrations! Typically, a pilot will unconsciously increase the collective, or power control, and accelerate the aircraft to some comfortable airspeed where vibrations are lowest. However, in the 222/680, there was lower-than-normal vibrations affecting the pilot's seats and that created a problem. There was virtually no increase in vibration, which resulted in an incredibly smooth ride, regardless of airspeed! Consequently, since there was virtually no perceptible increase in vibrations as the pilot pulled in more power to increase airspeed, the pilot essentially lost his cue for a normal comfort zone and that resulted in guest pilots continuously pulling in more power, thus getting extremely close to an over-torque condition (By way of explanation, torque is a means to indicate power in a helicopter, just as engine RPM indicates power in an automobile. There is a red line limit to what you can safely use; if you exceed that limit, or over-torque, you'll be damaging various parts).

That occurred repeatedly during every flight demo! The solution was to adapt a Boeing 737 stall warning stick-shaker to the collective control. Much like a rattlesnake, it would begin a low magnitude vibration of the collective when the pilot was approaching an over-torque condition, then increased in intensity the closer the pilot got to an impending over-torque. It worked well and most pilots adapted to that new "invention", and responded accordingly to avoid an over-torque. In fact, it worked great! Well, almost.

I remember one memorable demo flight with a guest pilot who bore a slight resemblance to Godzilla, at least physically. Unfortunately, he was every bit as strong as he looked! He also had very large, ham-hock sized hands that enabled him to get such a vise-like grip on the flight controls that they seemed to be encased in cement! Since there were no vibrational cues to alert him, he continued pulling up on the collective until it was shaking and buzzing like a rattlesnake! He had such a death-grip that I swear it resulted in the stick-shaker vibrating the entire helicopter rather than the collective! It was a long, stressful flight during which I spent the entire time trying to override his grip on the collective by continuously pushing the collective down to prevent an over-torque condition! As much as I had always enjoyed giving demo flights, I was never so happy for that flight to end!

After demo flights began, there seemed to be a steady stream of invited guest pilots that flocked in from all over the U.S., all of them being both anxious and excited to fly the 222/680 helicopter.

We not only accomplished demo flights at Bell's Flight Test Center located at Arlington, Texas airport, but we also accomplished demo tours throughout the U.S. and Europe, to include the Paris Air Show in 1985. The experience level of pilots ranged from "not long out of flight school" to high time military test pilots, Generals, Admirals and even a few elected congressmen. During all that time, I can't recall a pilot who left the cockpit disappointed. Everyone was highly impressed with the remarkable helicopter with the jet smooth ride!

Later in the program, Bell's engineers designed and built a new set of rotor blades that were referred to as Advanced Light Rotor (ALR) blades. They really didn't look too fancy, but they were a good replacement for the old, modified 412 rotor blades that had served so well for the first portion of the program. Designed primarily to evaluate rotor suitability for high-speed helicopter applications, it was hoped they would allow us to expand our aerobatic envelope. After extensive ground tests, the first flight with the new blades was accomplished on February 2, 1987. Performance of the new ALR blades met expectations and were so good, they were utilized for the remainder of the 222/680 program.

Following the successful introduction of the ALR blades, in late 1988 Bell's Program Management folks decided it was time to "raise the bar" once again by expanding the flight envelope to include aerobatic maneuvers such as hammerheads, rolls and loops. I had never flown those maneuvers, even in an airplane, so once again, it was off to aerobatic flight training! We had also decided a second pilot would join me as part of the flight crew even though we still had concerns about emergency egress. We felt like we had more confidence in the aircraft itself, plus a second set of eyes was needed to assist in monitoring all instruments. Jim McCollough, another Bell experimental test pilot, joined me for the next step in envelope expansion and we began training in a small Decathlon airplane. It was a small, tail-dragger aerobatic airplane and, whereas the previous training was somewhat lacking, the new training program turned out to be excellent! That was because both of us were signed off to practice aerobatic maneuvers either solo, or together. Reflecting back, I think that was the single, most important element of the advanced aerobatic flight test program.

After completing all training in the Decathlon, on November 17, 1988 we performed split-S maneuvers and right-hand rolls; hammerheads and loops on November 18; and left-hand rolls on November 22. It was a very successful program and of all the maneuvers we attempted, the loop was the most difficult to accomplish. I think it was because after so many years flying helicopters, pulling the cyclic way back into my stomach to get the helicopter over the top of the loop, then dropping down the backside of the loop was a bit unnerving for me, especially on the first one! Loops aren't something you normally do in a helicopter, but everything was well within airspeed, aircraft loads and G limitations just as we had planned. Once again, everything had been well documented by utilizing a cameraman and photographer to film all maneuvers from the 206-chase helicopter. The 222/680 also looked a little different since it had been given a new paint scheme to support a Bell/McDonnell Douglas Helicopter Company joint venture to win the U.S. Army Light Helicopter Experimental (LHX) program.

In January 1989, I ferried the 222/680 helicopter to Falcon Field in Mesa, Arizona to support the LHX media days on January 19/20. Our LHX entry was to be a mix of the 680-rotor system matched with the NOTAR (NO TAIL ROTOR) so part of the photo missions was me flying the 222/680 in formation with a U.S. Army Apache gunship and a small NOTAR helicopter. The appearance of the 680-rotor as an LHX contender garnered a lot of attention and later that same month I flew with a guest pilot from the Aviation Week & Space Technology magazine. Flying with the guest pilot was an enjoyable experience, and the popular photo of me rolling the inverted 222/680 was on the cover of the January 30, 1989 edition of the Aviation Week & Space Technology magazine. It was wonderful testimonial to the 222/680!

Many times, when we were on the road conducting flight

demos, the maintenance personnel at some of their facilities requested a closer look at the 680-rotor system. Knowing how the simplicity of the system intrigued them, we were always glad to comply with their requests. I think some of them actually thought we had some secret piece of hidden equipment inside the cuffs because more than once, I heard softly murmured "ahhs" when the cuff was removed from the extended arm of the yoke. I can also recall with some amazement how we could remove the cuff and blade, let everyone inspect it, reinstall it, then roll the helicopter out the hangar door, start up and launch on the next leg of our scheduled destination.

It was purely coincidental, but during the same time I was ferrying the 222/680 around the countryside, there was a very popular television series entitled AIRWOLF. It was a very popular series, especially among young viewers, since one of its stars was a handsome young pilot played by Jan-Michael Vincent. He flew a high-tech helicopter known as AIRWOLF and the overall theme was a "renegade pilot goes on special missions for the good guys with an advanced, high-tech helicopter capable of many features that included high speed, silent-mode flight and many weapons at its disposal". It also bore a close resemblance to the 222/680 helicopter, and that's understandable, since they were both Bell Model 222 helicopters. So, it's easy to understand why the 222/680 was always being mistaken for AIRWOLF! We did have a different paint scheme, but it was very sharp looking. Actually, it was quite fun zipping into various airports in the sporty looking 222/680 and more than once I had to admit I was indeed "flying AIRWOLF!" I felt as though I had to say that, rather than disappoint a crowd of admirers who were always so excited to see the "real" AIRWOLF! What a fun time for me!

In May, it was back on the road as we ferried the 222/680 to Fort Rucker, Alabama to garner support for the LHX program, but we lost the program to the Boeing/Sikorsky team shortly afterward. Their entry became known as the RAH-66 Comanche. On the return ferry flight to Texas, I noticed a slight vibration increase that was barely noticeable. In fact, it was so imperceptible that flight test engineer flying with me couldn't feel it at all. But I had so much time in it, I knew when something changed – and the ride had definitely changed! Soon after our return to Bell's facility at Arlington, Texas airport, further inspections revealed my worst fears. We had separation of the laminations in the yoke itself. Actually, it really wasn't unexpected because the original program was only expected to last 200 flight hours. By the time the program ended, we had more than 800 flight hours, but it was still a shock when Bell management decided they no longer needed the 222/680 helicopter as a test aircraft. The ferry flight from Fort Rucker, Alabama to Arlington, Texas on May 22, 1989 proved to be the last flight for the beautiful 222/680 helicopter.

The 222/680 helicopter had a very short lifespan. Its first flight was accomplished on May 27, 1982, and I made the final flight on May 22, 1989. Shortly afterwards, the 680-rotor program was officially over. In much the same manner a shooting star streaks across the sky, the 222/680's short existence was also filled with flash and fanfare. Then, in the blink of an eye, it was gone – leaving nothing but a distant memory of what it was like to fly in the helicopter with the jet smooth ride.

Well-known aerobatic pilot and instructor Duane Cole and me beside his Decathlon we used during aerobatic training (06/09/1984).

Barney Barnoski, one of my favorite Bell mechanics, is performing work atop the Bell 222/680 helicopter's main rotor system.

The Bell 222/680 helicopter in Paris, France for the Paris Airshow; H332 on side of aircraft was assigned number for the Paris Airshow (05/29/1985).

Me, flight test engineer Dan Forester and test pilot Jim McCollough standing beside Bell 222/680 helicopter on flight ramp (11/1988).

CHAPTER 16

The Tiltrotors

XV-3, XV-15, V-22 Osprey

I don't want to get too deep into the history of the tiltrotor aircraft, but I do think I need to include a quick story about Bell Helicopter's first tiltrotor, the XV-3. Even though it didn't make its first flight until 1955, the idea for a convertiplane had actually begun in 1943 when Larry Bell first saw Arthur Young's new helicopter design. That's when he began to visualize an aircraft that would be a compromise between the helicopters hovering and vertical flight capabilities, and the speed and cruise characteristics of an airplane. When Bell finally got a government contract to construct the XV-3 in January 1953, 90 hand-picked employees assembled it at Bell's new facility located in Hurst, Texas. Conducted under a strict shroud of secrecy, the official roll out had occurred at noon on February 10, 1955. Actually, the term "roll out" was a misnomer since cold, blustery winds forced an assembled crowd of invited guests inside the experimental hangar to witness the ceremony.

The first ground run wasn't accomplished until June 23, 1955 but everything went smoothly enough for a first flight on August 11, 1955. At 9:00 in the morning, Floyd Carlson climbed into the cockpit, started the engine and was soon waved into the air for a short flight. Although it only lasted five minutes, Floyd had lifted the XV-3 to a 20-foot hover, then maneuvered about in every direction before an excited crowd of onlookers, all of them quickly declaring it to be a roaring success! It would go on to fly about another hour at Hurst before relocating to the Globe facility where follow-on flight tests would begin in earnest.

The first order of business at Globe had been for Floyd to check out Bell test pilots Dick Stansbury and E.J. Smith in the XV-3. That

was accomplished with the grand sum of .2 hours each! After seeing that in his logbook, I asked Dick Stansbury about it and his response was to chuckle, then go on to explain that the underpowered XV-3 could barely lift two good-sized pilots, so they basically lifted off the ground for three "take-offs and landings", so Floyd could sign them off as being qualified! The two new pilots conducted ground runs and flight tests at Globe, and on October 25, 1956, Dick Stansbury departed Globe on what was to be a routine run that ended up to be so much more than that.

He had departed to the north in a right climbing turn with a Bell Model 47 chase ship following closely behind in the event something went wrong. At an altitude of approximately 100 feet, Dick had begun slowly converting the pylons forward when a phenomenon called "rotor weaving" began whereby the blades began moving up and down with increasing intensity, thus creating an instability that became noticeably stronger. Since it had been an ongoing problem, Dick was familiar enough with it that he quickly returned the pylons to the 90-degree position, or helicopter mode, to dampen the instability. Unfortunately, the instability remained unchecked, and there was little Dick could do other than attempt to get the XV-3 back onto the ground as quickly as possible! Unfortunately, the rotor instability had increased to such an extent, the fuselage encountered an out-of-phase vertical vibration so violent that Dick blacked out. He regained consciousness just long enough to switch on the instrumentation, an act that provided critical data for the investigation board. He also got the aircraft into a level attitude prior to impact with the ground. It was an act that probably saved his life.

Unfortunately, just prior to Dick's crash, the aircraft had undergone a modification to the pilot's seat that literally rendered the airworthiness of the seat to be marginal, and it provided very little support for the pilot when the aircraft hit the ground. Sliding forward after impact, it literally rolled Dick under the aircraft, snapping the lower vertebrae in his back, leaving him paralyzed from the waist down for the rest of his life.

Ship No. 1 would never fly again but, on the positive side, the XV-3 project moved forward despite that setback and eventually, the project ended on a positive note when flight in airplane mode was accomplished. The project had only lasted from 1955 until both rotors came loose during wind tunnel testing on May 20, 1966. Although the program had been short, it had served as the catalyst for future Bell tiltrotor aircraft.

As a note, Dick Stansbury and I became good friends during the 1990's and I spent many wonderful days with him at his home where he held me in awe while listening to his stories. Despite his severe injuries, he continued to live a good, productive life and I never heard him complain about his disability. He also remained a much-respected employee of Bell Helicopter until his retirement.

The XV-15

In 1971, NASA and the U.S. Army agreed to finance a tiltrotor research aircraft in accordance with the "Tiltrotor Research Aircraft Program". Contracts were awarded to both Boeing-Vertol and Bell Helicopter and, in April 1973, Bell's concept had been declared the winner! On May 23, 1976, two Bell Helicopter experimental test pilots by the name of Ron Erhart and Dorman Cannon made the first flight on remarkable aircraft. It would change aviation history!

My introduction to the XV-15 began right after the XV-15 had blown everyone away with its performance at the Paris Air Show on June 4 – 14, 1981. My first job had not been too glamorous, since all I did was "outfit" guest pilots in a proper flight suit when they arrived to fly the XV-15. It became so because one of my "extra" assigned duties had been to maintain all pilot's flight equipment, a job that had been assigned temporarily when I first transferred to Bell's Flight Research Center in December 1980. I still had the job when I retired in 2005!

During that period, I had the pleasure of meeting numerous guest pilots invited to fly the XV-15, but one of them stood out

above all others. His name was Barry Goldwater, and he had been an iconic Senator from Arizona who not only flew combat missions in World War II, but had also been instrumental in forming the Arizona National Guard. Not only had he been a real gentleman, but he also had been a big man, even though he was only listed as being 6 feet tall. Perhaps he had only seemed big, but I do recall it had taken one of my larger flight suits to fit him comfortably! He had flown with one of our senior tiltrotor pilots who had been impressed with Senator Goldwater's piloting skills.

I progressed from being a "flight suit" guy to XV-15 flight operations in November 1981, after I received a checkout out in a Cessna C441 Conquest to become a chase pilot for the XV-15. I had been involved in numerous helicopter projects but getting involved in the XV-15 program had been exciting, regardless of how I was involved. If Bell Helicopter wanted me to fly chase on the XV-15 program, I was more than happy to oblige!

Our C441 Conquest was certainly hard to miss! It had been painted bright orange with green striping, and it quickly picked up the nickname of the great pumpkin! We didn't use it that long, but there was one time when we got in too close to the wake turbulence created by those huge 25-foot rotor blades, and it almost turned us upside down before we could break free from that effect! It provided a wild ride for a couple of seconds, but everything soon settled down and we were able to resume our job as chase pilots. However, we were damned careful to never get in that close again! Actually, it was always difficult for an airplane to even get close behind the XV-15 because of its speed!

Once it had been decided the C441 Conquest was inadequate as a chase aircraft, Bell management began thinking about utilizing a jet as a chase aircraft. It began in January, 1982, when Ron Erhart, Bell's Chief Experimental Test Pilot, stopped at my desk and asked if I would be interested in getting checked out in a Cessna Citation jet. I had been quite surprised by his question, but my response had been "Yes! Absolutely!" Bell Helicopter had never owned a corporate aircraft, but they selected a Citation I jet to be utilized primarily not only as a chase aircraft, but also as an executive jet when not being utilized for flight test activities.

I attended the CE 501 course at Flight Safety in Wichita, Kansas from March 11 through March 19, 1982, with Chuck Anderson, another Bell Helicopter pilot. After completing our check rides on March 19, we boarded Bell's new CE 501 (N60PR) and flew from Wichita, Kansas to Arlington, Texas. The old cliché "the ink's still wet on their license" certainly did apply to us! Neither of us had ever flown a jet before, but after completing the simulator check rides and accumulating 2.9 hours, we flew Bell's first corporate jet to Arlington without any problems. We had little time to get acquainted with our new CE 501 because on March 20, we flew our first XV-15 chase mission and it never stopped from that moment on! Bell Helicopter had finally entered the "jet age"!

The Citation jet had been a game changer in that we could keep up with the XV-15 as it accelerated through its flight test program, and it wasn't long before we set off on the first of several demonstration tours on June 24, 1982. Our first stop had been at Fort Huachuca, Arizona where the XV-15 participated in Nap-of-Earth (NOE) flight evaluations during simulated enemy ground-to-air missiles controlled from self-guided radar installations. In almost all cases the XV-15 could "break radar lock" by executing NOE evasive maneuvers. Interestingly, the XV-15 had been painted in a desert camouflage paint-scheme, utilizing water-based paint prior to our departure from Arlington, Texas.

We eventually ended up at the U.S. Navy's historic Naval Air Station (NAS) North Island just outside San Diego, California and upon our arrival, the water-based paint had been washed off, transforming the XV-15 into the Navy gray paint scheme; a miracle! Soon afterward, on August 2, 1982, the XV-15 made its first offshore landing aboard the USS Tripoli located about 20 nautical miles off the coast of San Diego, California. Pilots for the occasion were Bell test pilot Dorman Cannon, and Navy LCDR John Ball (who later became

a Bell pilot). It had been a historic moment for everyone associated with the XV-15 program.

The only interesting event with regard to flying the jet was having to make a Precision Approach Radar (PAR) approach into NAS North Island on July 31. We had just returned from Arlington with several Bell executives on-board and the entire area around San Diego had been zero-zero due to ground fog. The Navy's approach control asked if I would accept a PAR approach so they could train some of their young personnel (a PAR approach is when the controller provides constant verbal guidance to you throughout a precision approach). It was the first and only time I ever had the opportunity to make a PAR approach! It really wasn't that stressful except for several executives who had gathered up just behind the cockpit to observe me making the approach since they had never seen one before! They had been very impressed and maybe even appreciative, too, since we had broken out of the fog just above the runway and made a smooth, safe landing! It was the one and only PAR approach I ever made.

Our next stop was Las Vegas on August 6, where the XV-15 participated in selected U.S. Air Force flight operations at Ellis AFB. Finally, on August 13, we departed for Texas, arriving at the Lubbock, Texas airport just about lunch time. A memorable moment occurred there since we arrived just about lunch time on a hot August day. We borrowed two airport crew cars and about twelve of us crammed into them, then made a bee-line for the nearest café that served chicken-fried steaks and iced tea! The wisecracking waitress obliged our orders with large glasses of iced tea, then set several pitchers on our tables. Huge platters of chicken-fried steaks arrived soon afterward! Everyone gave a hearty toast, followed by a loud cheer! It was great to be back home in Texas!

The leased CE 501 had done so well that Bell returned it to Duncan Aviation in Lincoln, Nebraska on September 14, and ordered a brand-new Citation 550. We ferried a Bell 212 model to Duncan (as a trade-in) on November 3 and, after a quick checkout

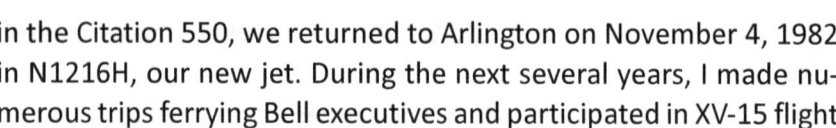

in the Citation 550, we returned to Arlington on November 4, 1982 in N1216H, our new jet. During the next several years, I made numerous trips ferrying Bell executives and participated in XV-15 flight test operations and demonstration trips all over the U.S.

My First Flight in the XV-15

My first flight in the XV-15 (N702NA) was on May 27, 1988 with Dorman Cannon, one of Bell's most respected tiltrotor pilots, as Pilot-in-Command (PIC). It was almost overwhelming when I first got into the seat, strapped myself in and had a chance to look around. One of the first things I noticed when sitting in the cockpit were four quick release "Cleco" pins with red flags attached to each of them. Located at various places around the cockpit, you pulled first one pin, then carefully placed it in a small metal receptacle that had four holes, one for each of the four flagged pins. It was all done sequentially and since removal of the last pin armed the XV-15's ejection seats, it was always the last to be pulled. Every time I pulled that last pin to arm the ejection seat, I remembered Dorman's words: "Whatever you do, don't reach down to adjust your seat belt and accidentally pull that round thing (the loop) because it will eject both of us at a one second interval!"

Once it had been stowed with the other three pins, the ejection seat was armed. Between my legs at the forward edge of your seat was a large cable loop that I didn't even want to breathe on, much less touch! Even though the ejection seat system was capable of a "zero-zero" ejection (meaning you could be successfully ejected while sitting on the flight ramp), I had no desire to test it, even though I have to admit, being propelled out of the cockpit while sitting on the end of an artillery shell would have been a hell of a ride!

Once we had gone through the startup, runup and all the required checks, Dorman called for takeoff clearance and I was ready for my first flight! I can still recall my first three impressions on that flight. My first impression was how easily it hovered! It seemed to hover like a rock with very little control input from me. Secondly, I

recall how surprised I was at how fast it accelerated! I was pushed back into the seat as those 25-foot blades clawed through the air, propelling me along with it! The third impression was how close those damned blades had been! Looking over my left shoulder, I saw the rotor disc and it appeared to be literally right behind my head, close enough for a haircut! But all apprehensions were quickly dispelled, and it was one of the most exhilarating flights I ever made!

My love for the XV-15 never diminished throughout the entire time I flew it! I absolutely loved flying guest pilots because it seemed to turn them into "a young child who had just been turned loose in a toy store!" The most common comment from an airplane pilot was how "they had always heard how difficult it was to hover a helicopter!" The XV-15 made it look easy! I've often wondered if they had tried hovering a helicopter since their flight, and what their comments would have been afterward!

Crash of XV-15 (N702NA)

Thursday, August 20, 1992 was a warm summer day, but it was perfect for flying! So much so, dust and cobwebs were dusted off the XV-15 in preparation for a demo flight for the Chief Pilot of Aerospatiale, a well-known French helicopter manufacturer. The XV-15 had been in the hangar for an extended period of time for inspections, and after a return-to-service maintenance test flight, it was prepped for its first demo flight in some time. Even among Bell personnel who had seen it fly so often, it had always been a special event to watch it fly! I think there was a song from the sixties entitled "Poetry in Motion" that best described the XV-15. The XV-15 seemed to thrive on that, even if it had been an inanimate object.

Ron Erhart, Bell's Chief Pilot, was the PIC for that demo flight and after starting it up, they took off and proceeded down range to dazzle the guest pilot. When he had approximately 30 minutes of fuel left, he returned and put on an impromptu air show at Arlington's municipal airport. Like everyone else, I stood and watched his mini-air show routine until they landed on runway 16, then began ground taxiing back towards the taxiway leading to Bell's flight ramp. That's when I walked back inside, and had just sat down when I heard our secretary let out a blood-curdling scream! I turned immediately to look out the window toward the XV-15 taxiing along Bell's taxiway and was stunned by what I saw!

The guest pilot had been so intrigued with how the XV-15 flew that he requested Ron to accomplish another lift-off to a hover while they were taxiing towards the Bell hangar. For some reason, the XV-15 came off the ground in a rolling motion and when I saw it, it was approximately 25- to 30-feet in the air and totally upside down! It seemed to hang for just a second before dropping vertically onto the ground, punctuated by a tremendous noise, but no explosion! I sprinted outside, then just stood there among the stunned crowd for a second or two just to take it all in! It was literally overwhelming! Both engines were screaming because there were no rotor blades to impede them! I could also see fuel vapor venting from damaged fuel cells and everyone was shouting to "stay back, it's going to explode!" It had happened so fast, there was no crash-rescue equipment in sight! My immediate thought was to either stand and watch it explode, or I could run over and at least try to get them out before it exploded! Neither one had been a good option! My next thought was how would I live with myself had I stood by and maybe had a chance to rescue them, rather than do nothing. I started sprinting toward the crashed XV-15!

I recall everyone screaming to stop, but I kept sprinting and as I approached, I could see the semi-conscious guest pilot who was still upside down, strapped in his seat. I immediately knew I couldn't assist him because the thick plexiglass dome was still intact, so I raced around to the other side and saw Ron was already unstrapped from the seat and sitting upright on the plexiglass dome of what was normally the top of the XV-15 cockpit. Still in shock, he had been trying to break the plexiglass by tapping on it with a small, heavy Cobra extraction knife. Running up to him, I asked if he was okay, as

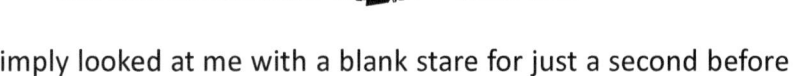

he simply looked at me with a blank stare for just a second before calling out my name, and saying how glad he was to see me!

Quickly looking around the edge of the plexiglass, I could see it had a crack just large enough for me to slide both hands inside, and with a mighty yank, I was able to split the plexiglass wide enough to reach inside, grab Ron and assist him out of the aircraft. I managed to cut both hands on the sharp edges of the plexiglass, and distinctly remember how amazed I was at the thickness of the plexiglass! I have no idea how I was able to break it apart to get the flight crew out! Adrenalin, I guess.

Just about the time I helped Ron out, Tom Warren, another Bell test pilot, came up and I told him to escort Ron away from the aircraft. As soon as they left, I crawled inside the cockpit and began to assist the guest pilot who had regrouped his senses enough to communicate with me. Telling me he was okay; I began unbuckling the seat belts just as something hit the plexiglass just above us with a mighty whump! It startled both of us and I looked up just in time to see a fireman swinging his crash axe in a huge arc to hit the plexiglass a second time! That was just what I needed; a crash axe to come crashing through the plexiglass and hit one of us on the head! It did seem to spur the guest pilot into a faster speed, and we were able to get out of the cockpit and sprint away from the aircraft! I also noticed the engines were quiet. They had used what fuel had been left in the lines, then shut down.

Shortly after exiting the crashed XV-15, an EMS helicopter arrived in response to the control tower's 911 call and medevacked both Ron and the guest pilot to the hospital. After ensuring we had done everything we could do, a couple of other test pilots and I drove over to the hospital to check on them. When we arrived at the emergency room, Ron was sitting up on a gurney and announced the doctor had already come by and told him everything seemed to be okay, with no broken bones. In the meantime, the guest pilot, who was French, was conversing with a nurse. Knowing his English might be lacking, we walked over to where he was telling the nurse he was okay except for a small "bimp" on his leg. The nurse was clearly puzzled by the word and, quite frankly, we were, too! In exasperation, he pulled up his pant leg, pointed to a goose-egg sized bump and proclaimed, "See, I have a 'bimp'!" It had been a Peter Seller's moment right out of "The Pink Panther" movies! Everyone started laughing, even the French pilot! I think we all needed that to dispel the heartbreak we all felt about losing the XV-15!

The next morning, the Chief Pilot had called the entire pilot's staff together to explain what had happened. He then announced he was to admonish me for running up to the XV-15 before the crash/rescue equipment had arrived. Then he said, "Now that I've admonished you, I would like to thank you! I thought you had been an angel when you ran up!" Probably not too many people get admonished, then a thank you in the same meeting!

I have to say, it had been very sad to look at what had once been a beautiful flying machine! The culprit turned out to be man-made, rather than mechanical failure. Simply put, a nut had been installed on a bolt connecting the rotor control tube to the rotor, but the safety wire had not been properly installed. Consequently, after a couple of years, the nut eventually screwed itself off the bolt, allowing the bolt to come completely out. At that point, the pilot no longer had control of the left-hand rotor system, and when Ron attempted to raise the XV-15 into the air one last time, only the right-hand rotor responded, thus putting the XV-15 into a left roll as it came off the ground, and turned upside down before crashing into the ground! Had it happened earlier while the XV-15 was still flying at altitude, or in the traffic pattern, it could have been even more catastrophic! Bell's engineers eventually determined it was not worth rebuilding so it never flew again. It was a sad demise for the XV-15 that had introduced tiltrotor to the world at the Paris Air Show, and all around the U.S., accumulating over 840 flight hours. To my knowledge, it had never missed a "curtain call".

Ironically, just prior to the loss of such an iconic aircraft on August 20, 1992, I had accomplished a notable first flight of an iconic helicopter, myself. The previous week, on Tuesday, August 13, I had made the first flight of the Huey II, thus extending the life of the legendary Huey! It was also made special by the fact that my Flight Test Engineer was a fine gentleman by the name of Leo Norman. When he hired in at Bell in the 1950's, his first job had been to work on the original Huey. His last job prior to retiring at the end of August was making the first flight with me on the Huey II. That was always a special moment for me.

Explosion in XV-15 After Takeoff

In the spring of 1999, Bell Helicopter's marketing department had arranged a very important flight demonstration for the U.S. Coast Guard (USCG) in Key West, Florida. It had been one of those types of operations that involved numerous individuals from both Bell and the Coast Guard, plus several ships, referred to as cutters, plus the XV-15 itself taking center stage. Bell test pilot Roy Hopkins and I departed on April 13, 1999, for Key West. In typical fashion, a Cessna Citation II, carrying our flight support crew, accompanied us on the entire flight. After spending the night in Mobile, Alabama, we arrived in Key West on April 14.

We began testing with the USCG the very next day, when Roy and a USCG pilot began conducting takeoffs and landings from a Coast Guard cutter, working with the USCG's Master Chief Rescue Swimmer in the process. It was a very professional operation that went as smoothly as any test program in which I've been involved. The debriefs were awesome, in that everyone, including the cutter's crew, pilots, mechanics and the Master Chief swimmer, all had their say as to what went well, or didn't. The swimmer was in the water while the XV-15 hovered directly over him, and he was amazed at how calm it was in the center of what was two mini-hurricanes churned up by the two counter-rotating, 25-foot rotors. In fact, he said it had been so calm that he could have probably communicated with the flight crew had anyone been standing in an open doorway! That was an important piece of information

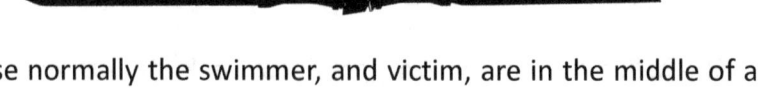

because normally the swimmer, and victim, are in the middle of a maelstrom. It had been a great trip and we departed for Opa-Locka, Florida, where we spent the night.

The next morning, we were soon ready for takeoff, but when we called for taxi/takeoff clearance, ground control insisted we taxi out for takeoff in the same manner as their departing airplanes were. Try as we might, we could not change their minds even though we normally took off like a helicopter! Consequently, we had to taxi the entire length of taxiway prior to departing on the main runway. Seeing as how it was a very hot day with high humidity to boot, Roy and I suffered mightily as we taxied. Little did we know the XV-15 itself was also suffering from the heat just as we were!

After finally reaching the end of the taxiway, we were cleared onto the active runway before being cleared for takeoff with our chase aircraft taking off in close pursuit just behind us. It had been a beautiful morning, albeit a hot one, and soon we were climbing to our normal cruising altitude of 12 – 15,000 feet. After all the frustration of dealing with the control tower, we were relaxed, cooling off and heading towards home when a huge explosion right behind us sent our adrenalin soaring! Roy and I quickly looked at each other and were thinking the same thing – midair collision with another aircraft! I think both of us were afraid to even breathe for a few seconds before finally realizing we were still flying so we wouldn't have to rely on our ejection seats to get us to the ground!

Very cautiously, we began a slow turn back towards Opa-Locka, while I advised our chase aircraft what had happened. I then called the tower and explained our situation. We didn't declare an emergency, but I did keep them advised of our progress. Since we were still very close to the airport, we lowered our landing gear and had our chase aircraft get close to check everything out. They reported everything looked normal, including the landing gear. We were relieved to hear that, but we were also puzzled about what could have caused such an explosion! Wanting to keep our taxiing to a minimum, we asked the tower for a direct approach to the same

ramp on which we had departed, without having to make a normal approach to the runway, followed by a long taxi. Fortunately, they approved our request!

After an uneventful landing and engine shutdown, we clambered outside and met up with our very concerned ground crew who had jumped out of the chase aircraft and sprinted over. Looking at the landing gear, all our jaws dropped when we saw the problem! One of our main tires had exploded! Fortunately, the XV-15 had tandem landing gear, meaning each of our three struts had two tires on them and the blown tire was on the right main gear. Had we tried to taxi, we would have also blown out the remaining tires and could have caused severe damage to the aircraft! We had been very lucky!

We finally figured out that the long, extended taxi in such hot weather had heated up the tires much higher than normal. Consequently, when we took off and stowed the gear inside wheel-wells and closed the doors, the exorbitant amount of heat built up with no way to dissipate it. The result was one of our main tires blowing up with the force of a small bomb, or at least that's what it sounded like! Fortunately, our support crew had planned ahead and after a couple of hours, we were soon back in business. After explaining to the tower what had happened, we were given clearance to take off directly from the ramp! The remainder of the flight home was uneventful.

Delivery of XV-15 (N703NA) to the Air & Space Museum September 16, 2003

When XV-15, N703NA, was running out of flight time, Bell Helicopter and NASA agreed to donate it to the Smithsonian's Udvar-Hazy Center near Dulles International Airport. Even though the pilots hated to let go of such a wonderful little flyer, we all agreed it could not spend its "retirement" in a more perfect location! It was to be quite an event, so we began planning for it well in advance, and that included deciding which pilots would fly it to Washington, D.C.

The two pilots who had made the first flight, Dorman Cannon and Ron Erhart, had both retired, as had Tom Warren, another Bell tiltrotor pilot. Surprisingly, the list of XV-15 pilots who held FAA Letters of Authorization (LOA's) designating them as PIC's was very short. Since Roy Hopkins and I had been the last of the XV-15 PIC's on Bell's pilot staff, we decided to do the honors! Bell also paid for numerous Bell retirees who had developed the XV-15 to attend the ceremonies as well. Ron Erhart and Dorman Cannon were two of the invitees, and I've always thought that was a classy thing for Bell Helicopter to do!

We had planned to depart on Friday, September 12, but the latest weather forecast predicted bad weather along our route of flight. In fact, I had been home on vacation when I got a call from Roy, advising me they had changed our departure date to Wednesday, September 10, which was the very next day! It took some scrambling, but I managed to get everything together. Suitcase in hand, I arrived in the early morning hours of September 10, at Bell's FRC ready to depart for Manassas, Virginia. Our route took us through El Dorado, Arkansas; Huntsville, Alabama; Charlotte, North Carolina; and, Lynchburg, Virginia. It was our last stop before arriving at Manassas, our final staging area prior to delivering the XV-15 to the Udvar-Hazy Air & Space Museum, located very close to Washington's Dulles International Airport.

We arrived at Manassas the next day, on September 11, and began preparations for the delivery on September 16. While Roy and I were planning for the delivery, the Smithsonian sent a photography team to Manassas to photograph the XV-15's cockpit for publication in a planned book. It had been interesting to watch, but I don't know if they ever published the book. While that had been going on, Roy and I drove over to the Udvar-Hazy museum to discuss the schedule of events with the Air & Space Museum's supervisor. Having never seen the new facility, Roy and I were stunned at how massive it was! After donning hard hats, the supervisor escorted us around the facility. It was fascinating to see the Enola Gay, the SST

and the Boeing 707 prototype that had started jet-age commercial transportation around the world! It was a fascinating visit!

The supervisor was a very pleasant gentleman who proudly escorted us to the north end of the huge facility where the big event would happen. They had seating for a hundred invited guests, and it looked fantastic! Then we went outside to look at where we would land, then taxi up and park the XV-15 after putting on an airshow. Roy and I looked at each other and shook our heads. The facility was still a semi-construction area and there were all kinds of junk metal, paper and other debris laying around that entire area! We then had to explain to that very nice supervisor we could not do the airshow as planned! We recommended relocating everything to the south end, where there was a very big ramp that was relatively free of junk.

It had been one of those "deer-in-the-headlights" moments! The supervisor just silently stared at us with wide-open eyes in hopes he hadn't heard us correctly! It was almost as if he could just shake his head, wake up and everything would be okay! Except it wasn't okay! After a moment of silence, he said, "Follow me, we've got a phone call to make." We had to drive back to his office where he picked up the phone, dialed a number and began explaining how Roy and I had just dropped "poo-poo in their punchbowl!" We couldn't hear the other end of the conversation until the supervisor activated the speakerphone function. The voice on the other end was aghast at how we were requesting a change in plans after everything had been set up! We patiently explained how we understood his situation, but it would be much safer to switch ends for the planned event. Sighing, he finally acknowledged "it would happen". We certainly understood their dilemma and hated to change things at such a late date, but we had no choice. All things considered, both gentlemen had been gracious about it and everything had been flip-flopped for our scheduled arrival.

Another factor we had to contend with involved the aftermath of 9/11/2001. The entire Washington, D.C. area had basically become a "no-fly" zone and could only be penetrated when in direct contact with the Air Traffic Controller (ATC) who was in control of that sector. Consequently, Roy called ATC every day to ensure nothing interfered with our scheduled September 16th arrival date. It would require close coordination and split-second timing because Washington's Dulles International Airport planned to shut down one of their major runways while we performed the last airshow ever in the XV-15. There would also be about a hundred invited guests on hand to watch. All we wanted was perfection for our last flight!

On the morning of September 16th, Roy and I carefully stowed our suitcases and travel kits on-board the XV-15 because we planned to go directly to Dulles Airport after the event to catch our flight home. Exactly on time, we started the XV-15, took off and arrived overhead the Udvar-Hazy facility at the appointed time. The airshow went off without a hitch, and we were welcomed by a large group of invited guests when we taxied in. The title had been handed over to the Smithsonian officials and all things considered, it could not have gone any better! Roy and I had been guests of honor and the food was delicious! What could go wrong? Well, there's always someone who never seems to get the word!

As soon as Roy and I had seen a window of opportunity to make a graceful exit, we scurried out to the ramp where we had left the XV-15, but it was nowhere in sight! Seeing a security guard, we quickly asked about the XV-15. "They towed it to the north end of the hangar", he said! Now, that was a very long way from the south end to the north end of the Udvar-Hazy facility! In fact, by the time we got to the north end, we were both nearly out of breath since we had jogged most of the way, still dressed in our flight suits. Seeing the XV-15, we approached it to get our luggage when a security guard demanded to know what we were doing there!

"Oh", we said. "We're the pilots who flew it in and we're here to get our suitcases".

"That's the property of the U.S. government and you can't have access to it"!

Are you kidding me?! Finally, after arguing and pleading our case,

we were finally able to get our bags, but had no time to change out of our flight suits! Flight suits that had metal zippers all over them!

After catching a ride to Dulles International Airport, Roy and I checked in, picked up our boarding passes and proceeded through security where we set off every alarm bell known to man! Our many-zippered flight suits created havoc, especially being that soon after 9/11! However, in the end, we prevailed and finally made it to the gate just in time. Sitting back and relaxing, I explained to Roy our fifteen minutes of fame had been too fleeting! We had been welcomed as heroes at the Smithsonian, but the folks at Dulles security really didn't care about such trivialities! I guess it put everything back into perspective for us.

In the end, the XV-15 had been the real hero, and leaving our favorite tiltrotor behind had been difficult! It was the aircraft that started it all, and throughout its life-span, it never let us down! It had even performed flawlessly on our last flights from Arlington to Manassas, Virginia, then it's final airshow at the Udvar-Hazy Air & Space museum. That XV-15, N703NA, had accumulated 680 hours in its lifetime, and had less than one-hour of flight time left on its airframe when we left it. Our favorite aircraft had given all it could give, and was totally used up in the process. Nothing was left in its tank! I'm very proud and honored to have been one of the pilots who flew it to its forever home. I think the XV-15 is in good company sitting there alongside so many legendary aircraft. I also hope it has forgiven us for leaving it there.

The V-22 Osprey

All the hard work put in to demonstrate the XV-15 paid off when John Lehman, Secretary of the Navy, directed NAVAIR to sign a contract in the spring of 1983! He had taken a ride in the XV-15 and, like everyone else who had flown it, he loved the unique capabilities it offered. The new program was called "Joint Services, Vertical Lift and Experimental", or JVX. The scope of the contract was massive since it called for the Marines to purchase approximately 550 and the

Navy 50, plus the Air Force requested 200 for special operations and search and rescue. In a nutshell, to accommodate such a huge order, Bell Helicopter and Boeing Vertol joined forces to make it happen.

I still recall the basis of what the military wanted the new tiltrotor aircraft to be capable of. First of all, it had to be able to carry a crew of three, plus 24 battle-equipped Marines, off the deck of a carrier to a shoreline some 50 nautical miles away and be able to make two complete roundtrips without refueling! It also had to be able to cruise at 250 knots, have a "dash" speed of 300 knots and it had to be capable of operating on the Navy's existing carriers. In other words, the new aircraft would not require development of new carriers from which to operate. It also had to be capable of deploying some 2,400 nautical miles with special fuel tanks that could be rolled into the fuselage. And oh, by the way, it also had to utilize the latest technology, such as computerized "fly-by-wire" flight controls, instead of standard bell-cranks and push-pull rods, such as those on the XV-15. One of the grizzled old test pilots suggested the "fly-by-wire" lines be strung directly through the control tubes, but he was ignored. It also had to be able to fly in any weather, day or night, plus be able to operate in poisonous gas dispersion by being slightly pressurized. All those requirements were clearly a quantum leap from capabilities of the little XV-15 demonstrator!

The XV-15's flight controls had been developed by a genius who must have been locked up in a "rubber room" at night to prevent him from pulling his hair out, or escaping the clutches of Bell! I'm still amazed at the complexity of that system, but it worked, and it worked very well! As far as being all-weather, forget about it! We flew only in good weather, but that's how concept demonstrators are designed! Suffice to say, to move from the archaic XV-15 to a new state-of-the-art tiltrotor was a massive challenge for the newly formed Bell/Boeing company, but they rolled up their sleeves and went to work!

Somewhere along the way, the new JVX tiltrotor also picked up a new name. In typical military fashion, it was given a letter, "V" for

vertical takeoff and landing, and the number 22, or V-22. Later in the program, I can recall Bell/Boeing had a competition for what the new V-22 would be called. Bell test pilot Mort Meng and another gentleman received a neat model of the V-22 for their entries of Osprey. I've also heard John Lehman himself favored that name, but I guess it really doesn't matter who selected it. The V-22 had become the Osprey!

March 19 is an easy date to remember for me since that's my birthday! It's also the date of the V-22's first flight at Arlington, Texas, on March 19, 1989. Since the V-22 had been a joint effort between Bell Helicopter and Boeing, the pilots had been a split crew: Bell pilot Dorman Cannon and Boeing pilot Dick Balzer. Both had been superb pilots and excellent choices for the program. The first V-22, Ship 1, had an ejection seat just like the XV-15, so we always had to keep that in mind when doing chase flights. We only flew up close when the test crew had asked us to get in to look at some specific object, then quickly pulled away into a comfortable position. You could never relax during those chase missions! It was a constant symphony of coordinated work between both pilots in the cockpit. In fact, all the pilots I flew with were superb and we worked together as one; it was awesome just to have been part of that! All that coordination was essential because you always had to be prepared for the "unknown unknowns" that could occur in a heartbeat!

My first chase flight for the V-22 didn't occur until September 29, 1989 and it was in a Bell 214ST helicopter, not the Citation 550! In the flight test world, you always crawl before walking, so that's what that flight represented. However, we began stepping out a little faster on October 3 when I flew chase in the Citation 550. We began flying more and more, and I cannot recall how many chase flights I made during the entire V-22 program. It had been a fantastic journey just watching the growth of the V-22 and observing how much potential it represented! Sadly, my last flight in the Citation 550 was on August 3, 1994; my time as a jet pilot had ended. During that time, I accumulated well over 1,000 hours just flying chase missions and executive transportation. It had been quite a ride!

In my discussion about the tiltrotor, I'm aware of the crashes that occurred during the development of the V-22. In fact, I should say painfully aware, because I lost several friends who had been crewmembers in those crashes. In the eyes of many, the "fork should have been plunged deep into the V-22 turkey" program long ago, but that is an entirely different story to be told by others. Fortunately, the Marines never lost faith in the V-22, and ultimately, they persevered to keep the program alive. I think their faith in the program resulted in a battle-tested aircraft that changed their entire battle doctrine and seems to be on the must-have list of other countries as well! I take great pride in my contribution to that program. Also, working alongside the Marines left me in awe of what being a U.S. Marine is all about! It was my great honor and privilege to have worked alongside them for so many years.

Bell test pilot Chuck Anderson and me standing beside Bell Helicopter's first company jet, a Cessna Citation I that we had just flown down from Wichita, Kansas (03/19/1982).

I'm flying a typical chase mission for the U.S. Marine V-22 Osprey in Cessna Citation II jet airplane (1991).

Bell test pilot John Ball and me standing beside Cessna Citation II jet at U.S. Navy's Patuxent River, Maryland base after delivery of V-22 Osprey (12/19/1993).

Ron Erhart (Bell's Chief Test Pilot), me and Bill Duncan standing beside Citation II jet chase aircraft used for V-22 flight test operations.

I'm strapping into Bell's XV-15 Tiltrotor in preparation for flight testing at Bell Helicopter's Flight Research Center located at Arlington, Texas airport.

We're taking a break during refueling stop on demo tour in Bell's XV-15 Tiltrotor (04/26/1990).

I'm with XV-15 test pilots Dorman Cannon, Don Borg and flight demo support crew at Meigs Field Airport near Chicago, IL; Sears tower is in background (08/03/1987).

Crash site of Bell's XV-15 Tiltrotor (N702NA) at Arlington, TX municipal airport (08/20/1992).

Bell's XV-15 Tiltrotor (N703NA) and flight test crew at Key West, FL for demo to U.S. Coast Guard (05/19/1999).

Bell test pilots Roy Hopkins, Ron Erhart, Dorman Cannon and me after Roy and I had just delivered Bell's last flyable Tiltrotor (703NA) to the Smithsonian's Steven F. Udvar-Hazy Center at Dulles International Airport. Ron and Dorman made the first flight of the XV-15 on 05/23/1976; Roy and I made the last flight on 09/16/2003.

CHAPTER 17

Interesting Flight Tests

As an experimental test pilot at Bell Helicopter from November, 1980, until my retirement on July 31, 2005, I flew over 4,100 hours of experimental test flights, and had my share of funny, scary and downright horrifying moments, such as the ones described in this chapter.

120% Over-Torque Test in a Bell 212

A pilot is always acutely aware of what his torque and engine temperature limits are because closely monitoring them will prevent him from exceeding specified limits set by the manufacturer. Simply put, torque is a measurement of transmission power in a helicopter and temperature is measurement of engine power. I don't recall what the purpose of the tests were, but many years ago, I had been one of several pilots assigned to a Bell 212 over-torque test; in fact, it had been a 120% over-torque test! We had all been quite shocked because a pilot was always limited to 100% torque. I still can't recall what the purpose of those tests were because pulling such high power could be very risky!

To accomplish such a test, the helicopter had to be restrained to the ground by large chains linked to massive iron tie-downs anchored in tons of concrete. They also had to pump pure nitrogen, a frigid inert gas, directly into the engine to prevent over-temps of the engines. Actually, all things considered, none of the pilots had been too keen on participating in such a test, but our intrepid Bell engineers felt it had to be done; consequently, the pilots' concerns were overruled!

The site selected for such an unorthodox test had been very close to Bell's Electromagnetic Compatibility (EMC) "shack", a facility where they tested a wide variety of avionics equipment for

Electromagnetic Interference (EMI) to ensure electrical impulses emitted from them didn't interfere with each other when being used. For some reason, it had been built like a blockhouse that resembled some World War II bunker. I don't know the reason for that, but I would have to say it had been a safe place to be during those over-torque tests!

It was probably on the third or fourth day when I found myself in the cockpit dutifully grinding away, eagerly waiting for my assigned hour of ground run to be over. Ground runs had always been boring, since all a pilot did was start the engines, ensure everything was "in the green" as far as the instruments were concerned, then pull up on the collective until you felt like you were pulling the guts out of those engines! It was definitely not normal sitting in there while those engines shrieked in opposition to such torture, plus the main rotor and tail rotor were howling in their own disgust at the whole thing. It really hadn't been a good place to be in the first place, then suddenly, it got worse!

I had just looked over at the flight test engineer to make sure he was awake, when a tremendous explosion went off directly behind me! I immediately glanced at the instruments and about the same time a fire warning light began flashing! As soon as I focused on the engine instruments, I knew immediately the number 1 engine had blown up! It couldn't have been more than a couple of seconds before I rolled both number 1 and 2 throttles to the off position, flipped off the fuel switches, then pulled the fire handle for number 1 engine! It had all occurred in just a matter of seconds and that's when the flight test engineer sitting in the copilot seat "came to life" and began screaming to shut down the engines just about the time our control tower was telling me we were on fire! While trying to digest all that, the mechanic yanked open my door to tell me the same thing! That's about the time the young engineer and I decided it would be much safer watching all the action from a good distance away! Fortunately, it wasn't long before the fire truck roared up, the firemen bailed out, then set to work extinguishing the flames.

It was pretty hectic for a while, but soon everything calmed down to the point we could walk back up to the helicopter and discuss what happened. I don't think there had been any doubt about what caused the engine to explode, because we had abused the engines beyond the norm. It was far more interesting listening to the stories of those who had seen the entire episode. Heck, I even began to wish I had seen it myself! Everyone said they thought a bomb had exploded because of the tremendous explosion and resulting fire, plus shrapnel going in every direction, even making indentions into the blockhouse! The other pilots were pretty blasé about it, but they did thank me for putting an end to a stupid program.

"Explosive Nodal Beam Bottoming" in the 222

Beginning with the popular Bell 206-L model, several Bell helicopter models were built with what we call nodal beams. Their sole purpose was to provide pilots and passengers a smooth ride in much the same manner a suspension beam mounted on the axle of an automobile works. The suspension beam of an automobile is attached between the bottom of the car and axles, and its purpose is to provide a buffer between the automobile and all the potholes and ruts. Conversely, the nodal beam serves as a buffer between the roof of a helicopter and transmission to which the main rotor is attached. It provided a smooth ride by eliminating the bumps and vibrations of the main rotor as it flies through gusty air. All things considered, it did its job remarkably well, except for those few times it didn't. One such incident occurred in January 1991, when I was flight testing a Bell 222 helicopter.

I really hadn't liked the way the flight test engineer had decided to use an old rotor blade that had behaved in much the same manner as a tired old bull, wanting to be left alone! I say that, because It had been one of those stubborn blades that could never be properly balanced to provide a smooth ride, regardless of which helicopter it had previously been mounted on. Consequently, it had been removed from service and stacked in the corner of the hangar where it had gathered dust and spider webs! However, since any rotor blade used for flight test programs could never be sold to a commercial operator, the Bell folks really hated to put new blades on a flight test aircraft, and the only blade available was the old blade that had been difficult to balance. So, it was resurrected from its nest of rats and cobwebs in the corner, dusted off, and attached to the main rotor system of our flight test helicopter.

It had taken several flights, but the mechanic had finally forced the blade to behave as it was supposed to, and it provided a relatively smooth ride. Without getting into details of what it takes to balance a rotor blade, let's just say sometimes it was akin to forcing a stubborn bull to do something against its will. The kind of bull a cowboy might ride in the first round, but he always knew the bull remained alert, just waiting for a chance to pay him back when he least expected it! Rotor blades sometimes react in much the same way, especially when the mechanic had to force that difficult blade into a spring-loaded position it really didn't like. Kind of like it did on the day of my own experience.

Making matters worse was the fact we had to dive the aircraft to 1.1 Vne (Velocity, never exceed) to test a set of new nodal beams that had been mounted on the test aircraft. Since the helicopter had a Vne of 150 knots, we had to dive to 165 knots to achieve 1.1 Vne! By way of explanation, if your automobile had a maximum safe speed of 100 mph, that would be the Vne, or the fastest you could safely drive your car. Assuming you had to do drive to 1.1 of 100 mph, that would be 110 mph. That's the speed you would attain to ensure a safety margin for the 100 mph.

Our test aircraft was heavily instrumented, which simply means we had wires running everywhere, each attached to critical components so Bell's engineers sitting back in Bell's TeleMetry (TM) control room could monitor how everything was functioning. The control room itself looked very similar to NASA engineers sitting in the famous Houston Control Center while monitoring a rocket blasting off at Cape Canaveral. The only difference was the control

room at Bell Helicopter's Flight Research Center at the Arlington, Texas airport was much smaller than NASA's. However, in my opinion, those Bell engineers were every bit as good as NASA's, and pilots loved having them watching out for our well-being!

On the day of the tests, we took off from the Arlington airport amid gusty winds below 3,000 feet Above Ground Level (AGL), but much smoother at higher altitudes. In order to achieve a higher-than-normal speed in a helicopter, you usually had to climb to a higher altitude, then dive at a fairly steep dive angle to achieve such a speed. In that case, I climbed to a much higher altitude because my plan was to accelerate slowly. You always proceed slowly during flight testing, because you always knew there was a speedbump lurking out there somewhere, ready to blossom into a full-blown mountain of trouble!

At 8,000 feet, I pitched the nose over and began my dive to accelerate to 1.1 Vne. It's always exhilarating to begin a dive from such a high altitude, sort of like a ride on a roller coaster! Down we went, rapidly accelerating at first, then slowing down so I could sneak up on my target airspeed of 165 knots. You never want to blow past any target airspeed, because that's where bad things can happen! On that day, all went well until we hit 160 knots. As we slowly crept up to 165 knots, we hit gusty air that caused us to rapidly accelerate past the target airspeed, and that's about the time all hell broke loose!

The helicopter began vibrating violently and it sounded like a dozen angry men had attacked us with sledgehammers! It was a crescendo of noise so loud, it engulfed us to the point it was hard for me to even think, much less react! I also was having trouble pulling the cyclic stick rearward, because the flight controls had become hard to move! I was so involved in wrestling with the controls, I barely heard the engineers in the control center screaming for me to slow down, because everything was "going off band edge!" That meant instead of seeing small squiggly lines on the graph paper, they were seeing those normally docile lines gyrating wildly from one edge of the graph paper to the other! In the blink of an eye, we had gotten into the type of situation where luck had suddenly become more important than experience! The speed bump had suddenly become a mountain!

Desperately trying to get everything under control, I was finally able to pull the cyclic control rearward, thus forcing the nose of the helicopter back up to a level attitude! After what seemed like forever, the helicopter finally began decelerating to a normal airspeed and the hammering slowly diminished, then disappeared completely. That's when I finally heard the engineer in the control center ordering me to "land immediately!" I had been so focused on getting everything under control I had totally tuned out everything, including radio calls. When I finally found my voice, I asked what they had seen during our wild ride. His response was: "We may have over-stressed everything and you probably need to land!" Now that got my attention, so I began an immediate descent before I even responded to his comment.

When we had descended to 200 feet AGL, everything seemed normal, so I made the recommendation to continue inbound toward Bell's Flight Research Center since it was only a few miles away. I figured I could always land immediately at the first sign of trouble and they agreed with my decision. Fortunately, we made it back to Arlington airport with no further trouble. After landing and shutting down the engines, we exited the helicopter, opened the engine cowlings and were stunned at what we saw!

It looked like an eagle with giant talons had attacked the helicopter because the arms of the nodal beams had flailed upward with such tremendous force they had hammered against the transmission, gouging deep scratches into it! There were no oil leaks, but red hydraulic fluid was seeping from strained cylinders caused by high control forces. If truth be known, there might have been stains in the seats where the flight test engineer and I had been sitting, too! Suffice to say, the aircraft was rendered unflyable for many months after that.

During the post-flight inspection, they discovered the new nodal beams had been installed improperly, plus, they also determined the culprit blade had gone out-of-track as soon as we hit gusty air, resulting in a severe helicopter vibration! The combination of those two events caused the nodal beams to explode into severe vibration of near catastrophic proportions! The engineers casually referred to it as "explosive nodal beam bottoming". It had been a term I had never heard of, and "casual" didn't do it justice! I had absolutely zero desire to even hear of it again!

Later that afternoon, after everything had quieted down, the program manager walked by my desk and asked if I had heard the tape of the cockpit voice recordings of all conversations between the flight test engineer and myself. When I answered "no", he then motioned me to follow him to the TM room so I could listen to it. I could hear when the flight test engineer had keyed his "mike" and then let loose with a scream that sounded like a high-pitched Mickey Mouse! I think he was asking "what the hell is going on?" It was almost impossible to decipher because we could also hear the sound of the nodal beams pounding on the helicopter! It had been a sound I never wanted to hear again, and I really couldn't blame the flight test engineer's reaction, because that had been his first flight as a flight test engineer! Actually, if I hadn't been so busy wrestling with the flight controls, we probably would've been harmonizing in mouse-speak together!

AH-1W Marine Cobra Crash

May 1, 1992, was not only a beautiful spring day, it was also the day I witnessed one of the most bizarre accidents I have ever seen! It involved a twin-engine AH-1W Super Cobra, the primary gunship helicopter for the United States Marine Corp (USMC).

Even though I was assigned to Bell's Flight Test Research Center at Arlington, Texas airport as an experimental test pilot, I had flown to Bell's main facility earlier in the morning for a project meeting. After completing the meeting, I wandered down to the production pilot's office to hitch a ride back to Plant 6. At the time, the pilot's office still had a large V-shaped protrusion made of glass jutting out from the brick wall that had been utilized in Bell's early days as sort of a control tower. I say control tower because that's where Claude Goode, Bell's first aircraft controller, had sat with a clipboard in hand to keep track of takeoffs and landings while controlling helicopter traffic during the late 50's-early 60's. The V-shaped window was very rudimentary in design, since Claude had to twist and turn his head quite a bit just to monitor existing traffic. Despite its shortcomings, it had worked well until a control tower had been built. That's where I was standing, since it had a nice little shelf to lean on, just about waist high. It was also a good place to monitor helicopter movements on Bell's massive flight ramp.

On that particular morning, I could easily see a large AH-1W Cobra helicopter sitting on a pad located on the flight ramp not far from where I was standing. It's rotor blades were turning at high RPM, and I recognized Jon Honaker, Bell Helicopter's Chief Production Test Pilot, at the controls. I could tell he was preparing to lift off, just by the way the blades were beginning to cone upward, and the skids had begun to slide ever so slightly as he pulled the collective upward to lift off. I also mentally imagined John making an obligatory call to Bell's control tower for permission to lift off into a stabilized hover before taking off into forward flight. Just as the aircraft was about to lift off and break ground contact, one of the production pilots sitting behind me called out, and I had just started to turn when I heard a tremendously loud explosion on the ramp, closely followed by a secondary loud bang in the south hangar! My immediate reaction was to duck, even while turning my head back toward the flight line to see what was going on! What I saw was astounding!

The Cobra appeared to be about 10-feet off the ground, momentarily suspended in air, and there was no main rotor attached to it! All I could do was stare in disbelief! A split second later, the aircraft began dropping rapidly before slamming onto the concrete ramp and

rolled onto it's left side, finally coming to rest on it's left-wing stub. A brief second or two later, the huge transmission, with blades still spinning, hit the ramp no more than 50-feet from the crashed Cobra! The flight line mechanic and electrician who had been assigned to that aircraft started running toward the Cobra to assist John, who was already starting to unstrap his seat belt and shoulder harness.

Unfortunately, as they raced toward the damaged Cobra, an assortment of various-sized aircraft parts began dropping like hail all over the flight ramp! There was so much metal debris falling, the would-be rescuers quickly reversed course and began darting and dodging, not unlike a jackrabbit trying to escape the jaws of a hungry coyote! It was sort of amusing watching such "artful dodger" aerobics even though it had been a very unamusing situation. Since the aircraft cockpit was literally at ground level, John had little problem exiting the aircraft and started running for a safe place, but no one seemed to know where that safe place was! With so many metal pieces drifting down from the lower stratosphere, it was hard to tell. John later said he didn't know "Where in the hell he was heading! He just wanted to get as far away from the crash site that he could!"

After what seemed like an eternity, the last bit of twisted metal hit the ramp and the result was a deafening silence. You could've heard a mouse smacking on cheese from across a football field! It was as if everyone had quit breathing and were trying to comprehend what had just happened, even as they stared at the carnage strewn all over the flight line. Step-by-step, everyone slowly began making their way out on the flight line with heads cast upward to ensure there were no late incoming bits of shrapnel floating down from the sky. No one dared to breathe a word at first, but when they got closer to the aircraft, then saw the huge transmission laying some 50 feet away, there was a crescendo of murmuring, gasping and "oh my Gods!" from all the onlookers.

Jack Ater, a Senior Bell mechanic who had been working on a Bell 222 model helicopter inside the south hangar, called out and said "hey, you guys might want to come and take a look at this." As we walked into the south hangar, everything looked OK until he pointed across the way to the south wall of the hangar. Scrunched up against the wall sat his huge portable toolbox, the kind that must've weighed several hundred pounds with all his tools in it! It looked like it had just gone toe-to-toe with Paul Bunyan and his axe, and lost! The toolbox had been sitting directly alongside Jack while he worked on the helicopter, before some unknown object from the Cobra had cleaved in its entire front side, and driven it about 75 feet across the hangar where it had slammed into the wall! It had been a close call for both Jack and me. In fact, we both said a silent prayer of thanks, because had that as-yet-unidentified object flung from the rotor blade passed just a few feet either side of its trajectory, one of us would have surely been cut in half by whatever it was that slammed into his toolbox! I say that because I was standing just inside a window, while Jack was standing alongside his toolbox when the object smashed into it!

In the final assessment, there really hadn't been much damage to the surrounding environment. A piece of metal had actually travelled several hundred feet across Highway 10 before slamming through one of the Cobra Club's large windows, missing several patrons in the process! Fortunately, aside from a few pock-marked bricks, the Cobra Club's shattered window and Jack's toolbox, there had been very little damage. The really good news was the fact no one had been injured! That was amazing, considering all the metal shrapnel and debris that had been launched in all directions as if fired from a cannon!

A post-accident investigation revealed a 44-pound, mid-span lead weight was improperly installed in one of the main rotor blades, had broken loose, then flung from the blade like it had been fired from a cannon before slamming into Jack's toolbox. The loss of the weight then created an unbalanced blade situation that resulted in the main transmission literally ripping itself from the fuselage of the Cobra, then being propelled upwards by the spinning rotors.

As for John, he quickly passed his physical exam and post-accident check ride, and was soon back in the air as a production test pilot. All things considered, I think John sort of enjoyed his new notoriety because it gave him new bragging rights! He went around bragging that "there are lots of pilots who have flown and logged time in MD Helicopter's NOTAR (No Tail Rotor) helicopters, but I'm the only pilot who could log NOMAR (No Main Rotor) time!" He was, and still is, a great pilot and good friend who is now retired, but I think he still enjoys telling that story. I also think he's maintained those bragging rights because I don't know of another pilot in the world who can lay claim to that statement!

Caught Red-Handed

In November, 1996, I had been assisting a Bell Helicopter Textron Canada flight test crew in the process of conducting Category A flight tests in a Bell 430 Model in Albuquerque, New Mexico. We had already accomplished the same tests a couple of months earlier at Morgan City, Louisiana, and had planned to repeat the same tests at a higher altitude. The Canadians had arrived there a day or so earlier, so when I got there it was a simple matter of moving on with the same test plan we had used previously, in Louisiana. Having worked with the same crew for some time, I had total confidence in their abilities, plus they were just downright enjoyable to work with. Everyone worked hard and we thoroughly enjoyed our time off. It didn't get any better than that!

The pilot I shared the cockpit with was a test pilot by the name of Leo Meslin, an old friend of many years, dating back to when we had made the first flights of the Bell 230 helicopter. He was a first-rate pilot and good friend, and we had always worked well together. That, coupled with the beautiful New Mexico scenery, had combined to make it a very enjoyable trip, even though the development of Category A procedures is probably one of the most difficult tests an experimental test pilot can be tasked with.

Some of the more difficult flight tests dictate instrumentation to monitor critical parameters of various aircraft systems; consequently, a control center is required, and they also must have a radio and an assigned frequency from the Federal Communications Center (FCC). That is very important, because there might be multiple users in a specific area and you don't want one user to overlap another user on the same frequency; hence, the different frequencies.

We had a preflight briefing before the first flight, and I noticed the radio frequency the control center was using had been the same frequency they had used in Canada. Thinking that was rather odd, I spoke up and asked if that same frequency had been assigned to them in Albuquerque as well as in Canada. Their response was that they hadn't bothered to get a new frequency in Albuquerque, because they weren't going to be there long enough; consequently, they would simply use the same frequency they had been assigned in Canada. I accepted that, but was still a little concerned because there could be a local company who had the same frequency we would be using. But hey, what were the odds?

With that all settled, we got on with our flight testing and everything was going extremely well until one day, we had just finished flying when Leo and I noticed a big white van with dark windows pull up almost to the flight ramp, then stop. The fact it was just kept sitting there was ominous enough, but what really got our attention was all the antennas mounted all over it! It looked like it had escaped from a "Star Wars" movie set! And there it was, just sitting there, idling. Neither Leo nor I could take our eyes off it.

After what seemed like forever, the driver's side door slowly opened and out stepped a gentleman wearing white coveralls. He paid no attention to anything around him as he reached inside and picked up what looked like a hot dog & beer tray you see being carried around a baseball park and set it on the hood of his van. Damn! It even had a strap for looping around his neck to carry it! Surely, he wasn't going to try and sell us hot dogs or beer? But then we saw something else. Radio antennas sticking up! The gentleman then

looped the strap around his neck and began walking toward us. He appeared to be turning small dials on the top part of the box while intently watching something on top of the box. He walked right up to us and stopped. He then looked at us, and simply said "gotcha!"

Too stunned to say a word, Leo and I just stared at him, looking much like a deer staring at headlights. He then went on to say he had been hired to find the culprits who had been using the same frequency permanently assigned to a large operator of cement trucks. The same gentleman was also highly agitated because the mysterious "someone" had been blanking out his radio transmissions whenever he had tried to use the radio! In fact, it was bad enough that he was having our "Star Wars" fellow looking for us! But a cement operator? How unlucky could that be? With our luck, he probably looked like Hulk Hogan!

Regardless of his appearance, we were going to meet him, because the mysterious man in white told us he had already called him, and he had already left for the airport! Leo and I did the only reasonable thing we could do. We motioned for every member of the flight test crew to join us on the flight ramp. We remembered the old phrase about "safety in numbers." Shortly afterward, as promised, our unwanted guest arrived, driving up in a big, black Dodge Ram.

The pickup pulled up alongside the white van and stopped. Out stepped our latest guest, all dressed up in cowboy shirt, blue jeans and boots! All six-foot, five-inches of him. He wasn't as big as Hulk, but he sure could've been his little brother! He strolled up to us and without offering to shake hands, said: "Who are you guys and what the hell are you doing here? I want to know, because you've been making my life pretty damned miserable for a couple of weeks!" Just like that, he had gotten our rapt attention!

The test director for our project was a pretty cool guy and he began to explain in rapid conversation who we were and what we did. He went on to say we had been doing flight tests in the same helicopter sitting alongside us. Our visitor then explained he owned a large cement company that managed 62 trucks and for the past two weeks, we had shut him down completely between the hours of 5:00 AM to noon. Well, we certainly couldn't deny it had been us, because that was exactly when our technicians first powered up our equipment and we had always flown until noon, weather permitting. Instead of ranting and raving, our visitor politely asked us if we could work something out so we could both get our work done. Our test director assured him we would change frequencies that very day, and would never bother him again! Our visitor smiled and said "great!"

Gentleman that he was, our test director then invited our new friend if we could take him and his entire family to dinner some evening. The cement man said: "I'll take you up on that! My family and I would love that." He then said: "Reckon we could get a ride in that helicopter some time? I think we would all enjoy that, too!" "You got it!" said the test director. He arranged a ride with a local Bell operator, and we became pretty good friends with the man who could have made our life miserable. Instead, he had been a gentleman who had chosen to take the high road and we became friends during our stay there in Albuquerque. He even joined us a couple of times to share a beer or two. Don't get much better than that!

Hard-Overs at 12,500 Feet

Autopilots basically relieve a pilot from constant hands-on control of an aircraft by responding to his commands programmed into a flight computer. Every airplane in the airlines has that capability, as do smaller business jets and other comparable twin-engine airplanes in the marketplace today. In fact, many of the larger helicopters flying today also have them, but smaller helicopters usually do not. That's because the process of developing an autopilot is not cheap! Consequently, there has to be a good reason to develop one. That good reason for Premier Aviation had come about because the new owner of Bell 407 helicopter had decided he needed

an autopilot to assist him while flying from his home in Seattle to his ranch in northern Idaho.

Premier Aviation was a small company located at the Grand Prairie, Texas municipal airport. I had been assisting them on a part-time basis when they asked me to attend a meeting to discuss developing an autopilot for a Bell 407. My first response to their inquiry had been to "forget about it, because it would be too difficult and expensive to achieve!" They readily agreed with my comment, then went on to say they had signed a contract to develop an autopilot for a Bell 407 model helicopter and they intended to fulfill the contract, with or without me. Seeing how determined they were, I shrugged and said: "If you're crazy enough to do it, I reckon I'm crazy enough to help you." I think my hesitancy had been based on the fact I knew it wouldn't be an easy thing to do! I knew some of the required flight tests would be downright scary, and one of the first things that came to mind was hard-overs!

Generally speaking, autopilots fly aircraft by controlling actuators attached to the aircraft's flight controls. It only takes small movements to control even the largest aircraft, but in the FAA world of development, you always had to play the "what if" game. If actuator rods control flight controls with small movements, the obvious question is "what if" the actuator just happened to have a hiccup and drove the actuator rod to its maximum length? The FAA even provides guidelines for accomplishing such adverse events, and they're referred to as "hard-overs." The maneuver is accomplished by a flight test engineer utilizing a little black box similar to an autopilot to "inject" a hard-over to drive the rod to its maximum limit. Since our tests involved a helicopter, such an injection by the flight test engineer would direct the helicopter to pitch up or down; roll left or right; or yaw left or right. Making matters worse, the FAA also stated that any hard-over must also take into account pilot recognition time, and that's what made it so downright scary!

Once the flight test engineer injected a hard-over in any direction, the test pilot couldn't react until three seconds, plus an additional pilot recognition time (normally ½ second) elapsed. So, let's say you're doing a nose down hard-over. As soon as the flight test engineer injects the hard-over to direct the nose of the helicopter to pitch down, the aircraft reacts immediately by nosing over at a very fast rate! In the meantime, the pilot can do nothing until approximately 3 ½ seconds tick by! Imagine being in a rollercoaster as it pitches over the first precipice and you can't grab for a railing to hang onto for at least 3 ½ seconds! Then imagine what it must have been like if you had been doing it for the very first time, just like the flight test engineer that flew with me during those first flight tests in May, 1997.

The young flight test engineer had already flown with me to accomplish all the hard-overs at lower altitudes of 3- and 6-thousand feet, but those hadn't been so difficult because there were still adequate cues on the ground he could see. However, the FAA required us to accomplish hard-overs at the maximum altitude the applicant is requesting. Since the customer wanted to fly over 10,000-foot mountains along his route, we had to repeat everything at 12,500 feet! Doing hard-overs at lower altitudes had been stressful enough, but doing them at 12,500 feet was just downright scary, even for me!

During our briefing prior to takeoff, we discussed at great length how we were going to accomplish our mission. In fact, since we had been wearing parachutes, we also discussed how we would execute an emergency exit from the helicopter in the event the rotor should impact any part of the helicopter, causing it to come apart! I also explained the absence of any cues, and how flying at higher altitudes is akin to flying in a fishbowl, at least it was for me. I also gave him the opportunity to exclude himself from the mission, but he was adamant about going. He wanted to do it, and he assured me he would have no problems; consequently, based upon his past performance at the lower altitudes, I had no reason to doubt him.

Having to wear parachutes in such a small cockpit was always uncomfortable, but necessary in the event I made a mistake. It was

confining to both of us, but at least the engineer had room to place the all-important black box in his lap to inject hard-over commands. After departing the Grand Prairie airport, we had to fly about 35-40 miles to the south before initiating our long climb to 12,500 feet. Soon enough, we began our climb and were soon cruising along at our desired altitude. The thing is, I don't know of too many helicopter pilots who are comfortable flying at higher altitudes, myself included! I always felt as if I were flying in a fishbowl while trying to balance the helicopter on the pointed end of a needle!

Glancing over at the engineer, I could tell he was already nervous about just being up there, and I can't say I blamed him. I even flew around for a few minutes to acclimate ourselves to our new altitude environment before asking him which hard-over he wanted to do first. His response was to set up for a pitch down hard-over that would drive the nose of the aircraft down. I acknowledged that, then automatically checked all aircraft instruments, including a new one attached to the cockpit directly in front of me. It was a very large timer with a sweep-second hand, and that would be my focal point for approximately 3 ½ seconds before I could recover from the hard-over. If I didn't recover immediately after those few seconds, it would probably be too late; that's why we had parachutes.

After getting set up for the first test point, I advised the engineer to give me a countdown of three seconds before injecting the hard-over. He acknowledged me, then began his countdown of "three, two, one – now!" The helicopter pitched over immediately and with my eyes focused on the timer in front of me, I excluded everything else that occurred inside the cockpit! After a long 3 ½ seconds, I recovered from the maneuver and quickly returned to normal flight. Everything had gone well, and the young engineer had not uttered a peep throughout the entire maneuver! I had been both surprised and pleased at his silence, but then I turned to look at him.

His eyes were bulging outwards from a contorted face that

had turned greenish-blue from lack of oxygen! He was no longer holding the black box because both hands were wrapped around his throat as he gasped for air. The black box now lay perilously close to the flight controls! In a crowded cockpit 12,500 feet in the air, my young engineer had hyperventilated! I reached over with my left hand and began pounding on his back while yelling at him through the microphone. Getting no response, I quit pounding on him, grabbed the throttle and snapped it to a ground-idle position to enter autorotation. That was the quickest way to get back down to a lower altitude where there was more oxygen to breathe! Then I began slapping his back again while screaming "breathe, breathe!"

After what seemed like forever, he finally gasped and his head flopped back against his seat, but he was still out of it! After another few seconds elapsed, his breathing improved, and some color had begun returning to his face. The crisis was over, but he was still a bit wobbly when we landed, so we sat for a few minutes while he recovered enough to walk. After finding his voice, he informed me he was through as a flight test engineer. I didn't try to talk him out of it and I've always thought he made a good decision. As far as I know, he has never flown again as a flight test engineer, but he's still a very fine engineer that I hold in high regard.

Triple Hard-overs

A sharp young man by the name of Tonka Hufford replaced the engineer who decided he had enough of hard-overs. I've always respected the young man who had "opted" out of flight testing, and still hold him in high regard today. However, Tonka turned out to be the absolute best replacement I could have ever had. Not only was he a very bright and talented young man, he was also fearless. I say that, because he picked up on doing hard-overs and we never skipped a beat after that first attempt at 12,500 feet. It had taken a while to complete, but we finally completed all hard-over testing and submitted the results to the FAA in hopes of obtaining their approval. As it turned out, they really liked what we had done, but!

The "but" part turned out to be even more exciting than what we had already accomplished!

It seems we hadn't taken into account random Electrical Magnetic Interference (EMI) from such things as high-powered electrical lines, radars, etc., that could have possibly resulted in multiple hard-overs. The possibility of such an occurrence seemed rather remote, so we argued mightily against it; however, the FAA insisted we take their concerns into account. Losing our argument, we put on our thinking caps, and using their methodology, we said "what if" we accomplished multiple hard-overs at pre-selected test points we deemed critical to resolve our dilemma. When the FAA accepted our proposal, I think both of us immediately regretted proposing something so stupid as triple hard-overs, primarily because we then had to go out and accomplish them!

As far as I know, no one had even heard of triple hard-overs, much less performed them! However, since the FAA agreed that had been a good way to resolve the issue, we just had to come up with a plan to accomplish triple hard-overs! Actually, it had turned out to be quite simple. Premier's engineers rigged up the black box so the flight test engineer could dial in multiple, simultaneous hard-over "injections" for any axis! For instance, we could set up for a pitch down, roll left, and yaw right, or pitch up, roll right and yaw left. We could do any combination we wished to accomplish. We had the means, then all we had to do is decide which ones and how many to satisfy the FAA.

The FAA agreed they would require no more than 20 test points, provided we could prove they represented the worst scenarios possible, yet the pilot could still recover from them. In October, 1997, Tonka and I went to work and after several flights and multiple hard-overs, we finally selected 20 of the triple hard-overs we felt represented the worst case for a pilot to encounter.

Since everyone was satisfied with our selection of triple hard-overs, Tonka and I went about accomplishing each and every one of them. It certainly hadn't been easy, and more than once Tonka and

I both threatened to "jump ship" due to the stress of having to do triple hard-overs! Hard-overs causing the nose of the helicopter to pitch down, while simultaneously rolling left and yawing right, then riding through all that for 3 ½ seconds had not been for the faint of heart! I think the FAA folks agreed with us because they accepted data from company flight tests rather than conduct the maneuvers with us.

A good way to describe a triple hard-over experience is to watch bull riding events at a rodeo. The toughest bulls are those that launch themselves out of the chute, leap vertically, then spin while their massive back rolls either left or right to unseat the cowboy unlucky enough to have drawn his number! Tonka and I rode through something similar day after day, and throughout the entire flight test process, I never heard a peep from Tonka until we had to endure what had been the last triple hard-over we did.

He was sitting in the rear seat to replicate a heavy aft flight condition and when he injected a hard-over and we began contorting in multiple directions, he thought he saw a powerline flash by and let out a single, loud yelp! We both had a chuckle about it later, but the way I saw it, he certainly had earned the right to let out a yelp! He had stepped into the middle of an extremely difficult flight test program and never missed a beat! He still has my utmost respect for what he accomplished and if I had to do it all over again, he would be my first choice as flight test engineer.

Ghost in the Helicopter

In late July, 1997, one of our Canadian test pilots had just landed after ferrying a new Bell 430 model helicopter all the way from Canada to the main Bell plant in Hurst, Texas. After shutting down the helicopter, he walked directly into the production pilot office and announced in a shaky voice that "something is wrong with that helicopter!" He then went on to say not only did it feel funny, it had also let loose with a loud mournful cry like a ghost moaning! Before laughing at such a story, the staff pilots had taken one look at his

ashen-faced appearance and knew immediately he was not kidding around with them!

The chief production test pilot assigned one of the more senior test pilots to go and fly with the shaken pilot, hoping they could repeat what he had just described. It hadn't been a comfortable flight, due to the hot and gusty conditions, but sure enough, when they accelerated to a cruise airspeed, the helicopter's main rotor blades began weaving up and down in a strange manner. The helicopter began vibrating and they could hear what began as a rumbling sound that grew in intensity until it became a loud roar! The startled pilots immediately reduced airspeed until the mournful cry dissipated, then executed a quick 180-degree turn and beat feet back to the Hurst helipad, an excellent decision!

The next call was made to Bell's engineering department to advise them of what the two pilots had just encountered. I ended up on the program, mainly because I had spent a lot of time in Canada flying the Bell 430 in the development phase and was pretty familiar with it. However, once I heard what had occurred, I had to admit I didn't recall encountering anything remotely like what had been described. In fact, I began having second thoughts about even getting involved with an aircraft that made moaning sounds like a ghost!

Early the next morning, the production test pilot ferried the "haunted" Bell 430 from Bell's main plant to Bell's Flight Research Center at the Arlington, Texas airport. Since it was early morning, it was relatively cool, plus the wind was virtually calm. I climbed on board while the helicopter was still running so we could accomplish our flight test before it got too hot. After lifting off and flying out to our flight test area, I began diving, turning and climbing in an attempt to repeat what the two pilots had experienced the day before.

Surprisingly, the helicopter remained very docile, and the vibration level was virtually non-existent, a characteristic the Bell 430 was renowned for, and we did not hear anything that remotely sounded like a loud roar, or a ghost moaning! We could have declared success at that point, but two experienced Bell test pilots had personally observed the strange phenomenon. They had also been somewhat shaken by the experience, so we knew something had happened. The problem at hand was to simply find out what it was. In the meantime, Bell Helicopter issued instructions to reduce the Vne (maximum speed, never to exceed) of the Bell 430 to 120 knots, a big-time hit for the owners and operators of the helicopter.

The first step was to instrument various parameters of the Bell 430, so we could monitor critical components while doing flight test maneuvers. While that was going on, the flight test engineer and I began working up a flight test plan that we hoped would ferret out the culprit causing all the trouble, but it proved to be akin to finding a needle in the haystack! It wasn't long before the helicopter was rolled out of the hangar, fully instrumented and ready to go. We were ready to begin searching for the ghost hidden somewhere within the confines of the helicopter.

Whenever we conducted flight test operations, we always began flying around sun-up to take advantage of cooler temperatures and calm winds. That way, if the aircraft began to "wiggle-waggle" when flying in calm winds, you knew something on the aircraft was making it happen, not hot gusty air. That's what we had hoped for, but unfortunately, we flew several mornings with nothing to show for our best efforts. The Bell 430 helicopter had flown just as I had remembered, and the ghost remained happy and hidden somewhere deep inside the bowels of the helicopter!

Then, on August 12, 1997, we had a minor maintenance problem on the BH430 that delayed our departure and instead of departing at sun-up, we didn't get airborne until much later in the morning. Consequently, it was much warmer, plus hot gusty winds were starting to swirl, as we flew toward the flight test area, then climbed to an altitude where it was cooler and not as gusty. Again, we dove, climbed, turned and descended in hopes of detecting anything that might expose our problem, but everything was normal!

After another disappointing flight, I called the control center and advised them we were returning to the airport.

After spiraling down from altitude to about 500-feet above the ground, it quickly became much hotter, and gustier. Hoping to get back to the airport in time for a late lunch, I had accelerated to an airspeed of about 135 knots in the hot, gusty air when the main rotor began weaving in a strange fashion and we began to feel a strange vibration I had never felt before! Just as I was trying to get a grasp on the situation, we heard a loud rumbling sound that soon became a loud roar, or moan! Looking upward toward the main rotor, I was horrified when I saw how much the tip path plane had split apart, an indication the main rotor blades were out of synch with each other!

I knew immediately we had a serious, not-so-good situation on our hands, so I lowered the collective in a desperate attempt to slow down, just as I heard the control center screaming: "Land! Land!" I really didn't need to hear that, especially since I already had my hands full trying to control an aircraft that had gone from a tabby cat to wild beast! Fortunately, I was able to get everything under control and we literally hovered back to Arlington airport and landed! I could tell everyone was concerned, because a large crowd was on hand to welcome us home! After doing a postflight inspection, we discovered the short incident had been severe enough to actually have cracked one of the main rotor yokes! We had found the beast with the ghostly moan! Somehow, we had to figure out what kind of beast it was.

It had taken a while, but the culprit turned out to be a combination of malfunctioning main rotor dampers, hot temperatures and gusty air. Those same dampers had been on the Bell 430 during the entire development phase, but the aircraft had been built and tested in Canada where the temperature was always chilly, cold, colder or frigid! The dampers functioned as advertised in cooler weather, but on hot, 100-degree days, the dampers lost their ability to control the blades when flying in gusty air. Then, when it hit the gusty air, the useless dampers allowed the rotor blades to run amok, eventually emitting that horrifying mournful roar, just as it had on our test flight!

Every engineer at Bell Helicopter had been stunned to discover the dampers had lost their damping ability in hotter temperatures, because it had never happened before. Imagine what it would be like driving down a hot, Texas highway in the middle of August and the shock absorbers suddenly stopped functioning. I'm sure it would be a horrifying experience, because the car would careen down the road while you were doing your best to find the brakes, just as it was for me when I met up with the beast on that hot August day!

In the end, we were able to resolve the issue by installing new dampers that were impervious to cold or heat, and we never had another problem after every 430 had been retrofitted with the new dampers. To my knowledge, no one was ever hurt, nor was there ever any damage to a Bell 430 helicopter due to that problem. We owed our success to the Canadian test pilot who had recognized a problem, and hadn't been shy about reporting it. It had been an accident waiting to happen.

Airspeed Calibration Tests in a Bell 214 ST

In late 1999, I was contacted by Petroleum Helicopters, Inc. requesting my services to conduct a series of flight tests in a large Bell 214 ST helicopter destined to go to some Middle Eastern country. They had sent photos to show they had added what looked like a small stubby wing located just below the right-hand pilot's door where they could attach a high-powered searchlight. After looking closely at the photos, I had become concerned about the small wing-like structure being attached close to a static port, a very important element of what is referred to as the pitot/static system. That's the system that provides the pilot with airspeed and altitude, and it actually dates back to the barn-storming days. It has basically remained unchanged for the simple reason airspeed and altitude

are critical to a safe flight, and we needed to ensure the new structure had no effect on that system.

Based upon my observations, I advised the PHI folks we probably would need to perform an airspeed calibration with what we call a "trailing bomb", a small, cylindrical tube that looks very much like a standard pipe closed at both ends, but it was a very expensive one! Unlike a pipe, it's solid, heavy and aerodynamically clean so it can obtain accurate data. It hangs below the aircraft, attached by means of a 75 – 100-foot cable, and it also has a clear, plastic tube leading to an independent airspeed indicator in the cockpit for comparison with the aircraft's indicator. Use of the trailing bomb also required a chase aircraft that can follow along behind the test aircraft to keep the test pilot advised of what the trailing bomb is doing. I had known from experience an aerodynamically clean trailing bomb can do strange things in certain conditions, some of them quite dangerous!

When I finally arrived at PHI's main facility in Lafayette, Louisiana, the first order of business was to have a preflight meeting to discuss what we had to do to ensure a safe flight test program. My first concern was the size of their trailing bomb, because it was much smaller than what I had been used to. A trailing bomb tagging along behind the 17,500-pound 214 ST helicopter (the largest Bell ever produced) producing hurricane-type winds could easily be transformed into a rocket if the right conditions occurred. I expressed my concern to the PHI personnel, but they informed me they had difficulty locating another trailing bomb and that was the best they could do. Reluctantly, I agreed to proceed with the tests and moved on to the next item we needed - a chase helicopter.

I was getting some pushback from the PHI test personnel, because use of a chase aircraft would be expensive, and so they asked "if I were sure we needed one." Out of all the folks in the meeting, I was the only one adamant about the chase aircraft requirement. Undeterred, I informed everyone at the meeting that a chase aircraft was a requirement I would not relinquish, and that not having a chase aircraft would be a no-go item for me! That broke the stalemate, and everyone agreed to a chase aircraft for our airspeed calibration flight tests. We were now ready to proceed with our first flight test!

It wasn't long before I was sitting in the Bell 214 ST requesting permission from the control tower to hover over and pick up the trailing bomb. After the ground crewman hooked up the trailing bomb, I increased power to climb vertically until the trailing bomb cleared the ground and dangled some 100 feet below the aircraft. Satisfied all was well, I called for takeoff clearance for a flight of two since the Sikorsky S-76 chase helicopter and I would essentially become a flight of "one" for the purpose of the flight test. With the chase helicopter close behind, we headed out over the swampy area that lay just east of the airport and quickly established 60 knots as the first test point. That had gone well, so after moving through 70, 80 and 90 knots in 10 knot increments, I advised the chase aircraft I was accelerating to 100 knots, and suddenly, that's when everything got very exciting!

The chase pilot began screaming "abort, abort" before I even had time to achieve 100 knots, but it's difficult to reverse course of a 17,500-pound helicopter once it gets moving in a specific direction! I was doing my best to slow down, but for the chase aircraft crew, it wasn't fast enough because they kept repeating their call to "abort!" After what seemed like forever, I finally managed to reduce the airspeed to 70 knots, where everything became tranquil; the chirping myna-bird calls to "abort" ceased, and the silence was deafening! Accepting that as approval to continue flying, I called the control tower and requested permission to return to the PHI ramp, because we needed to re-boot our thought process!

After exiting the aircraft, I met the entire chase crew in the middle of the ramp and their faces were still chalk-white from what they had seen. They all began talking at once, but soon one of them took the lead in telling me what had happened. It seems as soon as I had begun accelerating, the trailing bomb appeared to have been

launched by a booster rocket as it zoomed upward toward the Bell 214 ST, barely missing the tail rotor and main rotor before running out of energy, then dropping back down to re-energize itself for another attack on our rotor blade systems! Fortunately, I had been able to reduce airspeed in time for the trailing bomb to once again become a docile tag-along object, suspended 100 feet below the aircraft.

At the end of the day, everyone agreed we needed a larger, heavier trailing bomb prior to attempting another airspeed calibration test. PHI test personnel also agreed they would never, ever argue with me again when it had come to flight testing requirements. It seems that being on board the chase aircraft and watching that trailing bomb arch upwards toward the Bell 214 ST had given them all nightmares! I didn't have nightmares, but then, I never saw what was going on back there. I think it's just as well I didn't. On my next trip down to PHI, we accomplished all tests with no problems whatsoever. Gentlemen that they all are, those PHI personnel have never questioned me again when I laid out a test plan of what we needed to do, and how we needed to accomplish it. I also still love trips to Lafayette, Louisiana, home of all those wonderful restaurants with Cajun-cooked specialties, washed down with a cold beer!

Never Fly Over Prisons

In January 1996, I was conducting an airspeed calibration test of the Bell 407 Model's airspeed system using a long, cylindrical silver tube with a cone-shaped, stream lined nose we refer to as a trailing bomb. It's attached by a 75-foot cable, and without getting technical, it is utilized to get an accurate airspeed since it dangles well below and behind the helicopter, thus negating any disturbance of the pressure field around it. Although commonly utilized in the industry, usage of the trailing bomb is considered a high-risk type of test since the bomb has been known to wrap itself around the tail boom, or even strike the tail rotor in worst case incidents. Consequently, we always utilized a chase ship to fly close behind

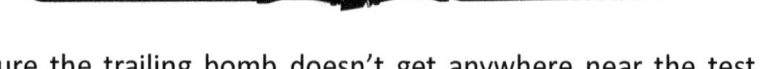

to ensure the trailing bomb doesn't get anywhere near the test helicopter.

The day of the test had been a typically clear, cold Canadian winter's day, whereby the ground was totally covered in snow and ice. I had departed Bell Canada's facility followed by a chase helicopter, and my intentions were to conduct all required tests by flying a large racetrack pattern while remaining within a 10 nautical mile perimeter of Bell's facilities. Everything was going splendidly, but unfortunately, the snow and ice had effectively removed all contrast from objects on the ground, thus rendering everything the same. In a nutshell, I had no idea what areas I was flying over, and that's what prompted a call from the Montreal Air Traffic Controller that began with "helicopter flying over XYZ prison, please say call sign!"

I was so focused on flight testing; I ignored the call until the pilot of my chase helicopter called over the radio to say "I think they're talking about you!" That broke the trance and when I looked below me, I was stunned to see the maximum-security prison directly below me! I had been aware of the prison and had always taken precautions to avoid it; however, airspeed calibration tests required a lot of concentration so I had failed to notice it due to everything on the ground being white. Flying a large racetrack pattern, I knew immediately I had probably flown over the prison numerous times, thus prompting their call. Realizing the error of my ways, I quickly called Montreal Center and gave them my call sign.

"Roger," said Montreal Center. The controller then proceeded to tell me they had received numerous urgent phone calls from nervous guards who were watching two helicopters flying repeatedly over the prison; making matters worse, one of the helicopters appeared to have a rope ladder suspended from it! Consequently, they were concerned we had planned to land inside the prison to whisk away one of their more notorious prisoners! Wishing to put everyone at ease, I responded to the controller that "we didn't have a rope ladder dangling below us. We simply had a trailing bomb since we were doing airspeed calibration tests."

You would have thought I had just told them we were planning to drop a bomb in the middle of Montreal! "BOMB! DID YOU SAY YOU HAVE A BOMB?! YOU MUST LAND IMMEDIATELY!" They heard the word bomb, and nothing else. I tried to explain but they kept screaming I had to land immediately! Finally giving up, I called Bell's control tower for landing clearance in hopes we could make a phone call to straighten everything out. But little did I know how much trouble I had created! It took several days and multiple phone calls to convince the Canadian authorities I had not been flying with a bomb hanging beneath my helicopter. After numerous apologies and promises that "I will never do that again", we finally prevailed. I had learned my lesson! I never again flew over the maximum-security prison, nor did I ever use the words "trailing bomb" again! At least, not in Canada!

The Vomit Comet

Another pilot and I had been working together on a Bell project in July, 2004, and had been having problems with what we call "handling quality issues." Simply put, the helicopter was not flying exactly as it should. Sometimes when you had such problems, the engineers would request a "tufting" test, whereby they would attach 6-inch strands of yarn, or tufts, all over the aircraft. It was sort of funny looking, because sometimes it looked like the aircraft was sprouting fur, kind of like a Chia pet. However, it was fairly efficient in showing how the airflow travelled over the entire fuselage of the helicopter.

When doing such a test, it always required a chase aircraft to assist the pilot in the test aircraft in watching out for other aircraft, plus engineers have to be on board to observe, and video the tufts so they can be evaluated. On the day of our tufted flight test, I was assigned the duty of chase pilot, so during the preflight briefing I asked who would be flying with me as copilot. Immediately the hand of the young engineer who had written the test plan went up. I knew he had been excited to be on the program, but I didn't know

if he had ever flown before. When I asked him that question, he responded "No; this will be my first." He was quite happy, but I was always a bit apprehensive about someone making their first flight.

The day of the test had been one of those miserably hot August days where I would have much rather been somewhere in air-conditioned comfort rather than in a hot cockpit! After I pre-flighted the helicopter, I then began briefing the young engineer copilot, as well as the cameraman who would ride in the rear passenger seat to film the tufts. It turned out that the cameraman had been an old pro who knew exactly what to do, so I focused on the engineer who would serve as my copilot on his first flight. I knew he was eager to see his tufting expertise in action, but I also needed him to assist in watching out for other aircraft that might be in the area. "Okey-dokey"; thumbs up! No problem!

Both helicopters cranked up and were soon hovering into position for takeoff. The engineer had a large legal pad and pen in his hands so he could write down notes about what he was seeing during the course of the test. No problem; however, I did advise him to avoid making rapid head movements back and forth between the test aircraft and his pad because that could induce air sickness, especially on such a hot day. Again, thumbs up! No problem!

After receiving clearance for a flight of two, I quickly moved into position alongside the test aircraft to ensure both the engineer and cameraman had a good view of the tufts. I was occupying the right seat, and since the test aircraft was flying off my left side, I could easily keep an eye on the engineer while he took notes. Unfortunately, he started doing exactly what I had cautioned him about. His head was moving rapidly so he could see the tufts, then write down his notes. That went on for a several minutes before I noticed his head movements had begun slowing down. After several more minutes, he was just looking, and not writing. Then, before too much longer, he had just been sitting there with a glassy look in his eyes. Things had definitely not gone well for my young copilot!

Sensing a problem, I asked if he was feeling sick. His response

was to turn to look at me, then slowly nod his head up and down. Oh boy; not good! I told him to turn the vent on the left door window toward him so he could at least get fresh air blowing over him, even if had been hot. He responded by very slowly raising his hand to adjust the vent as I advised him. I was thinking about my next move when the pilot in the test aircraft suddenly advised TM Control we were aborting our tests due to gusty air quality and returning to Bell's facilities. I had been both surprised and relieved, given my copilot's dilemma. I keyed my microphone to "Roger" his intentions, then took another look at my now ashen-faced copilot who was also sweating profusely! I asked the obvious, "Are you sick?" He nodded his head but could only mouth: "Yes". Definitely not good!

At that point, I told TM control I was breaking away from the test aircraft and landing since I had a sick passenger. Before they could respond I asked my copilot: "Are you going to throw up?" Again, the head nod! Oh boy! Things had rapidly gone to hell in a hand basket, and that's when Bell's TM control requested me to repeat my last transmission. They also added: "Did you say you were going to land because someone's sick?" I keyed the mike to respond while looking directly at my copilot, and just as I said "oh", my copilot upchucked everything he had eaten for the past week! All I could get out was "yeeeeeeeeahhhh!" because whatever had been in his tummy a scant few seconds ago, had suddenly been distributed all over me, courtesy of that damned vent I had just told him to adjust! I was coated down the entire front of my flight suit! It even puddled in my seat! All I could do was grit my teeth and focus on flying the helicopter! I also tried my best to have an out-of-body experience, just to keep from up-chucking myself!

I could hear, rather than see, the cameraman scrambling around in the rear passenger area in his desperate attempt to avoid flying chunks of spaghetti and meatballs, or whatever it was that had been percolating in my copilot's tummy! In spite of the cussing and gagging emitting from the back seat, I remained calm because I had to focus on flying! Meanwhile, my poor copilot didn't have the foggiest notion of where he was, nor did he care! Since it soon quieted down in the back seat, I assumed the cameraman had either passed out, or leaped from the helicopter! Either scenario was OK with me! I also thought it rather odd that I didn't receive another radio call from TM control center, nor did I attempt to make one all the way back to Bell's flight ramp. I was afraid to with so many unidentified food groups dripping down my face, not to mention the smell!

As soon as we landed, the mechanic sauntered up, opened the door and said; "Oh my gawd! I ain't cleaning that s**t out! No effing way!" That's about the time I returned from my serene mental place and the smell was overpowering, so I simply ignored the mechanic's loud rants, and said something real cool like "You ought to see it from my effing seat!" The cameraman had survived, because I could hear him scrambling out of the back seat with all his gear and probably running for the closest exit! My poor copilot couldn't even speak as he climbed gingerly out of the helicopter and squatted down alongside it. I have to admit, at that point, I could have cared less what anyone else was doing. All I wanted was to get out of that flight suit!

After exiting the helicopter, I walked rapidly to Bell's main building at Plant 6 and soon as I stepped in the door, I saw Bell's chief flight test engineer standing directly in front of me. He had been directing the Telemetry (TM) control center, and had actually been manning the radio; in fact, I had been talking to him when I let out with the loud "yeeeeaaah"! He was just standing there, leaning against a wall partition. I looked directly at him and said: "I don't want to hear a damned word from you!" He never responded. He just burst out laughing and actually had to hold onto the wall partition he was laughing so hard! The man was actually crying from laughing so hard! He was standing there laughing uncontrollably when I turned and stomped away, grumbling about getting no respect from the hired help!

After I had taken the only shower I ever took at Plant 6, I got dressed in a new flight suit, then met up with him again and he explained what had happened. When he had asked about someone being sick, my response was going to be "oh, yeah." However, my copilot let fly at that exact moment so all they heard was "Oh, yeeeeeaaahhh!" He then went on to say it had been as if everyone had been watching a movie! There had been twelve of Bell's brightest engineers inside the control center to monitor critical parameters of our test aircraft. When they heard my response, they had fallen over their desks, totally unable to monitor their squiggly lines because they were all laughing so hard! He also said that's why he hadn't attempted to call me while I was flying back in. He said if he had tried, everyone in the TM control room would have probably collapsed with laughter all over again! They all agreed that little episode was the funniest thing they had ever experienced in all their years spent in the TM control room!

In the end, my young copilot had offered to assist in cleaning the helicopter, but I told him to forget it! Stuff happens! The mechanic might not have liked it, but he had a job to do as well! The only real casualty was my flight suit because I threw it in the garbage! As for my young engineer/copilot, I never saw him again, but I never blamed him for the best entertainment anyone had ever provided for the TM control center personnel. Getting so sick inside a stifling hot helicopter cockpit had not been something he had chosen to do, or prepared for. It had just happened, and the only casualty was my flight suit!

Crash site of AH-1W Cobra helicopter at Bell's main heliport in Hurst, TX; the transmission and main rotor are laying directly in front of the fuselage (05/01/1992).

Crash site of AH-1W Cobra at Bell's main heliport in Hurst, TX; view is looking toward crushed left side of fuselage (05/01/1992).

Bell Canada test pilot Leo Meslin and me standing beside a Bell 430 helicopter at Albuquerque, New Mexico (12/02/1996).

Bell 407 helicopter with "fan-tail" tail rotor at Bell's Flight Research Center at Arlington, Texas municipal airport (07/21/2004).

Jeff Jones, Frank Everett and Tonka Hufford standing beside a Bell 407 helicopter that's ready for delivery (05/2000).

Bell Canada test pilot Leo Meslin and me receiving the American Helicopter Society (AHS) Frederick Feinberg award (04/20/1993).

CHAPTER 18

First Flight of (Bell Agusta) BA-609

Being able to make a first flight of a new aircraft was always an awesome event for me. It's something very few pilots have an opportunity to do, and I always felt it was such an honor just being part of such occasions. I've been privileged in my career to have been a pilot in numerous first flights, but I think being on board the BA 609 tiltrotor was a special privilege in itself. I say that because it was an entirely new category for civilian aircraft since the helicopter was certified in 1946!

The first flight of the BA 609 occurred on March 7, 2003, and it had been the culmination of many years of challenging work for a large, dedicated group of Bell Helicopter engineers, technicians and pilots who had all worked tirelessly to make it happen. Long anticipated, and despite the well-publicized problems that had plagued the Marine V-22 (more commonly known as the Osprey), more than 70 original BA 609 customers still had their refundable deposits on the table. Everyone was still supportive because we had conducted an enormous amount of both simulator and tied-down ground run tests to complete a long laundry list of "must do" items to satisfy even the most demanding engineer! Plus, ever since we had conducted the first engine start successfully on December 6, 2002, the aircraft's performance had been exemplary! After years of design reviews, systems development, aircraft buildup and more than 35 hours of ground run tests, the long anticipated first flight was close to becoming a reality.

Roy Hopkins, one of Bell Helicopter's most experienced tiltrotor pilots, was designated as the pilot-in-command, while I would perform the duties of copilot throughout the entire first phase of flight testing the BA 609. Both of us had spent countless hours in the BA 609 simulator to develop the fly-by-wire flight controls, as well as to develop our own skills in controlling the aircraft. As we slowly walked across Bell's large concrete ramp, there were many thoughts dancing about in the back of our minds. But there were no thoughts of failure! Our chatter was directed more toward what we hoped to achieve during that all-important event. Even the weather had cooperated because the day had been a flawlessly clear, blue sky with a light wind less than five knots. Perfect weather for flight testing! I might also add it provided a picture-perfect photo-op environment for literally hundreds of photographers that had migrated into the area surrounding the airport.

Even though we were mentally ready to fly, the first thing we had to accomplish was the most important part of any flight, experimental or not. It's called a preflight, whereby every square inch of the aircraft is thoroughly inspected by experienced technicians and pilots. Finally satisfied with the preflight, I accepted the aircraft logbook from the flight test engineer, read through several items, then signed it. My signature was the last in a long chain of signatures, but it was the most critical. With that final stroke of a pen, I was acknowledging the pilots were accepting responsibility for the aircraft from that point forward. Silently handing the logbook back to the flight test engineer, I acknowledged him with a quick nod of my head before stepping up to the fuselage door, pausing just long enough for one last look around before climbing inside.

As I entered the aircraft, I turned slightly, casting a cursory glance towards an apparatus located just aft of the door, sitting in the center of the fuselage. It was a modification unlike any I had ever seen! It was an emergency egress shroud some 28 inches in diameter and was designed to blow a hole through the fuselage deck! It was an expensive safety feature we hoped we would never have to use, since it would have been utilized only in a worst-case scenario. There had been instances in the past where a flight test crew had survived a crash but were unable to egress the aircraft, and had

perished in the ensuing fire. Consequently, Roy and I had deemed the emergency egress shroud to be necessary, even though there had been another emergency escape hatch in the overhead just behind the cockpit. The only preflight check was to ensure it was there, cocked and loaded!

The cockpit itself reminded me more of a small business jet than a helicopter. However, it did resemble a helicopter in one aspect. The pilot-in-command sits in the right-hand seat rather than in the left seat, the normal pilot-in-command seat in fixed wing aircraft. Since the cockpit was rather small, I had to step over the center console before settling into the left seat, then hit the release handle to slide forward, thus allowing enough room for Roy to settle into the right-hand seat. After we were both strapped in, I picked up the checklist and began running through it. At first, the checklist items I checked off sounded rather routine but soon changed to something more exotic sounding. Items such as ENGINE NACELLES - 90°; FLAPS – AUTOMATIC; FCS AUG – ON, all of which weren't normally found in helicopters. I also ensured the Flight Test Interface Panel (FTIP) was getting power since we would utilize that test equipment during our flight. The final item in the checklist was not only the strangest, but probably the most important!

Even though it simply involved the insertion and turning of a key, it was the hardest part of the checklist to comply with, at least it had been for me. Sort of a mental speed bump to overcome. I say that because the simple act of turning a key armed the emergency egress system, a very docile looking apparatus utilizing an explosive charge encompassing the entire circumference of the emergency exit shroud! It operated in such a manner that when activated, it blew a 28-inch hole in the floor, thus allowing large springs to slam the shroud downward onto the floor where it locked into place. Then, instead of standing two-foot-tall inside the cabin, the 28-inch shroud protruded two-feet beyond the aircraft fuselage where it would serve as the pilot's escape hatch. Most concerning was the fact that should the emergency exit system be activated, the multi-million-dollar BA 609 was no longer an aircraft capable of flying! It would have become just junk; a very expensive piece of junk with a 28-inch hole blown through the bottom of the fuselage! In fact, there had been so much concern about a mishap of this nature that only one key to arm the system had been issued; to me! After carefully inserting the key, I held my breath, then switched it to the armed position. Ah, thank God! No explosion! Finally satisfied all switches had been properly set, Roy was cleared by Bell's telemetry (TM) control center to begin a functional check of all flight controls.

Roy advised the test director in the TM control center of his intentions to conduct a check of all flight controls through their full range of motions to their respective limits. Referred to as "control boxing", that was a routine check, and he was promptly given approval to continue. He responded by moving the center stick, power lever and pedals throughout their full range of motion to their maximum limits. The BA 609's flight controls were fly-by-wire and controlled by three Flight Control Computers (FCC). In fact, every system on the aircraft was triple redundant, which simply meant one would have to have three failures of any system before losing that system entirely, a remote possibility. Roy and I were very familiar with those systems because we had spent countless hours in a flight simulator and Bell's Vehicle Management Systems Integration Laboratory (VMSIL) while engineers went through the laborious process of developing and shaping flight control laws for the FCC.

The instrument panel itself utilized a glass cockpit which simply means it utilized digital flight displays rather than the old mechanical style instruments so familiar on older aircraft. Other instruments located on the instrument panel were common to any aircraft and included such things as a magnetic compass, standby attitude indicator, radio control panel, flap indicator, gear handle and digital clocks. Satisfied everything looked good, Roy depressed the instrumentation trigger switch, and then verified the instruments were ready to "capture" all data during the engine start. Verifying that function as operational was critical to all flight test activities since it provided engineers with real-time data in regards to critical flight parameters; it was considered a GO/NO GO item. After ensuring everything was operating as expected, Roy advised Bell's control tower "Tango Romeo" (aircraft's call sign) was ready for engine start up". The tower's response was a terse: "Roger, Tango Romeo; you're cleared for engine start".

Roy then reached up and grasped the left-hand Engine Control Lever (ECL) on the overhead panel, then pushed it forward into the START/IDLE detent position. Engine starts on the BA 609 were automatic, so from that point on, we simply monitored all engine instruments for any abnormal readings. The engine responded by spooling up very nicely, with no surprises or exceedances. Everything functioned normally, just as it had during the previous 35 hours of ground runs. The automatic starter automatically dropped off-line when the gas producer (Ng) passed through 50%. All instrument indications were residing comfortably in the green as the 26-foot rotors began slowly spooling up to their operational ground idle RPM. Inside the cockpit, the only indication the aircraft was "coming alive" was a slight increase in cabin vibration. Satisfied everything was functioning properly, Roy repeated the process with the right-hand engine. Noting all engine instruments were comfortably in the green, he simultaneously pushed both ECLs forward into the FLY detent. Both engines and rotors reacted by spooling up to 100% rotor speed (Nr), and the only indication to us was another slight increase in vibration. We began a check of all aircraft systems to ensure everything was operating normally before Roy called the TM (telemetry) test director to verify all test parameters monitored by telemetry were functioning normally as well. TM's response was: "Affirmative"!

That simple one-word response confirmed all preflight and ground run activities had been completed! After years and years of ground tests, simulator tests and even more ground tests, there was nothing else left to do. The BA 609 could now join the realm of flight. The silence was deafening as everyone seemed to hold

their breath! Now, it was up to Roy and me. Roy broke the silence by calling the tower: "Tower, Tango Romeo requests takeoff clearance". The response was a "roger Tango Romeo; you're cleared for lift-off".

Prior to lift-off, Roy depressed the all-important instrumentation trigger once again and the amber light began flashing, an indication "all is well". Roy then began pulling vertically on the power lever to get the aircraft light on its landing gear, made a few slight adjustments to ensure all controls were in the neutral spot, then ascended vertically into a hover. Finally, the BA 609 had its landing gear off the ground on our historic first flight! Roy and I could literally hear everyone in the TM room begin to breathe again! It had just been our imagination, but it seemed as if we could hear the roars and claps of the large crowd of Bell personnel and invited guests that had assembled to watch history being made. In reality, the constant roar of rotor blades beating the air into submission would have precluded us from hearing anything, even had we been standing among our large crowd of boosters.

Even though Roy and I were euphoric, we were experienced enough to know there was much yet to be accomplished, so we continued with our checklist. I had been primarily monitoring instruments, but I also had taken note of Roy's movements on the controls as he smoothly pulled upward on the power lever until we were stabilized at a 20-foot hover. His movements were minimal, a good sign; the control laws we developed in the simulator had worked well! The next step was for Roy to perform a series of relatively simple tests to ensure all systems were functioning as designed. Roy performed fore, aft and lateral stick inputs, plus climbs, descents and turns to exercise all flight controls, allowing him to get the feel of the aircraft. Everything checked out perfectly, so we began setting up for the next series of tests which allowed me to get into the loop by participating in rotor stability tests.

Rotor stability tests during flight were very important because even though we had completed many rotor excitation tests, they had all been performed during restrained ground run tests. We all knew there was a significant difference in how rotor blades respond to any kind of excitation. That's because main rotors act like a large gyroscope, and the last thing you ever want to happen is for the blades themselves to go divergent in a violent out-of-track condition. A situation of that nature would have almost certainly ruined that historic, exciting day!

While Roy maintained aircraft control by keeping us at a steady hover, I used the Flight Test Interface Panel (FTIP) to dial in various input settings that would provide rotor disturbances the engineers desired. Once I had the desired settings, I began a countdown from "3 – 2 – 1", then bam! I hit the switch to inject an input to the actuators which represented a strong gust of wind hitting the rotor blades. The purpose for such a test was to ensure the rotor blades will "damp" rather than go divergent, a really bad thing. We know rotor blades don't always dampen after such an input, and while confident, we weren't 100% certain! In that instance, the engineers in the TM room were pleased the rotor blade excitation was well damped because the rotor blades had returned to a stable condition almost immediately after my excitation had terminated. Satisfied the main rotor blades would be stable throughout flight testing, we were able to move on to the next flight test sequence.

Since everything went so well, our next step was to accomplish a series of tests to ensure our aircraft controllability was safe for flight, our nacelles were operating properly and we had no "shimmy" tendencies in the landing gear when the aircraft was rolling down the runway. Soon, all those tests were successfully completed, and we were ready to fly a traffic pattern around the Arlington, Texas airport. Squeezing the microphone switch mounted on the center stick, Roy made a call requesting clearance to taxi onto the runway, and the tower's response was immediate: "Tango Romeo, cleared to taxi onto runway 16".

There's an old pilot's saying that "the runway behind you is useless", which is to say an aircraft needs so much runway to takeoff, and any runway behind you is of no use to you. Heeding that old saying, Roy hover-taxied out onto the runway "on the numbers". That meant we were sitting atop the number 16, which indicated a runway heading of 160° since the prevailing winds were from the south. By wisely allowing a long runway for a first flight, Roy would have had time to react if anything had gone wrong.

While still stationary at a hover over "the numbers", both of us quickly scanned all instruments to ensure everything was "in the green", or normal. Satisfied that was case, once again Roy requested clearance for takeoff and the tower responded quickly with clearance to takeoff. With his left hand gripping the power lever, Roy responded to the tower's clearance by using his thumb to push the nacelle position switch forward to 85° nacelle angle. The aircraft responded by accelerating into forward flight at 10:35 AM (CST). A very exciting time for aviation!

As we began climbing out, a helicopter chase ship slid into position behind us to provide extra sets of eyes to watch out for us, plus provide a "birds-eye" level platform for several photo/video type personnel on board (that's to be expected on a first flight of any new aircraft)! We climbed to 1,500 ft. MSL, then accelerated to 80 knots, our maximum speed for that particular series of flight tests since we had been restricted to 75° nacelle angle. There really wasn't much conversation between us since we were both busy monitoring instruments while Roy flew, and I was busy writing down data on my clipboard. It was only when we turned onto our downwind leg, I noticed just how quiet it was. Almost too quiet! I was just about to make a comment when Roy broke the silence by asking "is it just me, or is this aircraft really this quiet and smooth?" Softly chuckling to myself, I responded by saying "that's the same question I was going to ask you!" We both agreed; for a first flight, it was exceptionally quiet, and smooth!

In flight test, it's not uncommon for engineers and pilots to expect something to go wrong. When everything seems to be working "as advertised" without a hitch, sometimes they start to get an eerie feeling. Kind of like "this is working too good to be true!" Roy and I had both experience, and thousands of flight hours between us, and we had instinctively known not to relax. That's because, should something go awry, it would probably be sudden, without warning! Yet, there we were, cruising on the downwind leg during our first flight and everything was operating flawlessly. Unbelievable!

All too soon, we were preparing to turn onto base leg when the tower informed us that we were cleared to land, number two behind an airplane that had sneaked into the airport traffic pattern, and was already on final approach. Since we were already in helicopter mode, Roy turned final to make an approach directly to Bell's Flight Test Center flight ramp rather than make an approach to the runway. We had planned to go to the runway but the aircraft had been flying so flawlessly Roy decided to fly directly to the ramp so the huge crowd could get a close-up view of the BA 609 on final approach. After terminating the approach at a hover, Roy made a quick 90° pedal turn to face the crowd, then tilted the nose forward to make a "bow" towards the audience! The response was an eruption of cheers, clapping and "high fives"! It had been a good day! A damned good day!

Bell test pilot Roy Hopkins and me standing beside the BA609 Tiltrotor after making first flight at Arlington, Texas airport (03/07/2003).

BA609 Tiltrotor in flight at Bell's Flight Research Center at Arlington, Texas municipal airport.

CHAPTER 19
Staying Busy with Other Companies

Heritage Aviation
August 1, 2005 – November 5, 2007

My retirement party, celebrating almost 32 years of employment at Bell Helicopter Textron, was held at 9:00 AM on Friday, July 29, 2005. It was a short retirement because I began my employment with Heritage Aviation, Inc. on August 1, a scant three days later. It was an unexpectedly short retirement, but I had an opportunity to continue doing what I enjoyed most, test flying. Heritage Aviation was a small company located at the Grand Prairie municipal airport, and it was well-known as one of the best companies in the industry when it came to aircraft completions, especially interiors. All that was made possible by employees who were specialists in their respective fields, plus it happened in a remarkable facility I always referred to as the "skunk works". That's because they had a machine shop, a wood-working shop, leather interior shop and a paint booth large enough to fit a huge S-92 helicopter inside. They had just signed a contract with Sikorsky to do completion work on new 28,000-pound S-92 helicopters, so they needed an experimental test pilot. Since I was ready for a change, the timing was perfect!

It had certainly been a change of pace, since I began S-92 flight training at Flight Safety International (FSI) at West Palm Beach, Florida on August 15, 2005. After so many years away from simulators, the training at FSI had been difficult enough, but the oral exam and simulator flight check at the end of the course had been over-the-top difficult. Accompanied by another pilot, I managed to make it through, completing my final check ride at 2:30 AM in the morning of September 19. I had just enough time to get back to the hotel, take a shower, get dressed, then head to the airport for a quick snack before boarding the early morning flight to Dallas-Fort Worth (DFW). After arriving at DFW, the other pilot and I were picked up by Chris Pierce, Heritage Aviation's shop manager, and he casually informed us they had a brand-new Sikorsky S-92 helicopter waiting on the ramp for us to fly. Up to that point, I had been a little groggy since I had very little sleep the previous night and had just arrived in Texas with plans to catch up on my sleep. However, that casual comment about "having a new S-92 to fly" woke me up, big time!

I say that, because even though the other pilot and I had just completed almost five weeks of intense training on the S-92, but we had never seen the actual aircraft! It sounds crazy, but the simulator was an FAA-approved, Level D simulator deemed to be an exact replica of the real S-92. Consequently, there had been no need to fly a real S-92. The simulator, for all intents and purposes, was the same thing, at least as far as the FAA was concerned. However, I still thought it was reasonable to expect our first flight would be with an experienced S-92 pilot rather than it being "solo", albeit with a copilot.

At first, I thought Chris was simply kidding with us, but after we parked and walked through the hangar to the large ramp in the rear, I realized his comments had been no joke! He was damned serious because sitting before us was a S-92, the largest helicopter I had ever seen, much less flown! Cosmetically speaking, it wasn't a thing of beauty since it was still painted in the familiar green, zinc-chromate paint undercoat, but it was imposing, simply because of its size. The pilot standing alongside me had attended the same course and, like me, he was uncomfortable with the prospect of having to fly such a beast, especially since we had never even sat in one. Even though I sympathized with him, I explained to him the folks who hired us had paid thousands of dollars for our training and I had the feeling they would not be happy with us should we choose not to fly the S-92. I also mentioned the cockpit couldn't be that much different than the simulator's cockpit, otherwise, the FAA wouldn't have approved it, right? I also told him I would assume the duties of Pilot-in-Command while he would be my copilot. He half-heartedly agreed with my logical, non-sensical, crazy-ass approach to our dilemma. Our intro to the "real" S-92 had begun.

The reason for our unexpected flight was that the Sikorsky flight crew who delivered it had written up a maintenance "squawk" for a tail rotor chip detector light that illuminated while on the short final approach into the Grand Prairie airport. The maintenance "fix" for such an anomaly was to change the oil, then accomplish a 30-minute ground run before flying the helicopter for one hour. I was quite happy about doing a ground run first, because we would at least have some time in the cockpit to familiarize ourselves with the everything prior to flying.

After entering the cockpit and taking our seats, we just sat there taking it all in when my copilot broke the silence by asking what I thought about the cockpit. My response had been "It looks just like the simulator except it's easier to see". That was because most simulators utilize dark cockpits, and ours was sitting outside in the sunlight. Satisfied with that, we proceeded with the checklist, finally reaching the part where the copilot started the Auxiliary Power Unit (APU), thus providing power for the hydraulic pumps and generators so we could continue with the engine startup sequence. My copilot dutifully flipped the start switch on and what followed could best be described as a loud whistling sound followed by marbles rolling around in a tin can! It was a sound totally unlike anything we had heard in the simulator and our arms became entwined as we simultaneously tried to grab the switch to turn it off! Fortunately, the noise diminished within seconds after the switch was off, but it had left both of us in a real quandary in regards to our next step. After thinking about it for a minute or so, I decided to call a Sikorsky test pilot I had met so we could question him about what had just happened.

After exiting the S-92 and getting him on the phone, I asked about the APU and what it sounded like when going through a start cycle. He laughed and asked if it had made a loud whining noise reverberating in a tin can when we first engaged the APU starter. My response was to say: "Pretty much!" His response was, "That's how it's supposed to sound." I thanked him, hung up, then trudged back to the helicopter where I rejoined my copilot in the cockpit and ordered him to start the APU. It went through the whistling, marbles-in-a-tin-can sound, then quieted down and began humming like a sewing machine. That's when we both decided, even though simulators were great training devices, they simply couldn't capture all the quirks a helicopter may actually have. Both the ground run and one-hour flight were successfully accomplished and just like that, our S-92 career had begun!

Although I worked on numerous projects during the time I was with Heritage, there was one project that stood out above all others. It was ordered by some Head-of-State of a foreign country and he had decided his helicopter must have a shower on board in which he would be able to take at least two ten-minute showers when sitting on the ground. That turned out to be quite an order, because none of us had even seen a shower in a helicopter, much less built one. But it was time to let your imagination soar! And the Heritage folks did, big time!

It had taken some time, but the finished product had been spectacular! In fact, I still think it may have been the most beautiful interiors I have ever seen, at least in a helicopter. The interior was pure, white leather and also had a dressing room, day bed and even though the S-92 had been large enough to seat 19 passengers, that particular helicopter had only four seats, three of which could be considered normal. The remaining seat had been especially built for the Head-of-State who had requested the shower. It literally looked like a throne and the vibration so typical in most helicopters was virtually non-existent, especially for the "throne" seat. You could literally set a glass of water on that cushion and not see a ripple!

The shower itself was equally beautiful, since it had been constructed of what appeared to be marble, but was actually polished acrylic. Topping it off were gold faucets and sprayers, but the crowning jewel was a large curved, sliding acrylic plastic shower door that encased the shower. It had taken some masterful thinking and design-work to construct. The system itself consisted of two forty-gallon tanks to accomplish the simple task of taking a shower. One tank had been filled with "fresh" water; the other tank remained empty until the shower was in use at which time it began to fill with "gray" water. All that was happening while the gentleman enjoyed his shower, totally unaware of what a masterpiece of engineering had been required to provide his luxurious shower.

It had also required a lot of flight testing, and some of us even pushed the idea of contracting a cute, young Playboy bunny to conduct a final check of the shower to ensure the proper functionality of it. In fact, we had even insisted it was a safety requirement because there had been the possibility of someone dropping a bar of soap while lathering up, then step on it and take a bad tumble! Unfortunately, our good idea had fallen on deaf ears since even the FAA didn't require a bunny for flight testing.

However, we should have taken it more seriously, because about a year after the helicopter had been delivered, the shop foreman walked rapidly into my office one morning and asked if I had heard the latest news. When I responded "No", he then proceeded to tell me the gentleman who ordered the helicopter had slipped on a bar of soap while taking a shower, then fell and hit his head, thus killing him! All I could do was stare at him because I had been struck dumb, unable to say a word! Seeing my stunned response and ashen face, he started laughing and said "Well, he did die, but I don't think he was on the helicopter when it happened. I've just enjoyed telling everyone that soap story because they all nearly passed out just like you did!" At the end of the day, Heritage did completions on five S-92's, all of them different, and all extremely nice!

They also did completions on other helicopters which basically entailed taking a new helicopter off the production line and installing avionics, special equipment, painting or anything else the customer requested. In fact, one day a Sikorsky test pilot and I were sitting at a table in the break area inside the hangar, enjoying a cup of coffee when he slowly looked around, and in a very nonchalant manner said, "So, what does it feel like?" Surprised, my response was "What does what feel like?"

"Well", he said, "I see helicopters from every major helicopter manufacturer sitting inside this hangar, and that includes Sikorsky, Bell, Aerospatiale, Agusta and MD Helicopters. I know you fly all of them, so I was just wondering what it felt like to be working for all the major manufacturers at the same time."

"Well, guess I never thought that much about it", I said, "but as long as I can start'em, I reckon I can fly'em, you think?"

He agreed. Actually, I had been qualified to fly all of them, and I have to admit, I enjoyed every minute of it! I was never bored.

I also thoroughly enjoyed flying the S-92, and managed to amass a couple hundred flight hours in it, all of them experimental flight testing. It was a joy to fly and I've often wished that program could have continued, but it wasn't meant to be. A businesswoman none of us had never even heard of appeared on the scene and everything changed, even the location of the remarkable company that could literally do it all. It was a great job and I never regretted my decision to join them!

MD Helicopters, Inc. (MDHI)
November 5, 2007 – July 31, 2011

In early spring of 2007, the folks who owned Heritage Aviation, Inc. informed me "the little company that could" had been sold to MD Helicopters, a well-known, legendary company with long roots going all the way back to a gentleman who was somewhat of a legend himself; a fellow by the name of Howard Hughes! He had a love for all things connected to aviation and had ventured into it as early as 1932 when he established a new aircraft division within his Hughes Tool Company. He didn't get involved in helicopters until he formed a new helicopter division in 1955 called Toolco Aircraft Division. A year later, the company developed a civilian model called the 269A, and the U.S. Army purchased almost 800 of them. In keeping with the U.S. Army's history of tagging their helicopters with Indian names, it was designated the TH-55 Osage training helicopter. Most of them, if not all, made their way to Fort Wolters, Texas where they became the mainstay of their training fleet. I can personally attest to the capabilities of that small helicopter since I flew almost 1,000 hours in it as a flight instructor when I returned from Vietnam in late 1967.

Their next model was a turbine powered helicopter. It won a U.S. Army competition and began production of what was called the OH—6A Cayuse Light Observation Helicopter. However, it would become more commonly known as the "Loach" by the flight crews in Vietnam who flew it, fought in it, and survived most of the gun fights they got into. It became well known and much respected in the battlefields of Vietnam. It was fast, maneuverable and when it was shot down, it crashed "well", a trait they referred to as having good "crash-ability" qualities. That was due to a main beam structure that literally encapsulated flight crews in much the same manner NASCAR racing vehicles protect drivers today. Like the Huey, it became a much loved, legendary helicopter in the "helicopter war" of Vietnam. Hughes was two-for-two regarding the success of their helicopters, and derivatives of both of them are still being manufactured today. Now, that's quite a legacy!

The company who built such a wonderful helicopter had become engaged in a game similar to musical chairs since it ended up being sold several times, first being sold to McDonnell Douglas in 1984, then to Boeing in 1997. Boeing kept the military aircraft, then sold the civilian production models to RDM Holdings, a Dutch investment company that took the name of MDHI. Hard times prevailed and it was finally purchased by Patriarch Partners, LLC in 2005. The founder

and Chairman of the Board of that company was a woman by the name of Lynn Tilton, a very wealthy, self-made millionaire, many times over. Surprisingly, most folks had never even heard of her, especially those of us who had been in the aviation business. Purchasing MDHI had been her first venture into the helicopter business, and she had become CEO after the transaction had been completed.

I had no idea what would happen to my position at Heritage, but that question was answered when MDHI offered me the position of Chief Pilot, Director of Flight Operations, which I quickly accepted on November 5, 2007. My wife and I didn't relocate to their Mesa, Arizona facility until May 15, 2009, since I had to manage flight operations in Texas as well as in Mesa. It was the beginning of a journey in Arizona my wife and I remember fondly. I also recall how nice those 95- to 100-degree days were after wringing out all that Texas humidity I was used to; but then July rolled around. I still remember the first week I sat staring at the TV screen in disbelief! It was hard to wrap my brain around those non-ending 117, 117, 116, 118-degree forecasts I was staring at. It was hard to comprehend! Humidity was no longer a factor; as they say, "An oven doesn't have humidity in it either, but it's still hot as hell!"

After "almost" adapting to the heat, flying all of MD's helicopters rolling down the assembly lines kept me and all the other MD test pilots very busy. Not only did we have to fly production flight tests, we also had to fulfill the duties of instructing, flight demonstrations and any other duties dreamed up by everyone from the upper management folks to what seemed like groundskeepers at times. But we were all doing what we enjoyed. Besides, I've often heard a pilot really isn't happy unless he's complaining about something. That being the case, boy, were we happy!

There was a wide variety of helicopters from which to choose, including the 500, 530, 600 and the twin-engine 900. I have to admit, I thoroughly enjoyed flying each of them! After flying the 500 and 530, I certainly understood why U.S. Army flight crews loved flying the Loach in the jungles of Vietnam. Its predecessor, the Model

500, was everything they said it was, and with its bigger engine, the Model 530 was like a Model 500 on steroids! It's still a favorite in the law enforcement world. But the model I always enjoyed flying the most was the 902. Cosmetically speaking, it wasn't the most beautiful helicopter to look at, but it had a smooth ride and was always ready to go to work! I still recall reading a note attached to a 902 someone had traded for another brand of helicopter that went something like this: "Please take care of her because she always took care of us when it counted the most!" In the aviation world, those are mighty powerful words!

Death-Defying Dive

Throughout my entire career as a test pilot, I have to admit diving helicopters to 1.1 Vne has never been one of my favorite flight tests to do. Unfortunately, on one fine morning in September, 2008, I had been summoned to an unscheduled meeting to discuss an upcoming flight test that required me to do exactly that in a MD 600 helicopter. I can't say I was too concerned, because the MD 600 was an approved aircraft already in production; consequently, a test pilot had already accomplished that test sometime during the development phase. But that still wouldn't stop me from treating it with all the respect of a rattlesnake, because complacency can lead to disaster in the flight test world!

The fellow assigned to fly with me was not only one of the sharpest mechanics I've ever worked with, especially on MD helicopters, but he was also a damned good pilot! His name is Earl Greenwall and I had been very pleased he offered to fly with me during those tests. He had worked on, and flown, the MD 600 since its inception, so he was very experienced in it. We then discussed in great detail how we would do the tests, and what each of our responsibilities would be. Even though we had planned to make just one dive, each of us had to be totally aware of what to watch for, and what our actions would be, should we have to abort. Dives to 1.1 Vne can never be taken lightly!

Soon we were on our way to a flight test area located just south of Grand Prairie, Texas airport where we could climb to a higher altitude, then begin our dive to 1.1 Vne. I made wide, climbing turns so I could check for any other traffic in the area before we began our dive from an altitude of 8,000 feet. After we levelled off, we went through our "to do" list before turning ourselves into a yard dart, screaming towards the earth! It was also "gut-check" time because butterflies were still swirling in our stomachs before commencing the dive. After all, high speed dives going beyond the maximum speed which "mere" line pilots cannot exceed are not for the faint of heart! Or nervous tummies!

I looked at Earl and said: "You ready?" His response was: "If we have to!" I then dropped the nose of the MD 600 and we began rapidly accelerating towards our target airspeed of 171 knots. We very much resembled a yard dart hurtling from the sky, but I wasn't looking for any particular spot on the ground to punch into! I was totally focused on the airspeed indicator and altimeter. The altimeter was really unwinding, but it was the airspeed I was totally focused on. As we screamed earthward, we went through 140, 150, 160, then I began slowing down because I definitely did not want to exceed 171 knots! Any speed beyond that was unknown territory! An area where bad things could happen!

I first heard it just as we crept beyond 165 knots. It had sounded sort of like a rattlesnake's rattle, the kind of rattle it makes just before it strikes! I think Earl must have heard it, too, because he shifted in his seat and said: "Oh, hell!" Before I could even respond to Earl's comment, the helicopter seemed to leap past 171 knots, even as I was pulling aft on the cyclic control in a desperate attempt to slow down! Unfortunately, my best efforts were in vain because the controls suddenly seemed to be set in concrete! Unable to move the controls, I was no longer a pilot; I had become just another passenger, like Earl. Feeling helpless, I keyed the microphone and said: "I'm sorry, Earl". I truly thought I had lost control of the helicopter and both of us were about to become just another statistic. It was a weird feeling I had never felt before, or since. I also hope I never feel it again.

If there's one thing I have learned in all my years of flight testing, it's been to never give up! Although I couldn't budge the controls that seemed frozen in place, my hands remained gripped on the cyclic and collective controls, while my feet rested firmly on the pedals. As the speed seemed to dissipate, the nose of the 600 suddenly began to pitch up and we began soaring vertically into the cloudless sky! I never loosened my grip on the controls and just before I thought the MD 600 was going to flip over onto its back, I felt an imperceptible movement in the controls. It was very slight at first, but as the speed kept dissipating, I could feel the controls slowly returning to normal. Thank God! I had finally regained control and was able to return to normal flight! Breathing a sigh of relief, we began our descent and headed back toward Grand Prairie airport. Neither of us spoke a word on the flight home. We were simply too drained to speak.

After we finally made it back to the airport and landed on the flight ramp, the lead mechanic walked up to the helicopter, then opened the door. He first looked at me, then at Earl. Without saying a word to us, he simply said: "Damn!", then abruptly spun around and walked rapidly into the hangar. In just a matter of minutes it seemed everyone from the hangar had joined us out at the helicopter, all wanting to know what happened. My only response was to say "I thought we had lost it. I even apologized to Earl for killing him." The lead mechanic who had first walked out, allowed as how maybe I did because he said: "Both of you look like ghosts!" We both climbed out of the helicopter and went inside for a debrief, but I don't remember too much about it. I think Earl and I both were just happy to be sitting there! I'm not sure if we ever figured out exactly what happened, aside from the fact I may have gone beyond what had previously been tested during the development phase. Without a doubt, Earl and I had been close enough to peer over the "edge of the cliff" we often refer to while doing flight tests.

We had found the cliff, but thanks to God and good luck, we didn't pitch over it.

Working for MDHI had its share of good times and bad, just like it does at every company; but overall, the good times far outweighed the bad. The folks at MDHI were great to work with, and the pilot staff was one of the best I've ever worked with! To this day, I have nothing but good things to say about them. I've often been asked what it was like working for Lynn Tilton since she was very involved in MDHI's operations, and the news media churned out a lot of stories; but I'm not a big media fan. What I will say is she is a very smart woman who became wealthy through hard work and perseverance, and I respect that! What I enjoyed most though, was how well we got along, and the fact she had a great sense of humor! Resigning was a tough decision to make, but it had happened rather suddenly one Saturday morning.

My wife and I had been enjoying a cup of coffee on the patio of our condominium when she took a sip, then turned to me and said: "It's time". She said it quietly enough, but it spoke volumes! I mean, that was a woman who had traveled around the world a time or two with me, and had always been my biggest supporter. She was telling me it's time to go home; to our home in Texas. So, I listen when she speaks. I resigned from MDHI the end of July, 2011, and my wife and I returned to our home in Texas in August 2011. I've returned to MDHI on several occasions since then to assist with various flight test projects, and always felt as if I were going home. Soaring over the Superstition Mountains, the Four Peaks, and then following the Salt River back down onto Falcon Field is still one of my favorite places to fly.

AeroDynamix, Inc.
September, 2011

Not long after I returned from Arizona, I received a phone call from Tonka Hufford, the same young man who had accompanied me on all those multiple hard-over tests, plus we also worked together at Heritage Aviation and MDHI. He was manager of a small company that specialized in designing and modifying cockpits to be compatible with Night Vision Goggle (NVG) operations. I have to admit, I was quite surprised by the phone call, because my experience with NVGs had been very limited. In fact, I had flown with NVGs exactly one time, and that had been a demo flight several years ago while I was still Chief Pilot at Bell Helicopter. It was a limited introduction, but even then, I had intuitively known NVGs would become the wave of the future.

It wasn't long after joining AeroDynamix that I found myself at the Denton, Texas municipal airport early one evening to prepare for my first flight of NVG training. The company was Longhorn Aviation, and I was assigned to an instructor by the name of Justin Munroe. He was personable, well-qualified and a great instructor – the perfect combination for fun training sessions! I think my personal expectation was for me to struggle through the first few training flights, then get more comfortable throughout the remainder; but I was wrong. It may sound like bragging, but I felt like those NVGs simply enhanced the skills I already possessed. I say that because I felt so comfortable flying with them! It was like suddenly having Superman's x-ray vision, but instead of seeing through walls, I could see through the darkest night so long as I had some form of minimal lighting from the stars, moon or cultural lighting (man-made) to power my goggles. It was a fantastic feeling, and since that first memorable night I began training with NVGs, nothing has diminished that euphoric feeling!

Most folks think those "Star War" type-of-inventions are new technology, and for us humans, they are; however, Mother Nature is always a step ahead of us. She created them long ago for a few animals and birds, but especially for cats; and that includes your standard run-of-the-mill tabby cat. Even though domesticated, today's house cats are very capable of surviving in the wild if they had to; but it's the big cats such as those we see on "Animal Planet" or "National Geographic" that showcase their prowess and uncanny

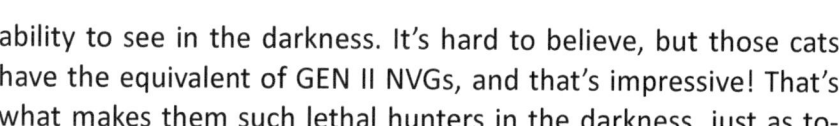

ability to see in the darkness. It's hard to believe, but those cats have the equivalent of GEN II NVGs, and that's impressive! That's what makes them such lethal hunters in the darkness, just as today's Special Forces warriors are.

The history of man's NVG capability actually dates back to World War II, but it was very primitive, at best. Germany's military attempted to use it on their tanks, but it was the Allies great fortune it didn't work; otherwise, it might have changed the course of the war, or at least prolonged it. I say that because there's an old saying in the military about "he who owns the night, controls the battlefield". Only a few primitive NVG-type weapons were utilized in Vietnam, but the invasion of Iraq in March 2003, really showcased the importance of NVGs! Literally every American soldier had access to small, single-tube NVGs, and use of NVGs has accelerated in the commercial world since then.

Throughout the years, NVGs have been developed, and strictly controlled, by the military; however, that once-strict control has relaxed to the point whereby the civilian aviation industry can also utilize them. The NVGs of choice are referred to as Generation III, or GEN III, for short. That simply means they are the third generation of technology dating back to the 60's, when NVG development began, and in today's military, unaided flight (flying without NVGs) is literally prohibited! I still recall Justin Munroe, my NVG instructor, telling me he had 2002 hours of night-time in the military, and only two were unaided. Without NVGs, the world would still be searching for Osama Bin Laden because I assure you, both pilots, and each member of Seal-Team Six wore them. Who knows, maybe even their dog was wearing NVGs when they ran Osama to ground, then eliminated him.

As you might imagine, I've traveled all over the U.S. to conduct certification testing for NVG compatibility on a wide variety of helicopters, and even airplanes. They've all been interesting, but one destination stands out above all others. It was the Faroe Islands, a small group of islands that lies almost midway between Denmark and Greenland. As Kevin Tran, an AeroDynamix engineer, and I approached them on an Atlantic Airways flight, they appeared to be nothing more than rocky projections of granite jutting out of the North Atlantic Ocean. A clear, blue sky seemed to magnify the small islands that lay surrounded by cold, dark water which stretched into infinity. The small white dots far below us were fishing boats, and simply flying over them gave me the shivers!

In that entire 540 square-miles of land, home to some 52,000-plus inhabitants, there was but a single airport. Built by the British in the dark days of World War II, Vagar International Airport is the lifeline between the Faroe Islands and the world. When our Atlantic Airways jet finally landed there, it became abundantly clear why there was only one airport on the islands. It had been built on the only spit of land where an airport could be built! Even then, the construction engineers had been forced to truck in extra stone and what little soil was available to expand the small, flat area. Just building it had been a remarkable engineering feat.

The island's rugged terrain also has weather to match, since it's described as being windy, wet, cool and cloudy. We saw "all of the above" on our orientation trip around the islands, moving through bright sunshine, a snowstorm, fog and high winds, all in a single afternoon! The sheep we saw atop some of the grass-covered roofs were not hallucinations, either. Why wouldn't they be there, since that's where the best grass was! Since we were on an island, we had fish for most meals, but there was also a choice of mutton, puffins and sometimes the "catch of the day" was penguin! Then there were the five hours of darkness we had to deal with to accomplish our NVG flight tests. The Faroe Islands is one of those places where you can "spend a month in one week", but the friendly folks who reside there certainly made it enjoyable.

Our team consisted of Kevin Tran and myself who flew in from Texas, a European Aviation Safety Agency (EASA) representative who flew in from Italy, a Bell 412 Captain who flew in from the UK, plus a project engineer who flew in from Denmark. We all came together on that small bit of land, solely to fly a total of two hours to certify the BH 412 for compatible flight utilizing NVGs. Surprisingly, it all worked out because in the early morning hours of April 24, 2014, we accomplished our mission. We took off at 2:00 AM in the morning and, true to form, the weather was intermittently cloudy, foggy, windy and sometimes even clear, but all in very short doses. The timing was fortuitous, too! We spent an entire week there for that one night, and it occurred on the same morning we were scheduled to depart. We completed everything by 4:00 AM, toasted our success with wine set outside our windows to chill, then caught our flight to Copenhagen at 8:00 AM. It had been perfect timing!

I'm standing with Bell pilots in front of Bell 47 helicopter at my retirement ceremony at Bell's Flight Research Center at Arlington, TX municipal airport (07/29/2005).

I'm flying a Sikorsky S-92 helicopter for Heritage Aviation at Grand Prairie, Texas municipal airport (07/25/2006).

I'm sitting in the cockpit of a Sikorsky S-92 helicopter (05/09/2006).

Group photo of MDHI pilot staff (left to right): me, Jerry Turchetta, Dave Salem, Angela Thomason, Rick Ponder, Mike Zale and Robert Rappoport at Falcon Field, AZ (2010).

Paul Peterson and me standing beside a MDHI 530FF helicopter at Grand Prairie, Texas airport (20007).

I'm sitting in cockpit of Bell 412 helicopter with Night Vision Goggles (NVG) still attached to my helmet after completing NVG test flight (08/26/2021).

CHAPTER 20

First Flight of Marenco Helicopter SKYe SH09

At Mollis, Switzerland Flughafen (Airport)
October 2, 2014

The first official sketch of his new helicopter had been drawn on the corner of a tablecloth when he had been trying to find investors to build his helicopter. Years later, in September, 2013, I was introduced to the gentleman who drew that sketch while having dinner at a small restaurant located in Leadville, Colorado, a beautiful, historic mining town sitting high in the Rocky Mountains. Martin Stucki was founder and CEO of a small company in Switzerland that was designing that new helicopter, and he was in a quandary, since he had no staff test pilot. We had chatted for a while, then after we had placed our order, I looked across at him and said: "Exactly, who are you?" He looked at me rather strangely, so I quickly added that I meant no disrespect at all. I had simply asked, because we have a saying in aviation that, "The best way to make a small fortune in aviation is to start out with a large fortune!" And there he was, amidst all the heavy-weight manufacturers in the world today, about to launch a helicopter with an entirely new design! Martin laughed and said: "I understand what you're saying, and I agree. It does sound crazy, doesn't it?"

He then went on to explain his rationale, and after we had completed our dinner, we retired back to the historic old Delaware Hotel, where we proceeded to discuss his new helicopter. I might also add we did have a humorous moment earlier that same day when checking in. I had pulled a muscle in my back a day or so earlier, so when I bent over to pick up my small suitcase, I let out a small grunt when my back muscles protested! Martin, ever the gentleman, looked over and asked: "What's wrong?" I explained my problem and he very graciously reached over, took my travel-bag and said: "I'll take it to your room." Martin had not climbed those stairs before, nor had I. That was the longest stairway I have ever seen! It ascended all the way up to the third floor, and Leadville, Colorado sits at 10,000 feet MSL! Poor Martin carried both our bags to the top floor, and he was really huffing and puffing by the time he made it to the third floor! He later accused me of taking advantage of his good nature, but in a fun way.

Martin Stucki was a mechanical engineer and commercial pilot who had long dreamed of designing, building and flying his own helicopter. He had not been happy with what he called "legacy designs" with no imagination. In 2002, he began working on his own design, a design he would call the SKYe-SH09 and we spent the remainder of the evening discussing his new helicopter design. It had been a very interesting and intriguing story! In fact, so much so, I agreed to join him on his journey. Shortly afterward, in December 2013, Wayne Barbini, a retired FAA engineer and close friend, and I made our first journey to their headquarters in Pfaffikon, Switzerland. After a "meet and greet" session there, we then drove to their flight test facility in Mollis, Switzerland, a small village located high in an Alpine valley. I had the feeling I was standing in the middle of a painting because it was one of the most spectacular sights I'd ever seen! It was everything I had always imagined Switzerland would look like, and so much more.

The village of Mollis is a beautiful little town full of "mom and pop" hotels, restaurants and shops, all of which are small and immaculately kept. There were no big-name hotels, motels, restaurants or even a McDonald's anywhere in sight. Just old, stately buildings that exuded a long history of a time gone by, and the warm personality of present-day owners. When we entered into the lobby of the Schuetzenhaus Hotel, we were greeted warmly by the proprietors who also doubled as chefs, waiters, waitresses and

maids. The rooms were small, clean and comfortable, the beer cold & frothy, and the food was excellent! It became my home away from home the entire time I worked in Mollis, Switzerland.

Early the next morning we drove out to the flughafen, or airport, and I was totally surprised by what I saw! The flughafen was formerly a small Swiss Air Force base before being abandoned and donated to the local townships that surrounded the airport. Nestled between majestic Swiss mountains, it had consisted of one main runway and one parallel taxiway that lay in the middle of lush, green grass that actually had cows grazing on it! Exiting the automobile and standing there on the flight ramp, I had the feeling I was standing amidst the most beautiful scenic view I had ever witnessed!

I also noticed a steady flow of pedestrians strolling nonchalantly down the taxiway as they made their way from one part of town to the other. I asked if that was normal and was surprised by the response, "of course!" I mean, there we were, on an active airport and a steady stream of bicycles, motorbikes, school children and adults were steadily making their way up and down the taxiway, even while airplanes and helicopters were starting up, then taxiing out for takeoff. When an airplane began taxiing from the ramp onto the taxiway, it was as if a silent warning went out and everyone cleared the way for the airplane, or helicopter, to navigate down the taxiway. Then, when it had passed, everyone moved back onto the taxiway and continued their journey to their destinations. It was business as usual and amazing to watch! They also told me they had never had any accidents or injuries! Amazing!

After taking all that in, we then entered Marenco's hangar and I was met with yet another surprise! They had been working on the SKYe-SH09 helicopter for a couple of years, so I thought it was very close to being complete; but I was wrong! It was still a virtual skeleton with no cockpit instruments or seats; in fact, there was very little that could even be considered complete. But the assembly crew in the hangar was a top-notch crew who were very enthusiastic and competent, and that counted for a lot!

The prelude to any first flight is ground runs. Lots and lots of ground runs and the SH09 had been no different. In fact, the SH09 program entailed even more than what I had been used to, thanks to the European Aviation Safety Agency (EASA). In their zeal for safety, nothing had been too small or insignificant to overlook!

Prior to making our first flight, I can still recall some of their concerns we had to address before being issued a Permit to Fly (Ptf). One of their concerns was based upon the lack of a magnetic compass, even though we could only hover over an assigned spot on the runway located approximately 50 meters from Marenco's hangar. There had been quite a bit of discussion before one gentleman (a local pilot) asked me if I thought there were adequate geographical cues available so that I would not get lost, even though I had no magnetic compass. When I assured him, I did, he responded with a "great" and checked that issue as being properly addressed. I still owe that fine gentleman a beer!

Everyone involved worked hard and, after a long and laborious series of multiple tests, we finally accomplished the first flight of the Marenco SKYe SH09 helicopter at 08:15 AM in the morning on October 2, 2014. It was a beautiful clear morning with no clouds and calm winds. However, it was rather chilly with temperatures hovering around 10° C (50° F). Since the doors had been removed for the first hover flight, I was bundled up in a heavy jacket and was sitting in the right seat while Martin Stucki occupied the left seat fulfilling the duties of copilot for the historical first flight. I might also add it was the first flight of a new helicopter in Switzerland – ever!

The first flight wasn't a thing of beauty, but most of the first flights I've been a part of normally are not. We were able to accomplish three take offs and landings before declaring it to be a successful first flight and shutting down. The crowd of admirers were limited mostly to the Marenco support crew, but that was enough! Everyone rushed over to congratulate us, and they bought a huge bottle of champagne to make it official. As soon as that was completed, the helicopter was towed back into the hangar, and the party began! It still rates high in my memory bank as the one of the best parties I've ever attended! Starting about 09:00 in the morning, it roared throughout the day and was still going strong when I departed for Zurich about 5:00 later that afternoon. It's probably just as well I left for Zurich when I did, because I heard the party didn't wind down until sometime the next morning! Those Swiss folks sure know how to build a watch, and how to party! They also know something about building a helicopter! I have nothing but the utmost respect for those wonderful people!

This was the first flight of Marenco's SKYe-SH09 Helicopter at the Mollis, Switzerland Flughafen (10/02/2014).

I'm shaking hands with Martin Stucki, CEO of Marenco, Inc., who was also copilot for first flight of SKYe-SH09 helicopter (10/02/2014).

I'm standing beside close friend and Bell pilot retiree Wayne Brown just after receiving our Wright Brothers Award at Bell Helicopter's Flight Training Academy (12/12/2015).

I'm standing beside well-known pilot and close friend Dub Blessing after being awarded the "Dub Blessing" award as HAI's "Flight Instructor of the Year" (01/2020).

Epilogue

Fulfillment of a Dream

Reflecting back in time, I still see myself as a young boy standing amid endless rows of cotton, once again watching never-ending streams of military aircraft flit about in that small sanctity of airspace directly above me, and dreaming of flying. Little did I know those childhood dreams would become the very foundation of my odyssey in aviation that began on November 11, 1964, the day I made my first flight. I still recall with great clarity taking off into a beautiful fall sky, then soaring in, around and over scattered puffy clouds that were so pure and white! I had never known such beauty had existed! That flight surpassed everything I had ever imagined, and from that moment on, the sky was my home. That was where I wanted to be. I often question how I got here, because the trip was not easy, nor was it a solo journey. I'm also quite certain luck played at least a small part in my career, because there are many young men who had dreams that were a mirror-image of mine, yet never made it.

Except for an Air Force Captain and a young U.S. Army Specialist 4th Class, my own dreams would have been unceremoniously terminated before even starting. A pompous, overzealous Air Force dentist would have ended any thoughts of flying helicopters in the military had not that young Captain stepped up and told him his job was to gather information, not pass or fail an applicant. Then, there was the young Specialist 4th Class at Fort Rucker, a draftee who had almost completed his two-year service obligation before volunteering to lead me through a minefield of government paperwork just to become a Warrant Officer Rotary Wing Aviation Candidate. Those simple acts seemed insignificant at the time, but thanks to them, my dream of flying prevailed and I met all requirements to begin flight school. Perhaps some luck had been involved, but I sure wish I could remember the names of those two guys who got involved! I still owe them.

After proving I had the "right stuff", I began flight training on December 10, 1965, and still remember sitting in a classroom at Fort Wolters, Texas when the instructor told everyone to turn and look at the students seated left, right, behind and in front of each of us. Somewhat puzzled, we complied. After a short pause, he told us we could all expect to go to Vietnam, a bit of news that certainly came as no surprise! Then he said one of us sitting in that tight circle would not be coming home from Vietnam alive. That was a sobering thought.

Another sobering moment came shortly after arriving in Vietnam when our platoon leader told us combat losses were just part of the risks we faced in Vietnam. He then went on to explain how all young aviators being rushed to Vietnam knew very little about flying helicopters since each of us had only flown a mere 210 flight hours in flight school. It seems the U.S. Army had compiled their own set of statistics that basically concluded all helicopter pilots were a liability until they had flown at least 700 hours. Putting it another way, we were all an accident just looking for a place to happen! That was in addition to flying combat missions! All were sobering thoughts to add to an ever-expanding list of such thoughts.

Surprisingly, 700 flight hours added up very quickly since the average flight time for helicopter pilots and crewmen in Vietnam was approximately 100 hours a month. I don't know how the loss of so many pilots and crewmen stationed at Vinh Long Army Airfield stacked up in the Army's carefully compiled statistics, but the loss of almost 30 airmen was tough on everyone. Each of those losses had been a brother airman, not just a statistic! I was fortunate to make it past the 1,000-hour mark with no visible scars, nor did I earn a purple heart; however, dark memories are constant reminders of long-ago days spent in the war-torn skies of Vietnam.

After spending the remainder of my military career at Fort Wolters, Texas as a flight instructor, my small family and I departed there on the morning of October 12, 1969, destination Lafayette, Louisiana, often referred to as Cajun country. Not long after we had passed through the gates of Fort Wolters for the final time, my wife began softly crying. Sensing they were not tears of joy, I asked why she was crying, and her response startled me! It seems she didn't want to leave the Army; in fact, she even told me she had loved being a military wife! Knowing how I had agonized about deciding what to do, I asked her why she didn't speak up and let me know how she felt. Again, her response startled me.

She said "I knew I would support you regardless of what you decided, so I wanted it to be your decision. My fear was for me to encourage you to stay in, then at some time in the future you might decide staying in had been a mistake and blame me. But my greatest fear was that you might have been ordered back to Vietnam and been badly wounded, or God forbid, even killed! That, I could not live with." Her logic was spot on, and I loved her all the more for it. After all, she was the young 18-year-old country girl I married on August 12, 1965, who had fit in perfectly with military life. Yet, she willingly left the military life she loved to support me, and my dreams.

Driving through the gates of Fort Wolter wasn't so much leaving the military behind as it was the entrance to our future, whatever that might be. I have no doubt the key to my accomplishments in aviation is based on how supportive she has always been when it came to my career. It was a career path that had no road map, nor blueprint. Nothing was planned. Everything seemed to have been more a matter of pure happenstance in a career that has lasted more than fifty years.

When I started flying on that November day so long ago, the aircraft I flew had old round gauges very similar to those used in cars of that time. Now, those same gauges are derided as "steam gauges" because of their antiquity. Shiny glass cockpits are all the rage today, again not unlike those in your sporty-looking car. Raised on a steady diet of computer games, iPads and iPhones, the younger generation of pilots fit right in. Now they're even heralding new flying machines without pilots, for God's sake! Suffice to say, none will ever have me on board as a passenger! I suppose you can consider me as being old fashioned, but when I read of such things, I just sigh and fervently wish for the "good old days"! I still miss those days, but I have little recourse except to accept the high-tech era in which we now live.

At the end of the day, my saving grace is knowing if there was a "golden age" of flying helicopters, I center punched the bulls-eye! It was a wonderful time to be in aviation, and it was my great privilege to have been a part of it. How cool and wonderful it was to have flown a wide variety of helicopter models all over the world, and see things most folks only dream about! Flying also seems to be a great equalizer regardless of race, color, creed or religion, and the list of folks I've flown with includes the folks who work in the trenches all the way up to CEOs, politicians, and movie stars. Regardless of one's status in life, everyone was soon visiting and carrying on like old friends, then parting ways at the end of the journey like two ships passing in the night, leaving behind nothing but good memories.

As I enter the twilight of my career, it's difficult to grasp I'm at that point in my life where I walk to aircraft on achy knees, then gingerly climb aboard before settling slowly into the pilot's seat so as not to disturb a gimpy back that's spent far too many hours in the cockpit! It seems like only yesterday that I was flying alongside some of the best test pilots in the industry! Now, no longer in demand to fly the more difficult experimental flight tests I once flew with ease, I have become just another "fly by night pilot", albeit with Night Vision Goggles (NVGs) strapped to my helmet.

Most of the flying I do now is in the early morning hours, on a moonless night when the sky is at its darkest. It's a time when the world below sleeps in silent slumber, and for an old man who has

spent most of his life in the air, it's a wonderful time to fly. It's also the fulfillment of a young man's dream that began so long ago in the cottonfields of west Texas. It's been a life full of cottonfields & copters.

CPSIA information can be obtained
at www.ICGtesting.com
Printed in the USA
JSHW050228110622
26826JS00001B/3